MW00610223

INTRODUCTORY BIOPHYSICS

PERSPECTIVES ON THE LIVING STATE

JAMES CLAYCOMB, PHD

Houston Baptist University

JONATHAN QUOC P. TRAN

University of Texas

Southwestern Medical Center

JONES AND BARTLETT PUBLISHERS

Sudbury, Massachusetts

BOSTON TORONTO LONDON SINGAPORE

World Headquarters
Jones and Bartlett Publishers
40 Tall Pine Drive
Sudbury, MA 01776
978-443-5000
info@jbpub.com
www.jbpub.com

Jones and Bartlett Publishers
Canada
6339 Ormindale Way
Mississauga, Ontario L5V 1J2
Canada

Jones and Bartlett Publishers
International
Barb House, Barb Mews
London W6 7PA
United Kingdom

Jones and Bartlett's books and products are available through most bookstores and online booksellers. To contact Jones and Bartlett Publishers directly, call 800-832-0034, fax 978-443-8000, or visit our website www.jbpub.com.

Substantial discounts on bulk quantities of Jones and Bartlett's publications are available to corporations, professional associations, and other qualified organizations. For details and specific discount information, contact the special sales department at Jones and Bartlett via the above contact information or send an email to specialsales@jbpub.com.

Copyright © 2011 by Jones and Bartlett Publishers, LLC

All rights reserved. No part of the material protected by this copyright may be reproduced or utilized in any form, electronic or mechanical, including photocopying, recording, or by any information storage and retrieval system, without written permission from the copyright owner.

QuickField™ is a trademark of Tera Analysis, Ltd. Copyright © 2009, Tera Analysis.
MATLAB® is a registered trademark of The MathWorks, Inc. Copyright © 2010, The MathWorks, Inc.

Production Credits
Publisher: David Pallai
Editorial Assistant: Molly Whitman
Production Manager: Tracey Chapman
Production Assistant: Ashlee Hazeltine
Associate Marketing Manager: Lindsay Ruggiero
V.P., Manufacturing and Inventory Control: Therese Connell
Composition: Lapiz Online
Cover Design: Scott Moden
Cover Image: Courtesy of Jacques Descloitres, MODIS Rapid Response Team, NASA/GSFC
Printing and Binding: Malloy, Inc.
Cover Printing: Malloy, Inc.

Library of Congress Cataloging-in-Publication Data
Claycomb, James R.
Introductory biophysics: perspectives on the living state / James R. Claycomb, Jonathan Quoc P. Tran.
p. cm.
ISBN-13: 978-0-7637-7998-6 (hardcover)
ISBN-10: 0-7637-7998-9 (ibid.)
1. Biophysics. I. Tran, Jonathan Quoc P. II. Title.
QH505.C53 2011
571.4—dc22

2009049892

6048
Printed in the United States of America
14 13 12 11 10   10 9 8 7 6 5 4 3 2 1

Dedicated to the memory of
John Vay

Introduction

Physics is the most basic natural science. It is the lowest-level science on a tree diagram representing the natural sciences. Other sciences such as chemistry and biology are higher-level sciences. Now it is possible to begin on one of the natural science branches and proceed quite a distance without knowledge of the more basic physical principles such as Newton's laws and gravitation. General physics, for example, is rarely a prerequisite for introductory biology and chemistry. To sprout new branches in our understanding of nature, and biology in particular, the tools of basic physics and chemistry are often required.

One goal of this textbook is to introduce biophysical theory concurrently with computer simulation. Another goal is to emphasize several important themes such as the role of entropy and the ubiquity of the statistical Boltzmann factor describing many aspects of the living state from the establishment of membrane potentials to the sedimentation of cell cultures. A key perspective of the textbook considers life in relation to the universe as a whole.

Additional computer simulation examples and exercises in the text and on the companion CD enable modeling of biophysical phenomena such as diffusion, establishment of membrane potential, bioimpedance, and the electrical response of cells and organelles to external fields. Dynamical models include the propagation of action potentials, nonlinear oscillations in population dynamics, and DNA vibrational modes. The formation of fractal structures, branching networks, and patterns in biology are also simulated. Contact with simulation methods further reinforces biophysical theory and strengthens the connection between physics and the life sciences.

Topics are arranged in increasing size scale and complexity. Chapters 1 and 2 give an introductory overview of biological structure and molecular forces. Chapters 3 through 6 cover heat transfer, thermodynamics, statistical mechanics, and diffusion, which play important roles in the biological domain. Chapter 7 covers fluid dynamics and the motion of bodies in fluid media. Bioenergetics, molecular motors, and the light absorption of biomolecules are discussed in Chapter 8. Chapters 9 through 11 discuss electrical and mechanical aspects of biophysics, including the passive response of cells and tissue to electric fields, field-induced stresses in biomaterials, and the propagation of action potentials in nerve cells. Biomagnetism and magnetic measurements are discussed in Chapter 12. Nonlinearity, chaos, and complexity in the life sciences are covered in Chapters 13 and 14. The textbook concludes with an introduction to astrobiology in Chapter 15.

Acknowledgments

We would like to acknowledge Vladimir Podnos for helpful suggestions for implementing QuickField™ for biophysical applications, software support from The MathWorks, Inc. and Tera Analysis. Thanks to Barbara Benitez-Gucciardi for sharing insights in computational biology and the F–N model. We are thankful for many helpful conversations with Saul Trevino and Gardo Blado on protein folding and statistical physics, respectively. Many thanks are due John H. Miller, Jr. for introducing us to many topics covered here from biomagnetism to torque generation in ATP synthase. We are grateful for helpful suggestions from Audrius Brazdeikis and for supplying fetal magnetocardiography data using SQUIDs. Thanks to Cindy Troung and Anum Umer for their investigations on the effects of microgravity on the human body. We are grateful for figure support from NASA and ADInstruments. J. R. C. acknowledges support of the Texas Center for Superconductivity at the University of Houston.

Table of Contents

1 Building Blocks and Structure

1.1 Atoms and Ions

1.1.1 Subatomic Particles

All matter essential to life is made up of atoms consisting of protons, neutrons, and electrons. The electron is a stable particle belonging to a class of subatomic particles known as fermions. The electron is elementary and without internal structure. Table 1.1 gives a list of elementary particles according to the standard model of particle physics. Protons and neutrons are composite particles made out of combinations of three up (u) and down (d) quarks. The proton is (uud) while the neutron is (ddu). The four other quarks: strange (s), charm (c), top (t), and bottom (b) do not form stable particles and therefore do not participate in the physics of the living state.

Another particle essential to life is the photon (γ) that is a massless member of the Boson class of particles. Without the photon, energy could not travel from the Sun to

Table 1.1 Particles of the Standard Model

Fermions			Bosons
u	c	t	γ
d	s	b	Z
ν_e	ν_μ	ν_τ	W
e^-	μ	τ	g

the Earth and warm our planet. Plants absorb photons to produce adenosine triphosphate (ATP) and to fix carbon that is consumed by more complex organisms on the food chain such as humans.

In Paul Dirac's theory of relativistic quantum mechanics, antiparticles were predicted that were later discovered in cloud chamber experiments. It turns out that for every particle species in nature there is a corresponding antiparticle, although antimatter is quite scarce, at least on Earth. Antimatter plays a key role in the strong nuclear force through the exchange of π mesons. These mesons are quark–antiquark pairs exchanged between nucleons binding the nucleus together. Antimatter is therefore essential for the existence of stable atoms and hence, life. Antimatter is also employed in medical imaging using positron emission tomography (PET) where the positron e^+ is the positively charged antiparticle of the electron. Gamma rays are detected in PET scans that are produced by the annihilation of matter and antimatter according to the reaction $e^+ + e^- \rightarrow 2\gamma$. From Einstein's mass energy relation and the energy of each photon $E = hf$ with $f\lambda = c$, the wavelength of the gamma rays emitted in PET are $\lambda = h/m_e c = 2.43$ pm.

The neutrino (ν), or "little neutral one," predicted by Wolfgang Pauli, is a participant of beta decay and other nuclear reactions. Roughly 10^{12} solar neutrinos pass through our bodies each second. Neutrinos are weakly interacting, so this rate does not change during the night as neutrinos pass straight through the Earth, only very occasionally interacting with matter. Without neutrinos, nuclear reactions essential to stars would not take place. The little neutral one is therefore essential for life in this regard.

The gluon (g) is a key participant in the strong nuclear force binding quarks in nucleons and mesons. Other particles of the standard model may not contribute directly to the physics of the living state but may be a natural consequence of a universe capable of sustaining life.

Highly energetic subatomic particles in cosmic rays can also influence life processes by breaking molecular bonds and causing genetic mutations. Fortunately, the Earth's atmosphere serves to greatly attenuate the flux of cosmic rays. The ozone layer is especially important in absorbing harmful solar ultraviolet radiation. Astronauts in space habitats are more vulnerable to solar protons that can cause health problems if received in excessive doses. Solar flares and coronal mass ejections can generate lethal doses of radiation, damaging eyes and vital organs. Cosmic rays passing though the gel-like vitreous humor in the eye produce Cherenkov radiation because the particles are traveling faster than the speed of light in a dielectric medium. Astronauts can perceive Cherenkov radiation in the eye as flashes of light.

High-energy particles can also provide lifesaving therapy by targeting cancerous tumors. Proton therapy is a recently developed technique that employs particle accelerators to target tumors with proton beams of specific energy.

1.1.2 Atomic Constituents of Life

Roughly 97% of living matter on Earth is made up of C, O, H, and N atoms. The remaining 3% consists mostly of S, P, K, Na, Ca, Mg, and Cl with ~0.02% of trace elements such as Fe, Cu, and Co. The elements highlighted in the periodic table (Table 1.2) have been found to contribute to life on Earth. The noble gasses in the last column are not capable of forming chemical bonds and therefore cannot form complex molecules necessary for life. Elements in the second row (B, C, N, O, and F) form strong bonds and can participate in chemical reactions. Notably, C and Si are capable of forming four bonds; however, the Si–Si bond is relatively weak so that only C forms polymer chains useful for terrestrial biology.

Living systems selectively assimilate certain atoms while expelling others; atomic species perform vital functions as shown schematically in Figure 1.1. An important factor determining the participation of a given element in the life process is its abundance. Scarcer atoms on Earth are less likely to form the basis for terrestrial biology.

1.1.3 Ions

Neutral atoms have equal numbers of protons and electrons. Positive ions, or cations, have fewer electrons than protons. Negative ions, or anions, have a larger number of electrons compared with protons.

Table 1.2 Periodic Table Highlighting Elements That Participate in Biology

H																	He
Li	Be											**B**	**C**	**N**	**O**	**F**	Ne
Na	**Mg**											Al	**Si**	**P**	**S**	**Cl**	Ar
K	**Ca**	Sc	Ti	**V**	**Cr**	**Mn**	**Fe**	**Co**	**Ni**	**Cu**	**Zn**	Ga	Ge	**As**	**Se**	**Br**	Kr
Rb	**Sr**	y	Zr	Nb	**Mo**	Tc	Ru	Rh	Pd	Ag	**Cd**	In	**Sn**	Sb	Te	**I**	Xe
Cs	**Ba**	Ln	Hf	Ta	**W**	Re	Os	Ir	Pt	Au	Hg	Tl	Pb	Bi	Po	At	Rn
Fr	Ra	Ac	Rf	Ha	Sg	Ns	Hs	Mt									

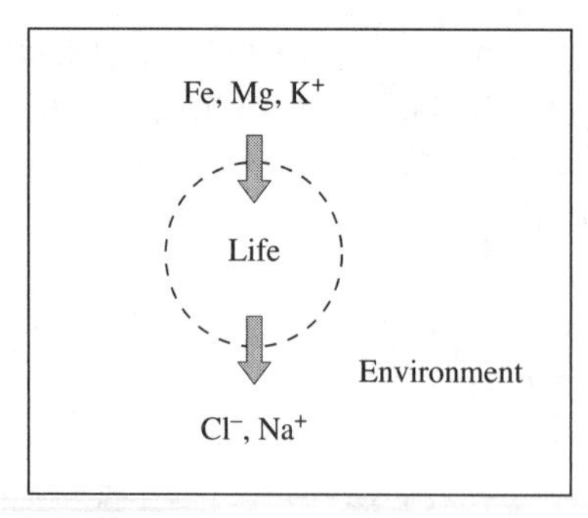

FIGURE 1.1 Living systems assimilate certain elements and expel other elements to establish concentration gradients and perform vital functions. (Adapted from Chopra A., and C. H. Linewater. 2007. *The major elemental abundance difference between life, the oceans and the sun.* Canberra, Australia: Planetary Science Institute.)

Ions are essential for terrestrial life processes. Their transport establishes membrane potentials and allows for the propagation of action potentials in nerve cells. Living cells rely on the transport of ions such as H^+ in the formation of ATP shown in Figure 1.2. The most important ions in biology are hydrogen H^+, sodium Na^+, potassium K^+, chlorine Cl^-, and calcium Ca^{2+}.

Less important for biological function are the isotopic differences between atoms with varying numbers of neutrons. However, various isotopes are important in nuclear medicine. Also, the isotope carbon-14 enables the radiological dating of biological materials.

1.2 Molecules Essential for Life

1.2.1 Water

Water (H_2O) is the most abundant small molecule comprising life on Earth, making up about 66% of the human body by number of H and O atoms. Bacteria such as *E. coli* are composed of about 84% by number of H and O atoms.

A water molecule consists of two hydrogen atoms and an oxygen atom, which interact through polar covalent bonds. The polarity of water results from the greater

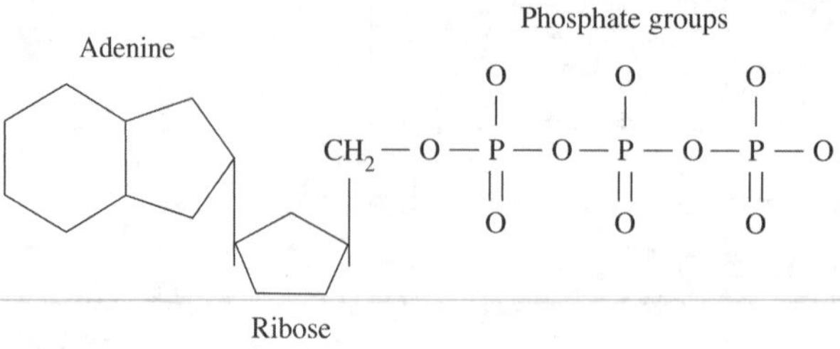

Phosphate groups

Adenine

CH_2—O—P—O—P—O—P—O

Ribose

FIGURE 1.2 Adenosine triphosphate (ATP) molecule.

affinity of oxygen for electrons compared to that of hydrogen. Thus, electrons in a water molecule tend to be concentrated near the oxygen atom.

When two neighboring water molecules come in close contact with each other, a hydrogen bond typically forms between the positively charged region (hydrogen) of one molecule and the negatively charged region (oxygen) of the other water molecule. These hydrogen bonds are weak relative to the intramolecular covalent bonds of water and are continually broken and reformed. Collectively, however, hydrogen bonds are strong and account for the many unique properties of water. These properties include its high specific-heat capacity, high melting and boiling points, and large surface tension. Interestingly, other polar molecules with similar densities to water evaporate at cellular temperatures. Water, however, exists as a liquid within biological systems.

The polarity of water also accounts for its function as a solvent in chemical reactions. Small amino acids such as glycine and polar substances such as sodium and potassium ions within the body easily dissolve in water. These molecules are deemed hydrophilic due to their "water-loving" properties of small size and charge. Hydrophobic substances such as triglycerides and aliphatic amino acids tend to associate together in a polar water solution and do not dissolve. It is thus the important function of amphipathic (both hydrophilic and hydrophobic) phospholipids within cell membranes to compartmentalize the cell and help maintain a highly organized biological system, in which water plays an indispensible role.

1.2.2 Proteins

Proteins are one of the primary building blocks of biological organisms and are comprised of small subunits known as amino acids. Approximately 20 amino acids are present in humans, 10 of which are essential. These essential amino acids cannot be manufactured by the body and therefore must be obtained from the diet. Amino acids are primarily composed of a central carbon attached to four chemical groups: an amine group, a carboxylic acid, a hydrogen atom, and a variable residue denoted R as shown in Figure 1.3. The –R group distinguishes one amino acid from the next and can be acidic, basic, hydrophobic, or hydrophilic. It can be as complex as a ring structure or as simple as a hydrogen atom. The many different properties of this variable –R group

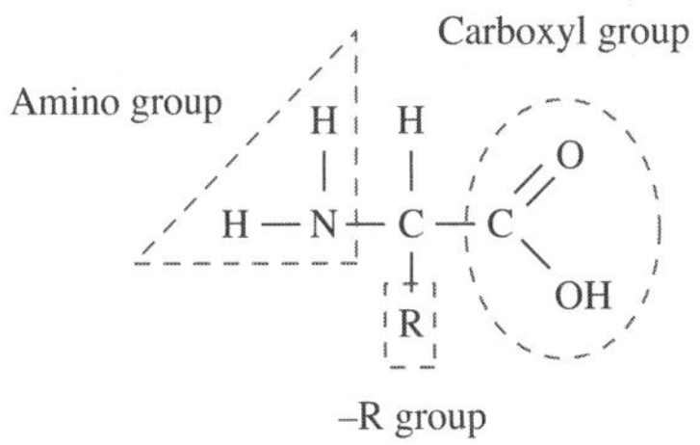

FIGURE 1.3 All amino acids are composed of an amino (NH_2), a carboxyl (COOH), and an –R group. Altogether there are 20 amino acids with 20 different –R groups.

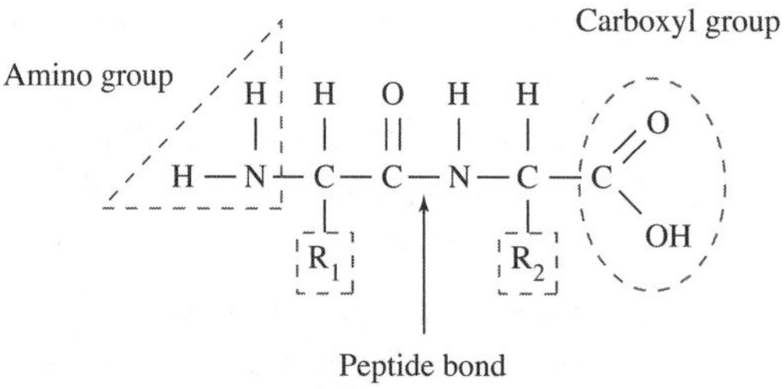

Carboxyl group

Amino group

Peptide bond

FIGURE 1.4 Amino acids link together to form proteins. The amino (NH_2) and carboxyl (COOH) groups link together to form a peptide (C–N) bond plus a water (H_2O) molecule.

allow for the many different structural and functional proteins throughout the human body. Moreover, the intricate relationship between different –R groups within proteins allows for the formation of specific geometric structures. These different protein structures result from a combination of electrostatic, hydrophobic, hydrophilic, and hydrogen-bonding interactions. The structures serve separate yet specific functions, such as sites for enzyme catalysis or channels for intercellular transport.

At the most basic level, proteins have a primary structure composed of a specific sequence of amino acids joined by peptide bonds as depicted in Figure 1.4. This amino acid sequence is crucial to the formation of a specific protein structure and thus is the basis for the protein's function. The second level of protein organization is the secondary structure. This structure consists of alpha helices and β-pleated sheets that form due to hydrogen bonding. At the tertiary level, a single protein can fold and form distinct geometric shapes due to the presence of hydrogen bonds, disulfide bonds, electrostatic or ionic interactions, hydrophobic interactions, and Van der Waals forces. Quaternary protein structures are simply complexes of multiple smaller proteins.

Proteins have many roles in biological systems. They provide structural support, facilitate in the transportation of other biological substances, and can even serve as hormones as a means to communicate from one bodily system to the next. Table 1.3 is a list of globular proteins, and Table 1.4 lists structural proteins and their accompanying function within the body.

In nature, proteins can be denatured by extreme heat or pH changes. Other denaturing agents include salt, urea, organic solvents, and mercaptoethanol that destroy the electrostatic interactions, hydrogen bonds, hydrophobic interactions, and disulfide bonds in proteins, respectively.

Table 1.3 Globular Proteins and Their Function

Globular Proteins	Function
Channels (e.g., Voltage-Gated and Ligand-Gated Channels)	Protein channels allow charged substances to cross the semipermeable cellular membrane. Voltage-gated channels are opened when a certain membrane potential is maintained. For example, Ca^{2+} ions are transported into the presynaptic neuron near the synapse via protein channels that are activated by the presence of intracellular Na^+. The influx of calcium initiates vesicular exocytosis and the release of neurotransmitters into the synaptic cleft. Ligand-gated protein channels open upon binding to a specific substance or ligand (e.g., acetylcholine).
Enzymes (e.g., Proteases)	Enzymes are biological catalysts that increase the rates of reactions without altering their thermodynamics. They are not consumed by reactions and therefore are neither reactants nor products. Proteases are enzymes that degrade proteins.
Hormones (e.g., Insulin)	Peptide hormones travel within the bloodstream from one organ to the next, are hydrophilic, and bind to a receptor on the surface of the target cell. Insulin, for example, is a large peptide hormone that lowers blood glucose levels.
Immune Response (e.g., Antibodies)	Antibodies in the blood adhere to the surfaces of foreign antigens (i.e., bacteria) and facilitate phagocytosis by macrophages.
Intercellular Transport (e.g., Hemoglobin)	Hemoglobin is a large, globular protein consisting of four smaller peptide units, two alpha units, and two beta units. It transports oxygen throughout the body via the blood.
Intracellular Transport (e.g., Myoglobin)	Myoglobin is the equivalent of hemoglobin in muscle tissue and transports oxygen to these tissues.
Membrane Pumps (e.g., Na^+/K^+ ATPase, H^+ Pumps)	Membrane pumps transport ions into and out of cells. The Na^+/K^+ pump helps maintain a resting membrane potential of -70 mV in neurons by transporting three Na^+ ions out of the neuron and two K^+ ions in. Hydrogen pumps maintain the acidic interior of lysosomes necessary for bacterial degradation.
Membrane Receptors (e.g., Neurotransmitter Receptors)	Acetylcholine receptors on the surface of postsynaptic neurons bind neurotransmitters and initiate the influx of sodium ions into the postsynaptic neuron via ligand-gated Na^+ channels.
Osmotic Regulators (e.g., Albumin)	Albumin is a protein that travels in the blood and regulates the osmolarity of the blood. Blood osmolarity increases with increasing blood albumin content.

Table 1.4 Structural Proteins and Their Function

Structural Proteins	Function
Collagen	Collagen is composed of a unique helical structure and helps provide support in tendons, ligaments, the skin, and bones. It is also the most abundant protein in the human body.
Glycoproteins	Glycoproteins are structural proteins composed of protein and carbohydrate components. They are part of the cellular membrane and play a role in cellular communication and recognition.
Microtubules	Microtubules are proteins made of tubulin. They play a role in axonal transport, ciliary and flagellar movement, and spindle formation during the cell cycle.
Proteoglycans	These structural proteins form the extracellular matrix and are composed of both protein and carbohydrate components.

1.2.3 Lipids

Lipids are hydrophobic molecules typically consisting of long hydrocarbon chains. Their hydrophobic nature allows them to form highly impermeable barriers in aqueous solutions like those shown in Figure 1.5.

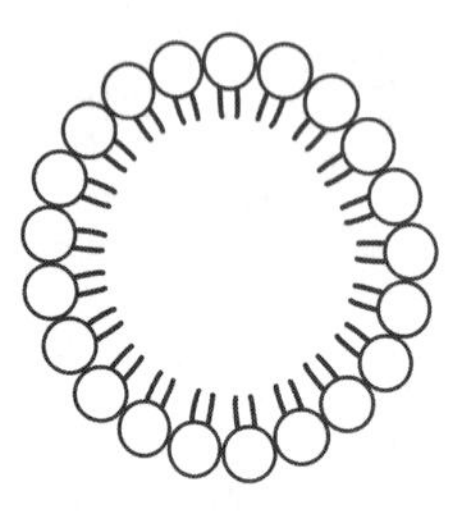

FIGURE 1.5 Micelle formed from a single lipid layer.

Cellular vesicles and cell membranes are mainly composed of lipids that separate one aqueous compartment from another. The semipermeable nature of the cell membrane is a result of pores and protein channels present between phospholipids. Lipids such as steroid hormones cannot freely dissolve in aqueous solutions and thus require vesicular transport throughout the human body. These hormones typically alter cell function at the level of transcription in the nucleus.

The fundamental building blocks of lipids are fatty acids. As their name implies, these molecules consist of a carboxylic acid group attached to a long hydrocarbon tail.

(a)

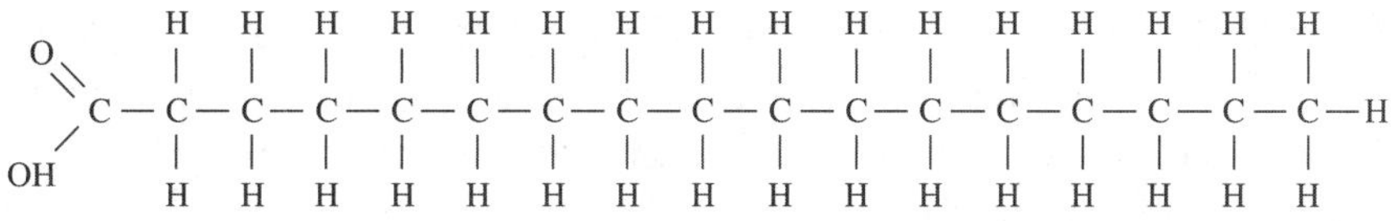

(b)

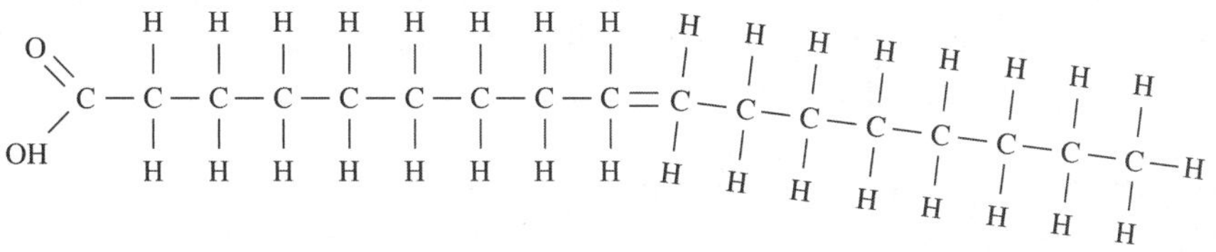

FIGURE 1.6 (a) Saturated and (b) unsaturated fatty acids. The nonpolar saturated fatty acid has a more uniform electronic charge distribution compared with the polar unsaturated fatty acid.

Fatty acids can be either saturated or unsaturated, depending on the presence or absence of carbon–carbon double bonds as shown in Figure 1.6. If a fatty acid contains at least one such bond, it is deemed unsaturated. When three fatty acids are combined, a triglyceride is formed, which primarily functions to store energy.

Lipid Digestion and Absorption

The digestion of lipids requires the action of pancreatic lipases that work in the small intestine as well as the action of bile acids that facilitate the emulsification of lipids. Micelles are formed by the bile acids, and they help transport lipids to the intestinal cell wall. Absorption of lipids occurs primarily through the jejunum of the small intestine. Previously, it was believed that lipids could simply be absorbed across the intestinal membranes via passive diffusion. Recently, however, it has been discovered that specific transport proteins assist in the transport of these lipids. NPC1L1, for example, is the protein that transports cholesterol and plant sterols into the intestinal cells. But because we have no essential use for plant sterols, another transporter known as ABCG5/8 is responsible for extruding these sterols back into the lumen of the intestines for excretion. Unlike other nutrient monomers that are small enough to enter blood capillaries directly from the intestinal villi, lipids are transported from the intestinal lumen to the lymphatic system through lacteals. Eventually, the lymphatic system drains into the thoracic duct, which joins the circulatory system in the upper left chest, and the lipids can be distributed to the tissues (adipose and muscle) or transported to the liver for processing.

1.2.4 Carbohydrates

Carbohydrates are the most common type of biomolecules that comprise most of the organic matter on Earth. The pentose sugar deoxyribose forms the structural backbone of DNA and RNA. Carbohydrates also store chemical bond energy used in metabolism. Simple sugars are described by the formula $C_n(H_2O)_n$ with $n > 3$. For example, $n = 6$ gives the formula for glucose $C_6H_{12}O_6$.

Blood glucose levels are maintained at relatively constant levels throughout the day despite increases in carbohydrate consumption during meals. This homeostasis is maintained through various metabolic pathways including glycolysis, gluconeogenesis, glycogen synthesis, and glycogenolysis. High levels of glucose in the blood can lead to the excess glycosylation of proteins in the body. Insufficient levels of blood glucose ultimately result in inadequate fuel for the brain.

Digestion and Absorption of Carbohydrates

Digestible carbohydrates include

- Polysaccharides–starch, amylopectin, and amylose
- Disaccharides–sucrose and lactose
- Monosaccharides–glucose, galactose, and fructose

Nondigestible carbohydrates include the plant sugars: cellulose, pectins, and lignins. Only monosaccharides can be transported from the gastrointestinal lumen into the blood. This transportation occurs through specialized membrane transporters on the intestinal membrane. Fructose enters the intestinal cells via facilitated diffusion while glucose and galactose monomers enter through a secondary active transport system. The active transport of Na^+ across the basolateral surface of the cells allows Na^+ from the intestinal lumen to enter the cells down its concentration gradient. The transport of glucose and galactose into cells is against a concentration gradient but is possible due to coupling with the more energetically favorable transport of sodium ions.

The breakdown of starch and disaccharides into absorbable monosaccharides occurs through amylases, lactases, and sucrases. Lactase deficiency results in a condition called lactose intolerance, which can be managed with lactase supplements or a diet low in lactose sugars. In the absence of lactase, the lactose that remains in the intestinal lumen is degraded by enteric bacteria and results in the accumulation of various gases that can cause cramps, bloating, and flatulence.

1.2.5 Cholesterol

Cholesterol is an essential component of the plasma membrane of living cells. It functions to regulate the fluidity of the membrane. It is also an essential component of the myelin sheath surrounding neuronal axons and serves as a precursor to bile acids, steroid hormones, and vitamin D.

It can be derived from the diet, formed via de novo synthesis from acetate, or from the uptake of low-density lipoproteins. Cholesterol is found predominantly in neural tissue but is also abundant in the liver and the adrenal glands. In the liver, cholesterol is used to make lipoproteins and bile acids. In the adrenal glands and gonads, cholesterol is used for hormone synthesis. Defects in cholesterol synthesis can lead to inherited disorders such as Smith–Lemli–Opitz syndrome, an autosomal recessive disease resulting in feeding problems, mental retardation, male pseudohermaphroditism, and polydactyly. Statin drugs are effective at lowering blood cholesterol levels by inhibiting an important enzyme in cholesterol synthesis called HMG CoA reductase.

1.2.6 Nucleic Acids

There are two main forms of nucleic acids: deoxyribonucleic acid (DNA) and ribonucleic acid (RNA). DNA is different from RNA in that the former is double-stranded, contains thymine instead of a uracil, and is present only within the nucleus, whereas the latter is single-stranded, contains uracil, and can be present inside both the nucleus and the cytoplasm. Furthermore, DNA contains a hydrogen atom on the 2′ carbon of its sugar moiety instead of the hydroxyl group in RNA.

DNA is composed of four specific nucleotides: ATP, CTP, GTP, and TTP. Each nucleotide in DNA is composed of a deoxyribose sugar, a phosphate group, and a nitrogenous base. These bases include adenine, thymine, guanosine, and cytosine, each represented by the letters A, T, G, and C, respectively. Hydrogen bonding in DNA allows for the formation of a double helix, a structure elucidated by James Watson and Francis Crick. Two hydrogen bonds connect thymine to adenine, and three hydrogen bonds connect guanosine with cytosine. In RNA, uracil replaces thymine and bonds to adenine. Individual nucleotides in both DNA and RNA are connected by phosphodiester bonds. In DNA, these phosphodiester bonds form what is known as a sugar–phosphate backbone.

Like proteins, DNA can be denatured by heating. This process by which the hydrogen bonds between paired nucleotides are broken is known as DNA melting. The rate of DNA melting can easily be monitored in the lab by measuring an increase in light absorption when double-stranded DNA denatures into single-stranded DNA.

1.3 What Is Life?

Erwin Schrödinger delivered a series of lectures in 1943 at Trinity College in Dublin, Ireland, before biophysics was an established discipline. These lectures addressed fundamental questions such as how space–time events that occur within the boundary of living organisms could be explained by the laws of physics and chemistry. Schrödinger formulated some questions concerning life from the perspective of a physicist based on these lectures in his book *What Is Life?* first published in 1944.

In a time before the discovery of the structure of DNA by Watson and Crick, Schrödinger postulated that genetic information was encoded in something like an aperiodic crystal, and he put forward the idea that organisms can create order internally by increasing the disorder of their environment. Schrödinger thus reconciled the tendency toward order in biological systems with the second law of thermodynamics that states that the entropy of an isolated system must increase.

1.3.1 Requirements for Life

- All terrestrial life processes occur in the presence of liquid water.
- Life therefore requires a range of temperatures where liquid water can exist.
- Energy in the form of chemical bonds or photons is required to maintain internal structure and to carry out processes associated with life.
- Conditions favorable for life must exist long enough for life to arise.

1.3.2 Domains of Life

The three domains of life on Earth include bacteria, archaea, and eukaryotes. Bacteria and archaea thrived for nearly 2 billion years of Earth's history before the arrival of eukaryotes.

- Bacteria are prokaryotic cell types lacking a cell nucleus, and they reproduce asexually through binary fission.
- Eukaryotes are cells with nuclei and membrane-bound organelles. The cells comprise most multicellular life including plants, animals, and fungi.
- Archaea are the simplest life form lacking any internal organelles. These single-celled organisms are the most ancient and can survive in extremely inhospitable conditions on the planet.

1.3.3 Characteristics of Living Cells

- Metabolism—Absorb nutrients and use chemical bond energy to carry life processes.
- Reproduction by cell division.
- Organization—Manufacture internal structure from elements assimilated from the environment.
- Sense environmental stimuli such as light, changes in nutrient level, temperature, and pH.
- Response to such stimuli by cell motility and growth toward light sources.
- Membrane potential—Characteristic of all living cells that facilitates the transport of ions and drives molecular motors essential to life processes.

- Signaling—Through propagating action potentials.
- Apoptosis—Programmed cell death.

1.3.4 Structure of Living Cells

According to cell theory, all living organisms on Earth are made up of cells that develop from parent cells. A eukaryotic cell is depicted in Figure 1.7, where key organelles are shown.

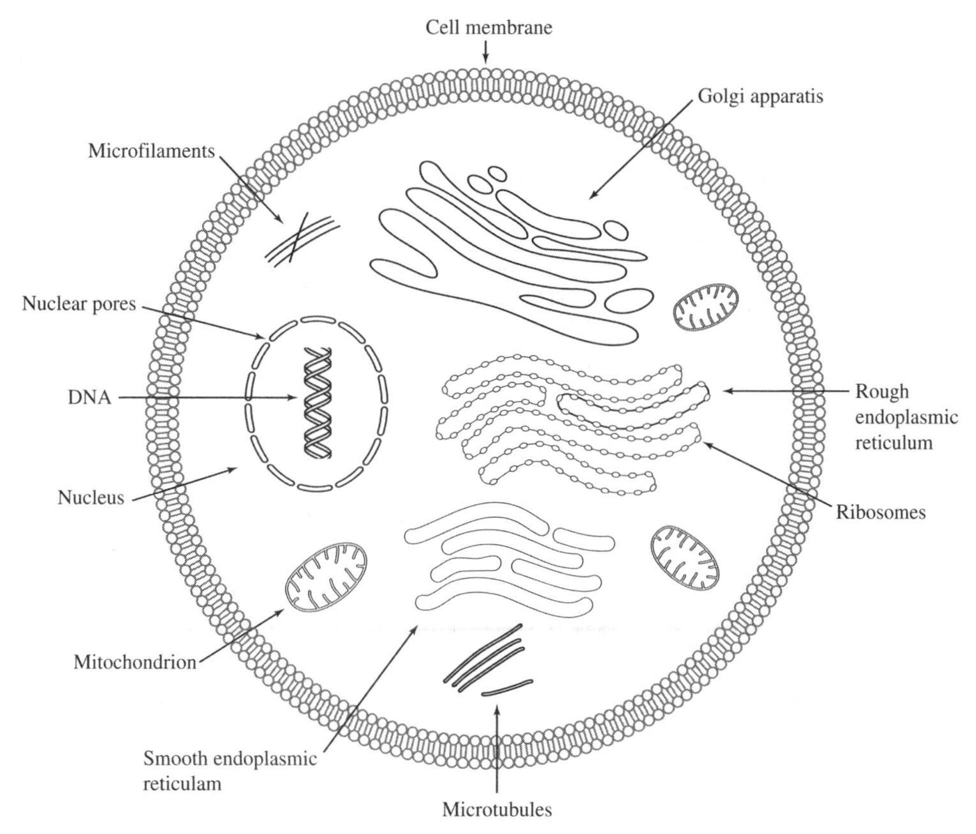

FIGURE 1.7 Schematic of a eukaryotic cell showing key organelles.

A list of organelles is given in Table 1.5, indicating the function and type of cell in which the organelle is found.

1.3.5 Boundary of Life

Viruses and viroids have no metabolism and are not classified as a part of any domain of life. They are made of nucleic acid molecules that can replicate by invading living cells and therefore play a significant role in biology. These entities represent the boundary between living and nonliving matter.

Table 1.5 Cellular Parts List

Organelle/ Structural Element	Function	Eukaryotes	Prokaryotes
Cellular Membrane	Phospholipid bilayer separating the interior of the cell from the exterior region as depicted in Figure 1.8. Each phospholipid molecule consists of a hydrophilic phosphate group that faces outward with hydrophobic fatty acid tails facing inward.	¤	¤
Cell Wall	Rigid cellulose layer enclosing the cell membrane in plants and bacteria.	¤	¤
Chloroplasts	Photosynthesis; chlorophyll; light-activated electron transport chain.	¤	
Cytoskeleton	Proteins including microtubules and microfilaments that allow for cell rigidity support the cell's shape and are used in cellular motility.	¤	¤
Cytosol	Gel-like fluid containing organelles inside the plasma membrane.	¤	¤
Golgi Apparatus	Packages macromolecules including lipids and proteins produced in the ER for transport and secretion.	¤	
Lysosomes	Garbage disposal vesicles of animal cells containing enzymes that digest expended organelles, food particles, and endocytosed particles.	¤	
Membrane Transporters	Proteins that facilitate the transport of ions and large molecules across the lipid bilayer. Membrane transporters can allow active or facilitated transport.	¤	¤
Mitochondria	Powerhouse of the cell; synthesize ATP; contain outer- and inner-mitochondrial membranes.	¤	
Nucleolus	Contained inside the cell nucleus. Involved in the synthesis of ribosomes.	¤	
Nucleus	Control center of the cell. Contains genetic information.	¤	

Organelle/ Structural Element	Function	Eukaryotes	Prokaryotes
Ribosomes	Organelles without a membrane that translate messenger RNA to synthesize polypeptide chains.	¤	¤
Rough Endoplasmic Reticulum (ER)	Rough ER membranes are covered with ribosomes where proteins are synthesized.	¤	
Smooth Endoplasmic Reticulum (ER)	Smooth ER consists of a network of membranes where lipids are synthesized.	¤	
Vacuoles	Large storage vesicles in plant cells containing nutrients.	¤	

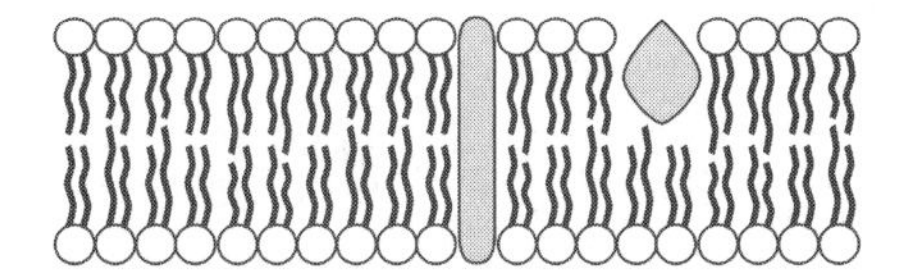

FIGURE 1.8 Section of plasma membrane with embedded membrane proteins. Cell membranes from a physical boundary between living and nonliving matter on Earth. The membrane encloses the cytoplasm containing organelles while remaining permeable to small molecules.

- Viruses have genes made up of RNA or DNA but can only replicate inside host organisms. Protective protein coats called capsids surround viral nucleic acids.
- Viroids are plant pathogens consisting of strands of RNA that are lacking a protein coat.

Figure 1.9 illustrates a λ-phage virus injecting a strand of DNA into an *E. coli* bacterium. The relative sizes of the bacterium and the virus are shown in this figure.

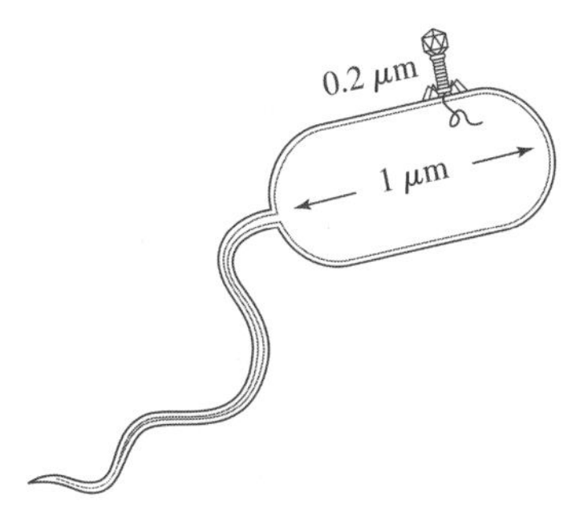

FIGURE 1.9 λ-phage virus injecting DNA into an *E. coli* bacterium (only one bacterial flagellum shown). (Drawing created using Motifolio.)

2 Living State Interactions

2.1 Forces and Molecular Bonds

In this section we consider molecular forces that form bonds and mediate interactions between biomolecules. We first consider the force between two neutral atoms. At short ranges $r < 0.4$ nm, the Pauli repulsion prevents the overlap of electron orbitals, while at slightly larger distances an attractive force results between fluctuating atomic dipole moments. The Lennard–Jones potential energy

$$U_{\mathrm{LJ}}(r) = \varepsilon\left\{\left(\frac{r_0}{r}\right)^{12} - 2\left(\frac{r_0}{r}\right)^{6}\right\} \tag{2.1}$$

contains a positive (repulsive) term proportional to r^{-12} and a negative (attractive) term proportional to r^{-6}. The negative energy term is due to the Van der Waal interaction between fluctuating instantaneous dipole moments. The force is obtained from the negative derivative with respect to r

$$F(r) = -\frac{dU_{\mathrm{LJ}}}{dr}. \tag{2.2}$$

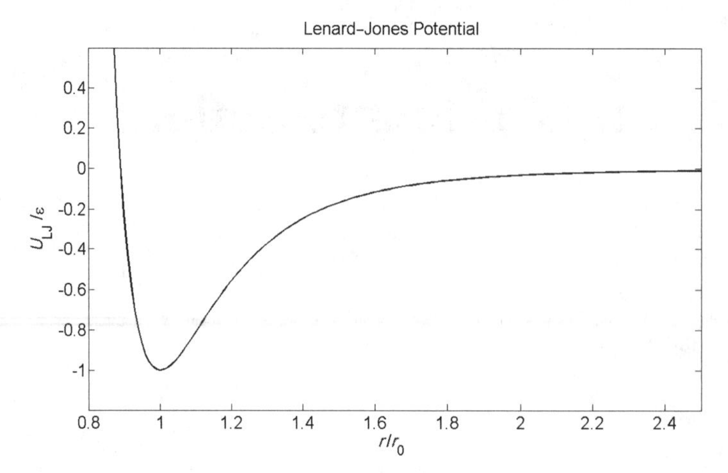

FIGURE 2.1 Van der Waals potential energy and force between two neutral atoms as a function of their separation.

The Lennard–Jones force between two neutral atoms is then

$$F(r) = 12\frac{\varepsilon}{r_0}\left\{\left(\frac{r_0}{r}\right)^{13} - \left(\frac{r_0}{r}\right)^{7}\right\}. \tag{2.3}$$

The atoms will be in mutual equilibrium when $F(r) = 0$, or when $r = r_0$. The force is maximal when $dF/dr = 0$. This occurs when $r = \sqrt[6]{13/7}\ r_0 = 1.11\ r_0$. The following MATLAB code plots the Lennard–Jones potential energy in Figure 2.1.

```
% define r range
r=.8:.01:2.5;

% specify L-J potential
U_LJ= (1./r).^12 - 2*(1./r).^6;

plot(r,U_LJ,'k','LineWidth', 2);

% specify graph properties
axis ([.8 2.5 -1.2 .6])
xlabel('r/r_0','FontSize',20);
ylabel('U_L_J /\epsilon','FontSize',20)
set(gca, 'FontSize', 20)
title('Lenard-Jones Potential','FontSize',20)
```

2.1.1 Ionic Bonds

- Ionic bonds form between atoms that are positively or negatively charged.
- Positive ions are atoms that have lost at least one electron, whereas negative ions are atoms that have gained at least one electron.

- Ionic bond formation is a result of atoms striving to attain a full outer shell of electrons.
- Sodium, for example, has 11 valence electrons and typically loses an electron to form a positively charged Na^+ ion.
- Positively charged ions are called cations, while negatively charged ions are called anions.
- When a cation comes in contact with an anion, an ionic bond typically forms between the two atoms.
- These bonds are easily dissociated in water due to the favorable interaction between charged ions and water.

2.1.2 Covalent Bonds

- Covalent bonds, such as the phosphodiester bonds between sugar and phosphate molecules in DNA, are stronger than noncovalent bonds.
- These bonds form due to the sharing of electrons between two atoms.
- Oxygen, with six valence electrons, can form two covalent bonds with hydrogen in order to complete its outer electron shell.
- Because carbon has four valence electrons and requires eight to complete its outer electron shell, it usually forms four covalent bonds with other atoms.
- The electrons that are shared between two atoms in a covalent bond form a cloud of negative charge that balances the repulsive forces of the positive nuclei.
- The bond length is the specific distance between two atoms at which the repulsive forces between positive nuclei are canceled by the attractive forces resulting from the shared electrons.

2.1.3 Hydrogen Bonds

- Hydrogen bonds are generally formed between hydrogen atoms and an atom of greater electronegativity such as oxygen or nitrogen.
- Unlike the electrostatic interaction between oppositely charged ions, hydrogen bonds are directional and are thus a polar interaction.
- A polar interaction occurs between an electropositive hydrogen atom and two electronegative atoms, one of which is directly bonded to the hydrogen via a covalent bond.
- When water is present, the atoms forming a hydrogen bond can form new hydrogen bonds with the surrounding water molecules, thus weakening the former bonds.

2.1.4 Van der Waal Forces

- Intermittent dipoles occur in the fluctuating electron clouds surrounding nonpolar atoms.
- Such a dipole can induce another dipole in a nearby atom such that the positively charged region of one dipole is adjacent to the negatively charged region of the second dipole.
- The attraction between the slightly negative region of one dipole and the slightly positive region of the next is known as a Van der Waals interaction.
- These interactions can be significant considering the collective interaction of many nonpolar atoms.
- The Van der Waal attraction is represented by the negative term in Equation 2.1.

2.2 Electric and Thermal Interactions

The presence of an electric charge q in a dielectric medium gives rise to an electric potential in volts

$$U_e(r) = \frac{1}{4\pi\varepsilon}\frac{q}{r} \tag{2.4}$$

where $\varepsilon = \varepsilon_r\varepsilon_0$, ε_r is the relative permitivity, and ε_0 is the permitivity of free space. The electric field $\mathbf{E}$ in a region of space is given by the gradient of the potential U_e or,

$$\mathbf{E}(r) = -\nabla U_e. \tag{2.5}$$

For a single point charge, we use $\nabla = \hat{r}\partial/\partial r$ in spherical coordinates

$$\mathbf{E}(r) = \frac{1}{4\pi\varepsilon}\frac{q}{r^2}\hat{r} \tag{2.6}$$

where $\hat{r}$ is a unit vector pointing in the radial direction. A point charge q located in a region of space with electric field $\mathbf{E}$ will experience a force $\mathbf{F} = q\mathbf{E}$ so that the magnitude of the electric force between two identical charges separated by a distance r

$$F(r) = \frac{1}{4\pi\varepsilon}\frac{q^2}{r^2}. \tag{2.7}$$

The force between point charges embedded in a dielectric medium is reduced from the free space value by a factor of ε_r. Water is thus a particularly good solvent, with $\varepsilon_r = 78$ at 25°C.

The electrostatic energy W_e associated with a point charge in a region with electric potential U_e is $W_e = qU_e$. The electrostatic energy stored in a system of two identical point charges is then

$$W_e = \frac{1}{4\pi\varepsilon}\frac{q^2}{r}. \tag{2.8}$$

The electrostatic energy is on the order of the thermal energy k_BT near the characteristic Bjerrum length with $q = e$

$$\ell_B = \frac{1}{4\pi\varepsilon}\frac{e^2}{k_BT}. \tag{2.9}$$

At room temperature $\ell_B \sim 0.7$ nm.

2.3 Electric Dipoles

Biological molecules such as water and many proteins have net dipole moments. The vector dipole moment is $\mathbf{p} = q\mathbf{d}$ for two equal and opposite charges $\pm q$ separated by a displacement $\mathbf{d}$. The units of dipole moment are charge times displacement. In Debye units 1 D $= 3.3 \cdot 10^{-30}$ C m. The dipole moment of water is $p_{\text{water}} = 1.85$ D. The dipole moment of proteins are between $p_{\text{proteins}} = 10^2 - 10^3$ D. The electric potential of a single dipole is given by

$$U_e = \frac{1}{4\pi\varepsilon}\frac{\mathbf{p}.\mathbf{r}}{r^3} = \frac{1}{4\pi\varepsilon}\frac{p}{r^2}\cos\theta \tag{2.10}$$

where θ is the angle between $\mathbf{p}$ and $\mathbf{r}$. The potential will be zero in the plane perpendicular to and bisecting the dipole moment. Also, the electric field will be perpendicular to this plane. The equipotentials surrounding a dipole embedded in the plasma membrane of a living cell are plotted in Figure 2.2. The electric field is perpendicular to the equipotential lines in this figure pointing from the positive end to the negative end of the dipole.

The torque acting on a dipole in an electric field $\mathbf{E}$ is given by the cross product

$$\boldsymbol{\tau} = \mathbf{p}\times\mathbf{E}. \tag{2.11}$$

The torque is zero when the dipole moment points in the direction of the electric field so that the torque will tend to align the dipole with the field. There will be a net force on a dipole in a nonuniform field given by

$$\mathbf{F} = (\mathbf{p}\cdot\nabla)\mathbf{E}. \tag{2.12}$$

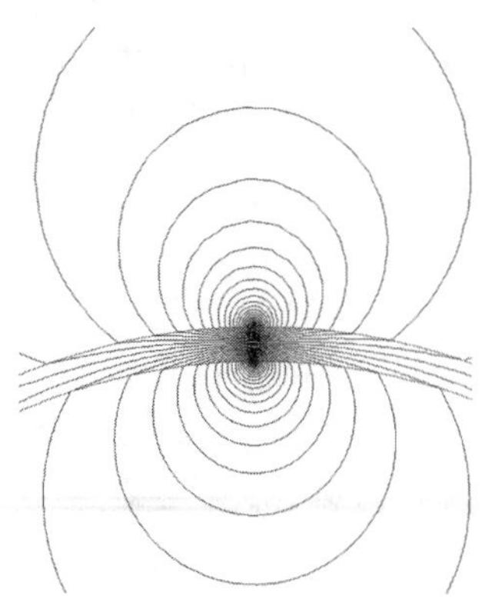

FIGURE 2.2 Finite element calculation showing lines of equal potential due to an electric dipole embedded in a plasma membrane with 7 nm thickness. The field is distorted because of the differing relative dielectric constant of the membrane ($\varepsilon_r \sim 10$) compared to the intercellular and extracellular spaces ($\varepsilon_r \sim 80$).

This dielectrophoretic force can be used to separate different types of cells and biomolecules with varying sizes and dielectric constants. However, the force on a dipole (with zero monopole moment) in a uniform electric field is zero since the positive and negative ends of the dipole experience equal and opposite forces. The interaction energy of a dipole in an electric field is

$$W_e = -\mathbf{p}\cdot\mathbf{E}. \tag{2.13}$$

Note that the minimum energy configuration corresponds to a dipole moment parallel to the electric field $W_e = -pE\cos(0) = -pE$. The maximum energy corresponds to the dipole antiparallel to the electric field where $W_e = -pE\cos(180) = -pE$. The interaction energy between two dipoles with moments $\mathbf{p}_1$ and $\mathbf{p}_2$ separated by a displacement $\mathbf{r}$ is

$$W_{12} = \frac{1}{4\pi\varepsilon}\left(\frac{\mathbf{p}_1\cdot\mathbf{p}_2}{r^3} - \frac{3(\mathbf{p}_1\cdot\mathbf{r})(\mathbf{p}_2\cdot\mathbf{r})}{r^5}\right). \tag{2.14}$$

Table 2.1 Dipole and Ion Interaction Distance Dependence

Interaction	Energy	Force
Ion–Ion	$1/r$	$1/r^2$
Ion–Dipole	$1/r^2$	$1/r^3$
Dipole–Dipole	$1/r^3$	$1/r^4$
Dipole–Dipole (fluctuating)	$1/r^6$	$1/r^7$

The first term in (2.14) vanishes for dipoles with perpendicular moments, while the second term vanishes for dipole moments perpendicular to **r**. Table 2.1 gives the distance r dependence of interaction energies and forces between ions and dipoles. The last interaction term in this table corresponds to the Van der Waal attraction.

2.3.1 Polarization and Induced Dipoles

A charge such as a sodium ion in aqueous solution will polarize the surrounding water molecules as illustrated in Figure 2.3.

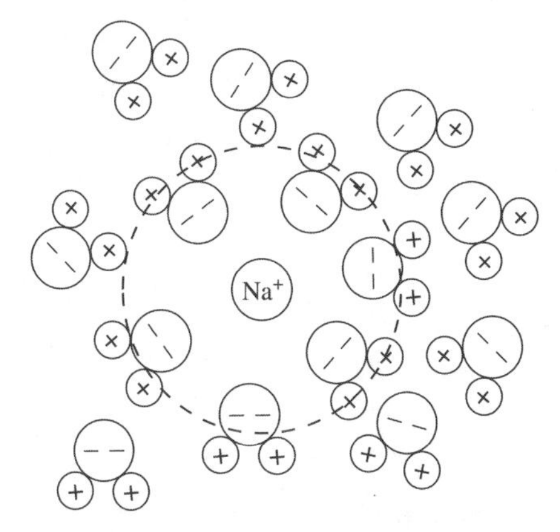

FIGURE 2.3 Hydration of a sodium ion by water molecules. The hydration radius of the dotted circle is approximately 1 Debye length Λ. If the sodium ion were removed, the water molecules would assume a random orientation in a relaxation time τ.

Water molecules are also polarized in a uniform electric field of magnitude E as depicted in Figure 2.4. The induced dipole moment **p** of an atom in an electric field **E** is proportional to the electric field for sufficiently weak electric fields

$$\mathbf{p} = \alpha \mathbf{E} \tag{2.15}$$

where **p** and **E** point in the same direction and α is the polarizability. The macroscopic polarization vector is proportional to the average dipole moment

$$\mathbf{P} = N\mathbf{p}_{\text{avg}} \tag{2.16}$$

where N is the number of atoms per unit volume so that

$$\mathbf{P} = N\alpha\mathbf{E}. \tag{2.17}$$

Polarization vector is also expressed as

$$\mathbf{P} = \mathbf{D} - \varepsilon_0\mathbf{E} \tag{2.18}$$

where **D** is the electric displacement

$$\mathbf{D} = \varepsilon\mathbf{E}. \tag{2.19}$$

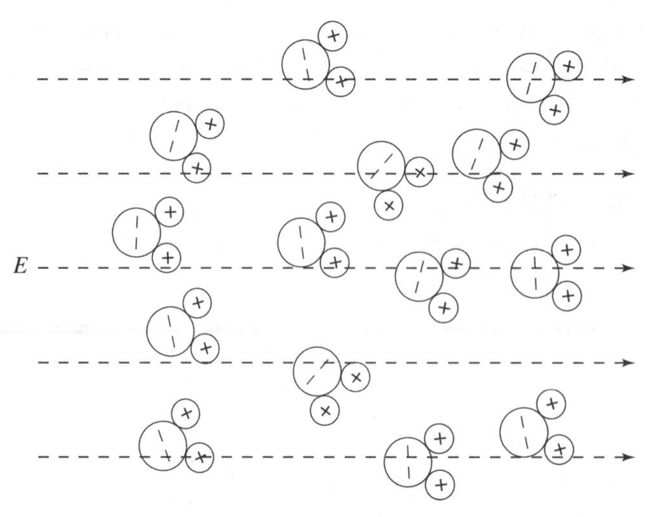

FIGURE 2.4 Polarization of water molecules in an electric field of magnitude E.

Thus we can express the polarization as

$$\mathbf{P} = (\varepsilon_r - 1)\varepsilon_0 \mathbf{E} \tag{2.20}$$

where the quantity $\chi = (\varepsilon_r - 1)$ is the electric susceptibility. Clearly $\mathbf{P} = 0$ if $\varepsilon_r = 1$ or if $\mathbf{E} = 0$. The Clausius–Mossotti equation

$$\frac{\varepsilon_r - 1}{\varepsilon_r + 2} = \frac{N\alpha}{3\varepsilon_0} \tag{2.21}$$

relates the relative permitivity to the polarizability α of a medium. The temperature dependence of the dielectric constant is accounted for by the Debye equation

$$\frac{\varepsilon_r - 1}{\varepsilon_r + 2} = \frac{N}{3\varepsilon_0}\left(\alpha + \frac{p^2}{3k_B T}\right) \tag{2.22}$$

where p is the dipole moment magnitude.

2.4 Casimir Interactions

- The Casimir force is similar to the Van der Waal force where, instead of fluctuating real dipole moments, the force is caused by virtual quantum fluctuations predicted by quantum electrodynamics.
- The presence of conducting boundaries, or capacitor plates, excludes vacuum modes with wavelengths longer than the separation between the conductors as illustrated in Figure 2.5.

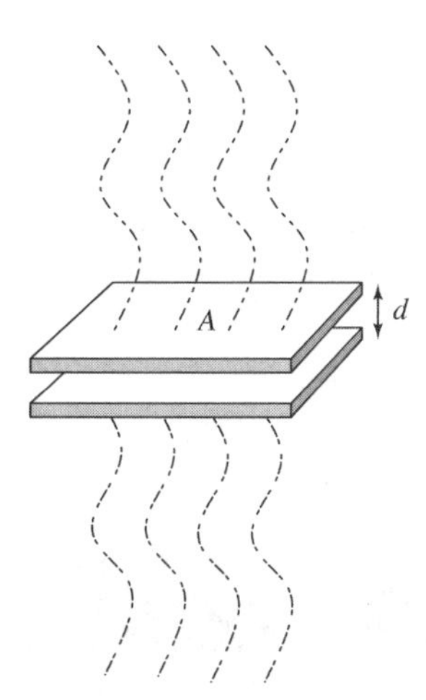

FIGURE 2.5 Two ideal conducting plates of area A separated by a distance d. Longer vacuum wavelengths are excluded from the central region between the plates, resulting in an attractive Casimir force that tends to push the plates together with a force that is inversely proportional to d^4.

- The exclusion of longer wavelengths results in a lower vacuum pressure between the plates than in external regions.
- The resulting pressure difference, or Casimir force, acts to push the conductors together.
- The Casimir force increases with decreasing separation and may exceed electrostatic interactions at very small distances.
- This tiny force has been measured experimentally in agreement with the predictions of quantum electrodynamics. The Casimir attraction between two parallel plates with area A and separation d is

$$\frac{F}{A} = -\frac{\hbar c \pi^2}{240 d^4}. \tag{2.23}$$

- Because of its $1/d^4$ dependence, this force has a very short range. For two plates of $A = 1\ \text{cm}^2$ separated by $d = 1.0\ \mu\text{m}$, the Casimir force is on the order of pN, roughly the weight of a red blood cell in the Earth's gravitational field.
- Casimir force can act between conductors and dielectrics and is highly dependent on the geometry of material boundaries.
- The Casimir force contributes to the formation of stacks of red blood cells known as rouleaux, and helps stabilize lipid bilayer tubes as discussed in Chapter 4.
- Short-range quantum electrodynamical effects may also play a role in protein folding.

2.5 Domains of Physics in Biology

Many life processes and characteristics fall under the major realms of physics, including statics, dynamics, fluid dynamics, thermodynamics, statistical mechanics, electrodynamics, and quantum mechanics (listed in Table 2.2). These are discussed in the following chapters.

Table 2.2 Domains of Physics in Biology

Domain of Physics	Life Processes/Characteristics	Governing Equations/ Physical Principle
Electrodynamics (Chapters 2, 6, 9, 10, 12)	Ionic bonds, membrane potential, action potentials, nerve signals	Coulomb's law, Laplace's equation, Poisson–Boltzmann equation, Hodgkin–Huxley equations
Thermodynamics (Chapters 3, 4, 5, 6)	Metabolism, heat transfer in living organisms, chemical reactions	Laws of thermodynamics, heat equation, Nernst–Plank equation
Statistical Mechanics (Chapters 4, 5, 6, 10)	Protein folding, diffusion, ion transport, membrane potential	Partition function, Fick's laws, Poisson–Boltzmann equation
Quantum Mechanics (Chapter 5, 6, 8)	Electronic structure, photosynthesis, light absorption in biomolecules, molecular electronic configurations, covalent bonds	Schrödinger equation
Dynamics (Chapters 6, 7)	Motility of organisms, transport of biomaterials, inertial effects	Newton's laws of motion
Fluid Dynamics (Chapter 7)	Breathing, circulatory systems, swimming and flight, Earth's atmosphere	Navier–Stokes equation, Bernoulli's equation, continuity equation
Statics (Chapter 11)	Elasticity and tensile strength of bones and tissue	Equilibrium conditions, Hooke's law, formulas for stress and strain
Chaos, Nonlinear Dynamics, and Fractals (Chapters 13, 14)	Population dynamics, biochemical oscillations, pattern formation, branching networks	Sensitive dependence on initial conditions, self-similarity, power law behavior

EXERCISES

Exercise 2.1 The Lennard–Jones potential between molecules is sometimes written without the factor of 2 multiplying the second term

$$U_{\mathrm{LJ}}(r) = \varepsilon\left\{\left(\frac{r_0}{r}\right)^{12} - \left(\frac{r_0}{r}\right)^{6}\right\}.$$

Compare the separation distance at which the two molecules will be in mutual equilibrium to the equilibrium separation obtained from L–J potential in Equation 2.1 with the factor of 2. At what distance is the force maximal in this model?

Exercise 2.2 Calculate the interaction energy between pairs of dipoles with configurations illustrated in Figure 2.6(a)–(e).

(a) **p** ↑ ↑ **p**

(b) **p** ← ← **p**

(c) **p** ↓ ← **p**

(d) **p** → ← **p**

(e) **p** ← → **p**

← r

FIGURE 2.6 Pairs of electric dipoles with various orientations.

Exercise 2.3 The Casimir energy between two uncharged plates with cross-sectional area A separated by a distance d is

$$U_{\mathrm{C}}(d) = -\frac{\hbar c \pi^2 A}{720 d^3}.$$

Calculate the value of d where the energy required separating the plates is comparable to the thermal energy $k_B T$ at room temperature.

Exercise 2.4 QuickField Electrostatics Simulation. Calculate the force between an electric dipole (modeled as two charges separated by a distance d) and the cylindrical dielectric membrane shown in Figure 2.7. Take the relative dielectric constant of the membrane $\varepsilon_r = 10$ with air $\varepsilon_r = 1$ outside the membrane. Compare the value of the force repeating the calculation with water $\varepsilon_r = 78$ outside the membrane. Perform the calculation using axial symmetry.

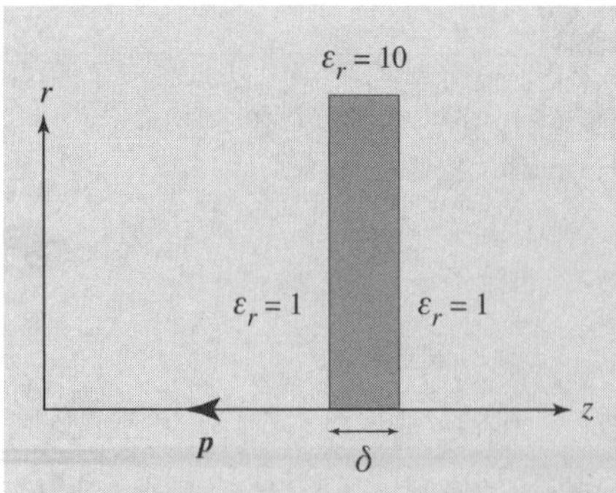

FIGURE 2.7 Electric dipole outside a cylindrical dielectric segment.

Exercise 2.5 QuickField Electrostatics Simulation. Calculate the dielectrophoretic force between an electrified needle and the dielectric sphere shown in Figure 2.8. Take the relative dielectric constant of the sphere to be $\varepsilon_r = 10$ surrounded by water $\varepsilon_r = 78$. Construct the model using axial symmetry.

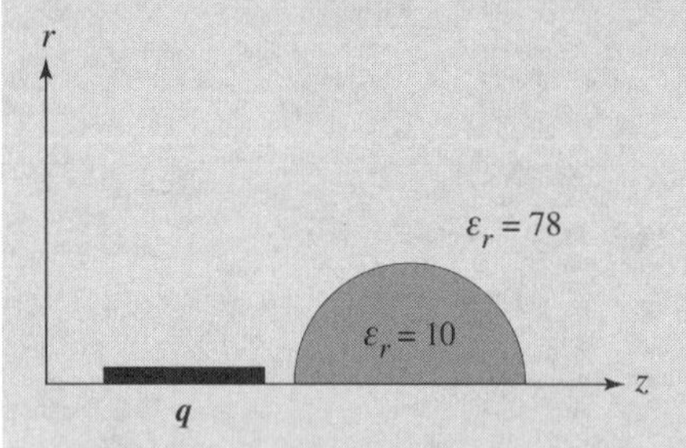

FIGURE 2.8 Charged needle outside a dielectric sphere.

Exercise 2.6 QuickField Electrostatics Simulation. Plot the equipotentials surrounding an electric dipole within a spherical dielectric membrane as shown in Figure 2.9. Take the relative dielectric constant of the membrane $\varepsilon_r = 10$ with water $\varepsilon_r = 78$ inside and outside the membrane. Construct the model using axial symmetry. Calculate the electric field energy inside the membrane.

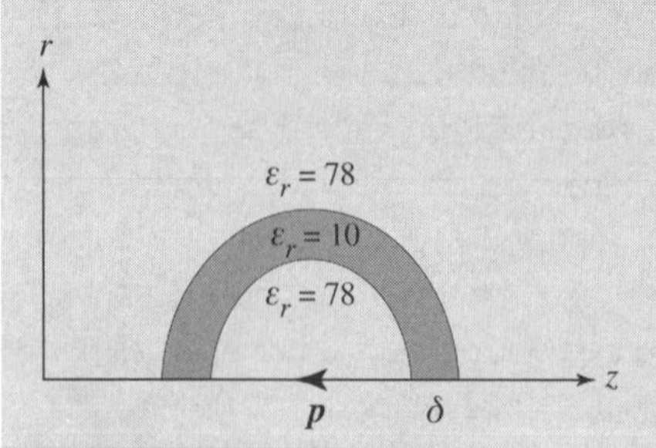

FIGURE 2.9 Electric dipole inside a dielectric sphere.

3 Heat Transfer in Biomaterials

3.1 Heat Transfer Mechanisms

Heat transport can occur between a living organism and its environment by thermal conduction, convection, and radiation. Figure 3.1 illustrates these processes at the skin surface of an organism.

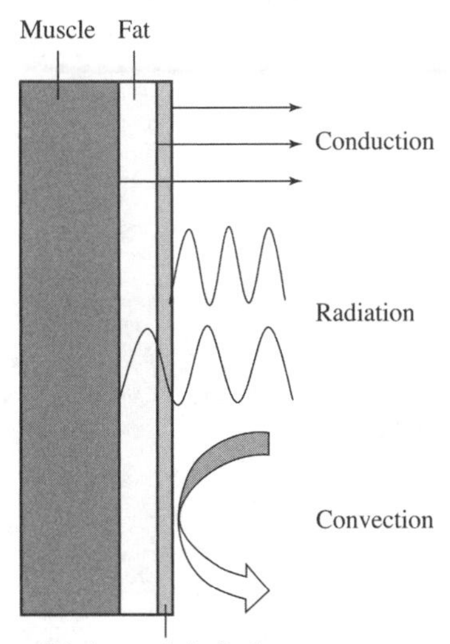

FIGURE 3.1 Heat transfer mechanisms between tissue and the environment. (Adapted from Glaser R. 2001. *Biophysics* Springer-Verlag Heidelberg, Germany.)

3.1.1 Conduction

Temperature gradients give rise to thermal conduction. The thermal conductivity of tissue can be regulated by the circulatory system. When the external temperature drops, capillaries can contract to slow heat loss and maintain the temperature of the body's inner core. This is because tissue with lower blood content has lower thermal conductivity.

3.1.2 Convection

Convection enables energy transfer by mass transport of fluids, including gases and liquids. This can occur via blood flow through circulatory systems and air circulation in the lungs. The circulatory system carries warm blood to the surface of the skin. Newton's law of cooling gives the time-dependent temperature of a body in contact with an environmental reservoir with temperature T_{env}

$$\frac{dT}{dt} = \beta(T_{env} - T) \tag{3.1}$$

where β is the convection coefficient. The following MATLAB code solves the differential Equation 3.1 and plots the time-dependent temperature.

```
% the parameter beta determines the rate of cooling

beta=1e-2;
dt=0.1;                              %time step
Tenv=33;

nmax = 1000;

T=zeros(nmax,1);
t=zeros(nmax,1);

t(1)=0;

T(1)=100; % specify initial temperature

for n=1:nmax-1

    T(n+1)=T(n)+beta*(Tenv-T(n))*dt;
    t(n+1)=t(n)+dt;

end

% create a plot of the time dependent temperature

plot(t,T)
xlabel('time (s)')
ylabel('temperature (C)')
```

Cytoplasmic Streaming

Convective transport of cytoplasm inside of cells results from the motion of membrane-bound organelles and actin filaments. Cytoplasmic streaming assists in the transport of

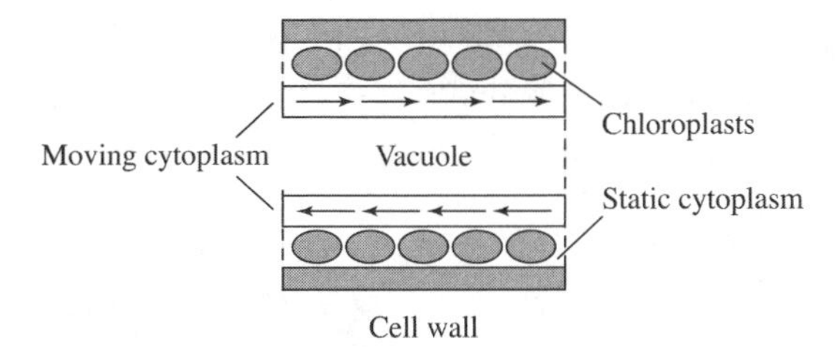

FIGURE 3.2 Schematic of cytoplasmic streaming inside a cylindrical cell such as in the green algae *Nitella.* (Adapted from Alberts, B. 1989. *Molecular Biology of the Cell*, Vol. 1, Courier Corporation, North Chelmford, Massachusetts.)

nutrients to different parts of the cell. For example, diffusion is insufficient to transport nutrients in the long green algae *Chara* and *Nitella*, where the streaming velocity of cytoplasm is as high as 75 μm/sec. A schematic of cytoplasmic streaming is shown in Figure 3.2.

3.1.3 Radiation

Radiation is the third mechanism by which living organisms exchange thermal energy with their environments. For an ideal emitter or absorber with surface area A, the power P per unit area emitted or absorbed is proportional to the fourth power of the temperature

$$\frac{P}{A} = \sigma T^4 \tag{3.2}$$

where T is in Kelvin and σ is the Stephan–Boltzmann constant

$$\sigma = \frac{2\pi^5 k_B{}^4}{15h^3c^2} = 5.67 \cdot 10^{-8}\frac{\mathrm{W}}{\mathrm{m^2K}}. \tag{3.3}$$

Such a body with ideal emission and absorption characteristics is known as a black body. The total power radiated by a nonideal body is

$$P = \sigma \varepsilon A T^4 \tag{3.4}$$

where the emissivity ε is a dimensionless number equal to one for a perfect emitter. The emissivity of skin is $\varepsilon \approx 0.98$, with most biological surfaces $0.9 < \varepsilon < 1$. A body at temperature T_{body} in an environment with temperature T_{env} will emit or absorb energy at a rate of

$$P = \sigma \varepsilon A\left(T_{\text{body}}^4 - T_{\text{env}}^4\right). \tag{3.5}$$

Note that the electromagnetic radiation emitted is a function of wavelength λ. No power is radiated or absorbed if a body is at the same temperature as the environment. Wien's law gives a relation between the temperature of a body and the wavelength λ_{max} that it emits the most power

$$\lambda_{max}T = 2.898 \cdot 10^{-3}\text{m} \cdot \text{K}. \tag{3.6}$$

All living things give off black body radiation with a peak wavelength inversely proportional to the temperature. The human body temperature is 37°C, or 310 K, corresponding to a peak wavelength of $\lambda_{max} = 9.35\ \mu\text{m}$, which is in the mid-infrared range.

Stars are nearly perfect black body emitters. The temperature of the Sun at the photosphere is ~5600 K, corresponding to a peak wavelength of about 500 nm. This is comparable to some of the characteristic absorbing frequencies of different biomolecules in Chapter 8. Sunlight supplies energy for all life on Earth (except possibly near deep ocean hydrothermal vents) with an intensity $I_{sun} = 1.37\ \text{kW/m}^2$ at the surface of the Earth. The Earth can be considered as a nonisolated system in steady state with a constant average temperature. The average power absorbed by the Sun must be equal to the average power radiated into space or

$$\underbrace{I_{sun}\varepsilon_{\oplus}\left(\pi R_{\oplus}^2\right)}_{\text{power absorbed}} = \underbrace{\left(4\pi R_{\oplus}^2\right)\varepsilon_{\oplus}\sigma T_{\oplus}^{\ 4}}_{\text{power radiated}} \tag{3.7}$$

where $\varepsilon_{\oplus}$ is the emissivity, $R_{\oplus}$ the radius, and $T_{\oplus}$ is the average temperature of the Earth. The surface area of the Earth is $4\pi R_{\oplus}^{\ 2}$, and the cross-sectional area of incident solar radiation is $\pi R_{\oplus}^{\ 2}$. Solving Equation 3.7 for $T_{\oplus}$ gives

$$T_{\oplus} = \left(\frac{I_{sun}}{4\sigma}\right)^{1/4} \approx 279\ \text{K}. \tag{3.8}$$

Note that this result is independent of the emissivity and radius of the Earth. This calculation does not account for the ~30% fraction of solar radiation reflected from the cloud tops that does not reach the Earth's surface. It also neglects the warming effects of greenhouse gases in the atmosphere.

3.2 The Heat Equation

3.2.1 Transient Heat Flow

Thermal conduction acts to equalize temperature differences between regions of higher and lower temperatures. The rate of thermal energy Q transferred between two reservoirs at temperatures T_1 and T_2 separated by an insulator of thickness Δx is given by

$$\frac{\Delta Q}{\Delta t} = \lambda A \frac{T_2 - T_1}{\Delta x} \tag{3.9}$$

where A is the cross-sectional area and λ is the thermal conductivity of the insulating barrier. The rate $\Delta Q/\Delta t$ is positive in the lower temperature reservoir and vice versa as energy is transferred from high to low temperature. Equation 3.9 may be written as a partial differential equation describing local heat flow in a material body with a one-dimensional temperature gradient

$$\frac{\partial Q}{\partial t} = \lambda A \frac{\partial T}{\partial x}. \tag{3.10}$$

Note that energy is transferred opposite the direction of the temperature gradient. We can write Equation 3.10 as a partial differential equation over a single scalar field $T(x, t)$ as derived in Appendix 3

$$\frac{\partial T}{\partial t} = \alpha \nabla^2 T \tag{3.11}$$

where $\alpha = \lambda/c\rho$, ρ is the mass density in kg/m^3, and c is the specific heat in J/kg·K. Equation 3.11 is a diffusion equation of the same form as Fick's second law encountered in Chapter 6, replacing T by the concentration C and $1/\alpha$ by the diffusion constant D.

The following MATLAB code solves the one-dimensional heat Equation 3.11 and creates a mesh plot of $T(x, t)$ for a body initially at 0°C immersed in a reservoir at 100°C.

```
r=0.2; % parameter r=dt*alpha/dx^2

imax=50;  % maximum number of spatial steps

nmax=1000; % maximum number of spatial steps

T_reservoir=100.0;

T=zeros(nmax,imax);

% specify boundary temperatures switched on at t=0;

T(:,1)=T_reservoir;
T(:,imax)=T_reservoir;

for n=1:nmax-1

      for i=2:imax-1

            T(n+1,i)=T(n,i)+r*(T(n,i+1)-2*T(n,i)+T(n,i-1));
      end
```

```
end

% create a mesh plot of the time dependent temperature

mesh(T)

xlabel('position (mm)')
ylabel('time (ms)')
zlabel('temperature (K)')
title('spatial-temporal temperature distribution')
```

3.2.2 Steady State Heat Flow

In steady state $\partial T/\partial t = 0$, so temperature distribution is described by Laplace's equation

$$\nabla^2 T = 0 \tag{3.12}$$

in regions without heat source and constant thermal conductivity. In regions with heat source q, the temperature satisfies Poisson's equation

$$\lambda \nabla^2 T = -q \tag{3.13}$$

and in regions with both heat source and inhomogeneous and/or anisotropic λ

$$\nabla \cdot (\lambda \nabla T) = -q. \tag{3.14}$$

Note that λ and q may also depend on temperature, resulting in a nonlinear heat flow equation. In general, a heat source will give rise to a heat flux with divergence

$$\nabla \cdot \mathbf{F} = q \tag{3.15}$$

where the flux vector $\mathbf{F} = -\nabla T$. Both the heat flux and temperature must be continuous at the interface between regions 1 and 2 with differing thermal conductivities so that

$$\left\{ \begin{aligned} T_1 &= T_2 \\ \lambda_1 \frac{\partial T_1}{\partial n} &= \lambda_2 \frac{\partial T_2}{\partial n} \end{aligned} \right\} \tag{3.16}$$

in the absence of heat source on the boundary where n is perpendicular. The analysis of steady state heat flow is similar to electrostatics with the potential U_e and electric field $\mathbf{E}$ analogous to T and $\mathbf{F}$, respectively.

Heat Transfer through a Living Cell

As an example we calculate the conductive rate of heat flow through the plasma membrane of a living cell. The radial heat equation is given by Equation 3.10 with $A = 4\pi r^2$ for a spherical cell

$$\frac{\partial Q}{\partial t} = \lambda 4\pi r^2 \frac{\partial T}{\partial r}. \tag{3.17}$$

Dividing by r^2 and integrating gives

$$\frac{\partial Q}{\partial t}\int_{r_a}^{r_b} \frac{1}{r^2}\,dr = \lambda 4\pi \int_{r_a}^{r_b} \frac{\partial T}{\partial r}\,dr \tag{3.18}$$

whereupon simplification gives

$$\frac{\partial Q}{\partial t}\left(\frac{1}{r_a} - \frac{1}{r_b}\right) = \lambda 4\pi (T_b - T_a). \tag{3.19}$$

The heat flow rate is then

$$\frac{\partial Q}{\partial t} = 4\pi\lambda \frac{(T_b - T_a)}{\left(\frac{1}{r_a} - \frac{1}{r_b}\right)} = 4\pi\lambda (T_b - T_a)\frac{r_a r_b}{r_b - r_a}. \tag{3.20}$$

Since $r_b = r_a + \delta$ with $r_b \approx 1\ \mu\text{m}$ and $\delta \approx 10$ nm, the quantity $r_a r_b \approx r_b^2$, so that

$$\frac{\partial Q}{\partial t} = 4\pi\lambda \Delta T \frac{r_b^2}{\delta} \tag{3.21}$$

where $\Delta T = T_b - T_a$. Using the thermal conductivity of lipid $\lambda \approx 0.17$ gives $\partial Q/\partial t \approx 2.1 \cdot 10^{-4}$ W with $\Delta T = 1$ K.

3.3 Joule Heating of Tissue

Hyperthermia treatment makes use of high-frequency electromagnetic fields to inhibit cancer by locally elevating tumor temperature. The technique is often used in conjunction with radiation and chemotherapy. The specific absorption rate (SAR) gives a measure of energy absorption in tissue

$$\text{SAR} = \frac{1}{\rho_m}\sigma E_{\text{rms}}^2 \tag{3.22}$$

where ρ_m is the mass density in kg/m^3, σ is conductivity in S/m, and E_{rms} is root mean square electric field in V/m. Figure 3.3 shows the Joule heating in a region of tissue surrounding a tumor under alternating electrical excitation. The SAR may be calculated by dividing the time-averaged Joule heat by the mass density ρ_m.

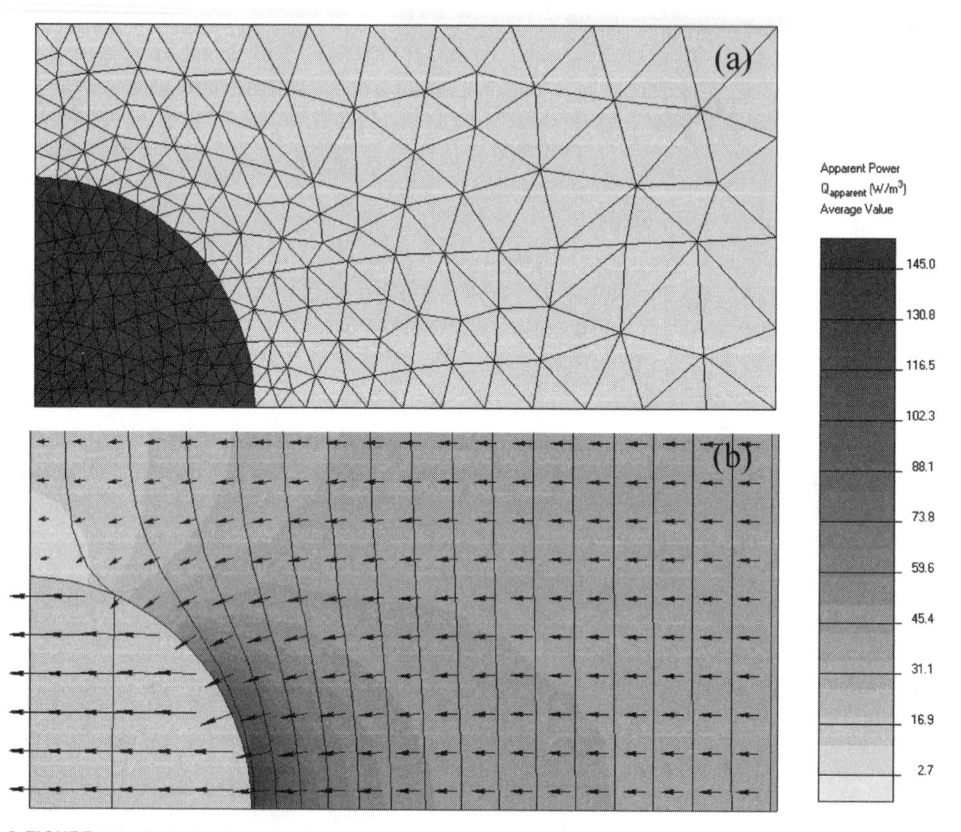

FIGURE 3.3 Joule heating of a tissue region containing a tumor.

It turns out that conduction currents provide the main contribution to the absorption. Figure 3.4 shows the temperature and heat-flux distribution resulting from the Joule heating calculated in Figure 3.3. Convection and radiation boundary conditions are applied to the body surface with convection coefficient $\beta = 8.6$ W/K $\cdot$ m^2 and emissivity $\varepsilon = 0.97$.

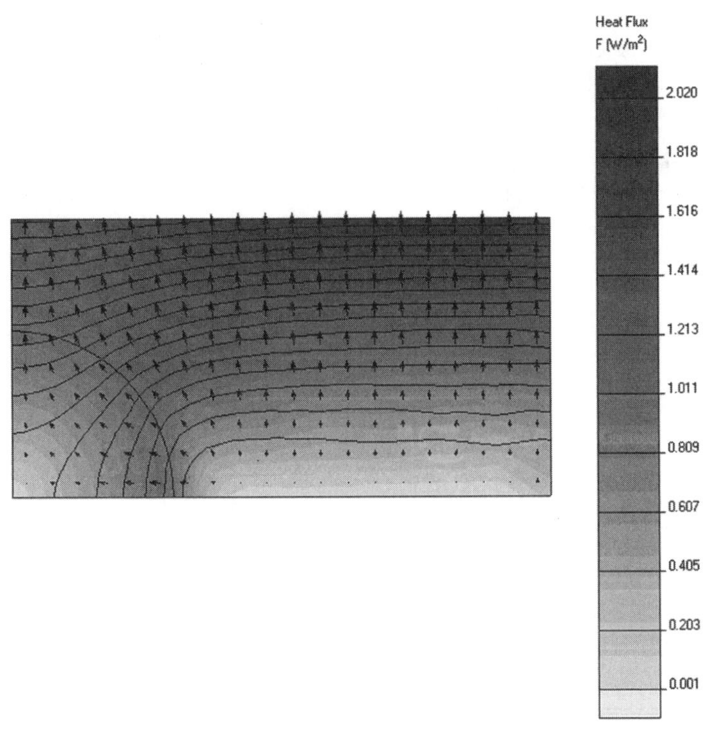

FIGURE 3.4 Temperature and heat-flow vectors resulting from Joule heating of the tissue region containing a tumor modeled in Figure 3.3. This calculation is performed using QuickField finite element software importing the Joule heat from the AC Magnetic module into the Thermal Analysis module of the software.

EXERCISES

Exercise 3. 1 Calculate the heat loss per unit length through a cylindrical membrane subject to a temperature gradient ΔT between its inner radius r_a and outer radius r_b.

Exercise 3.2 Use Newton's law of cooling to plot the temperature $T(t)$ of a body with initial temperature $T(t = 0) = T_0$. Plot $T(t)$ for $T_0 > T_{\text{env}}$ and $T_0 < T_{\text{env}}$.

Exercise 3.3 QuickField Coupled AC Magnetic and Thermal Analysis. Model the temperature distribution resulting from Joule heating in the human brain containing a tumor. Construct the model using axial symmetry. Provide the AC excitation using a single coil outside the brain surrounded by the cranium and tissue regions.

Exercise 3.4 QuickField Thermal Analysis. Model the heat flux through the human head. Take the outer surface of the brain to be at 310 K with the environmental temperature equal to 273 K. Specify the convection coefficient $\beta = 8.6$ W/K $\cdot$ m^2 and emissivity $\varepsilon = 0.97$ at the outer surface of the head. Compare heat flux for different convection coefficients simulating a top hat. Construct the model using axial symmetry with the radius of the brain 15 cm, thickness of skull 1.5 cm, and thickness of scalp 0.7 cm. Note that it is not necessary to mesh regions inside the brain because the outer temperature of the brain is taken to be constant.

4 Living State Thermodynamics

4.1 Thermodynamic Equilibrium

Thermodynamics describes the interaction of systems with their environments. The system in question consists of a subset of the universe, such as the living cell depicted in Figure 4.1. Living systems are not usually in equilibrium with their surroundings. Because of metabolism, the temperature of an organism can be higher than its surroundings.

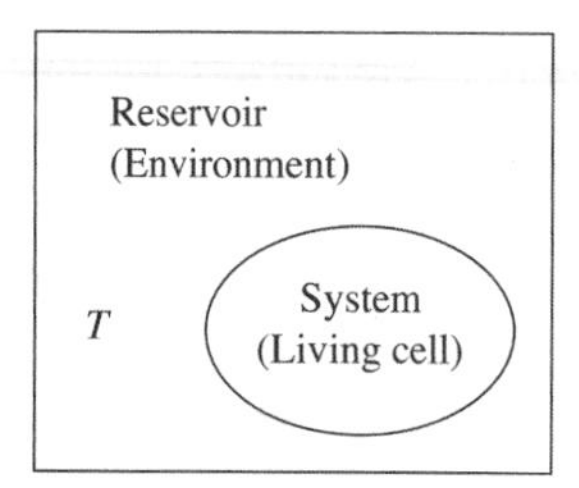

FIGURE 4.1 Thermodynamic system (living cell) in contact with a reservoir (the environment) at temperature *T*.

Equilibrium systems are described by state variables that can be extensive or intensive. Extensive state variables such as the volume V, entropy S, and energy E depend on the system size. Intensive variables such as temperature T, pressure P, and chemical potential μ are independent of the system size. The 0th law of thermodynamics states that two bodies in thermodynamic equilibrium will be at the same temperature. A given living body may not be in equilibrium with its surroundings but may be considered as a nonequilibrium system in steady state over sufficiently short periods of time. Over longer periods of time, living systems will gain or loose mass depending on whether their energy input is greater or lesser than their energy output. We often apply equilibrium thermodynamics to many biophysical problems such as protein folding and processes that occur on time scales much smaller than the life span of an organism or cell.

4.2 First Law of Thermodynamics and Conservation of Energy

The principle of conservation of energy is the foundation of the first law of thermodynamics. The change in total energy of a thermodynamic system dE is equal to the energy transferred across the system boundary. If we just consider the energy transfer mechanisms of heat flow and work, the differential form of the first law is expressed as

$$dE = dQ + dW \tag{4.1}$$

where dQ is the heat added or lost, and dW is the work done on or by the system. The sign convention used here is that if positive work is done on the system, then $dW > 0$ and E increases. Note that dQ and dW are not perfect differentials and are not referred to as "change in work" or "change in heat added" to the system. Some textbooks emphasize this by using a slash notation $dE = đQ + đW$ or by simply writing $dE = Q + W$.

4.3 Entropy and the Second Law of Thermodynamics

The conservation of energy as expressed by the first law of thermodynamics is not the only physical constraint determining whether a thermodynamic process can occur. Many processes that do not violate conservation of energy are not observed. For example, an object will not levitate off the floor by spontaneously converting its thermal energy into gravitational potential energy. The reason that this process is never observed is because the entropy S_f of the final configuration (when the block is temporarily airborne) would have to be less than the initial state entropy S_i. This would violate the second law of thermodynamics. Loosely speaking, entropy is a measure of the disorder of a system. According to the second law of thermodynamics, the entropy of an isolated system does not decrease. In its most common mathematical form, the second law is expressed as $\Delta S \geqslant 0$ where the change in entropy of an isolated system is positive or zero.

4.3.1 Does Life Violate the Second Law?

The second law of thermodynamics states that, for an isolated system, entropy can only increase or stay the same. Spontaneous processes occur only in the direction of increasing entropy. Living organisms are, however, highly structured systems consisting of a myriad of well-ordered molecular arrangements forming genetic material, cells, organelles, and networks of tissues and nerve fibers. One might wonder whether life is an exception to the second law. It turns out that the growth and structuring of an organism is accompanied by an increase in environmental entropy. In his 1944 essays *What Is Life?*, Erwin Schrödinger proposed a framework of thinking about the positive flow of entropy out of living systems as equivalent to an inward flow of negative entropy (negentropy). This would be analogous to how we conceptualize the flow of negative charges in an electrical circuit as an equivalent positive current flowing in the opposite direction.

Plants and animals, as well as their cellular constituents, are open systems that reduce their entropy by increasing the disorder of the environment. These systems require energy to create ordered structures while increasing the entropy of their surroundings. This energy can be in the form of chemical bond energy obtained from other organisms at various positions on the food chain, or directly from the Sun via photosynthesis.

Besides interactions between species, the tendency of entropy to increase may be one of the driving forces of evolution. Genetic codes stored in molecules of DNA represent states of low entropy. Mutations occur due to small random fluctuations that are irreversibly propagated by successive generations in accordance with the second law of thermodynamics.

4.3.2 Measures of Entropy

Entropy is not a quantity that can be directly measured in the laboratory such as temperature. There are no entropy meters in a catalogue of scientific instruments. However, the entropy can be determined from other thermodynamic quantities. In 1864 Rudolf Clausius defined the infinitesimal change in entropy of a system transitioning between two equilibrium states at constant temperature as

$$dS = \frac{dQ}{T}. \tag{4.2}$$

The change in entropy as a result of the reversible transition between two macroscopic states A and B is then

$$\int_A^B dS = \int_A^B \frac{dQ}{T}. \tag{4.3}$$

Thus the same amount of heat dQ causes a greater increase in S if added at lower temperatures. If an isothermal process occurs where energy dQ is transferred to the system whose sole effect is to increase the disorder dQ/T, then the energy input is not recoverable as useful work.

4.3.3 Free Expansion of a Gas

As an application of Equation 4.3, we consider the change in entropy of an ideal gas undergoing a free expansion from a volume V_i to a final volume V_f as depicted in Figure 4.2. We treat the process as isothermal so that there is no change in internal energy of the system

$$dE = 0 \rightarrow dQ = -dW \tag{4.4}$$

while work done on the system is $dW = -PdV$ so that $dQ = PdV$. The infinitesimal entropy change is

$$dS = \frac{dQ}{T} = \frac{PdV}{T}. \tag{4.5}$$

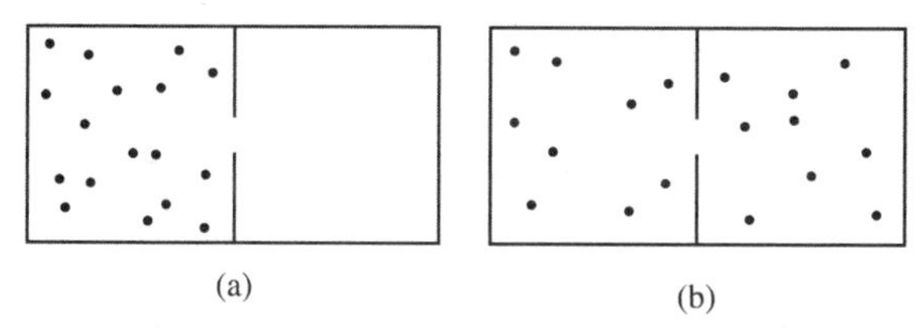

FIGURE 4.2 Free expansion of gas. (a) All of the gas molecules are located on the left-hand side of a thermodynamic system consisting of a box with a perforated central partition. (b) Gas molecules have leaked through the opening dividing the box thereby increasing the system's entropy.

Substituting $P = Nk_BT/V$ from the ideal gas law gives

$$dS = Nk_B \frac{dV}{V}. \tag{4.6}$$

Integrating both sides of this equation

$$\int_{S_i}^{S_f} dS = Nk_B \int_{V_i}^{V_f} \frac{dV}{V}, \tag{4.7}$$

we find that the change in entropy is proportional to the log of the final to initial volume fraction

$$S_f - S_i = Nk_B \ln\left(\frac{V_f}{V_i}\right). \tag{4.8}$$

Notice that entropy has the units of Boltzmann's constant k_B, or J/K.

4.4 Physics of Many Particle Systems

Boltzmann provided a statistical interpretation of entropy in terms of the number of ways that a system could be organized. The Boltzmann constant relates the entropy to the number of microstates, or possible arrangements of a system, denoted by Ω

$$S = k_B \ln \Omega. \tag{4.9}$$

For N particles that can occupy m states with occupation numbers $\{n_1, n_2, \ldots, n_m\}$ where

$$N = \sum_{i=1}^{m} n_i, \tag{4.10}$$

the total number of possible arrangements is given by the ratio

$$\Omega = \frac{N!}{n_1! n_2! \cdots n_m!}. \tag{4.11}$$

Consider an array of dipoles where each dipole can assume an up or a down orientation

↑↓↑↑↑↑↓↓↓↑↓↓

where $N\uparrow$ and $N\downarrow$ are the respective number of dipoles pointing up or down. The number of possible arrangements with $N = N\uparrow + N\downarrow$ is

$$\Omega = \frac{N!}{N\downarrow! N\uparrow!}. \tag{4.12}$$

For example, the number of possible ways to arrange $6\uparrow$ and $6\downarrow$ dipoles with $N = 12$ is

$$\Omega = \frac{12!}{6!6!} = 624, \tag{4.13}$$

corresponding to $S = 6.44\ k_B$, which is the maximum entropy of the 12-dipole array. There is only one way in which all the dipoles can point upward,

$$\Omega = \frac{12!}{12!0!} = 1 \tag{4.14}$$

where $S = k_B \ln 1 = 0$ is the minimum entropy of the array.

4.4.1 How Boltzmann Factors in Biology

In thermodynamic equilibrium, the Boltzmann factor

$$\exp\left(-\frac{E_i}{k_B T}\right) \tag{4.15}$$

is proportional to the probability of finding a system such as an atom or molecule with energy E_i among a spectrum of possible energy values at temperature T. This factor is central to statistical mechanics and threads through many aspects of living state physics. Table 4.1 lists a few appearances of the Boltzmann factor in biology.

Table 4.1 Boltzmann Factors in Biology

Topic	Formula	Application
Protein Folding (Chapter 4)	$P_c \propto \exp\left(-\frac{E_c}{k_B T}\right)$	Boltzmann factor is proportional to the probability that a protein will assume a conformation with energy E_c.
Reaction Rate Constant (Chapter 5)	$K_{eq} = A\exp\left(-\frac{E_a}{k_B T}\right)$	Slope of the Arrhenius plot determines the activation energy E_a.
Maxwell–Boltzmann Velocity Distribution (Chapter 6)	$n(v) \propto \exp\left(-\frac{mv^2}{2k_B T}\right)$	Boltzmann factor is proportional to the velocity distribution function.
Diffusion Constant (Chapter 6)	$D = D_0\exp\left(-\frac{E_a}{k_B T}\right)$	Describes diffusion across an energy barrier E_a.
Sedimentation of Cell Cultures (Chapter 6)	$C(z) = C_0\exp\left(-\frac{m^* gz}{k_B T}\right)$	Gives the vertical concentration of a colloidal cell suspension in a gravitational field g.
Centrifugation of Cell Cultures (Chapter 6)	$C(r) = C_0\exp\left(\frac{mr^2\omega^2}{2k_B T}\right)$	Radial concentration of a colloidal cell suspension with centripetal acceleration $r\omega^2$.
Concentration of Ions in an Electric Potential (Chapter 9)	$C_i = C_{i0}\exp\left(-\frac{q_i U_e}{k_B T}\right)$	Concentration of ions and electric potential described by the Poisson–Boltzmann equation.
Membrane Potential (Chapter 9)	$\frac{C_i}{C_0} = \exp\left(-\frac{q\Delta U_e}{k_B T}\right)$	Membrane potential proportional to the log of the concentration ratio inside and outside the cell membrane.

4.4.2 Canonical Partition Function

The partition function Z is simply the sum of all Boltzmann factors of a physical system. Once Z is obtained, all other thermodynamic quantities can be derived as depicted in Figure 4.3.

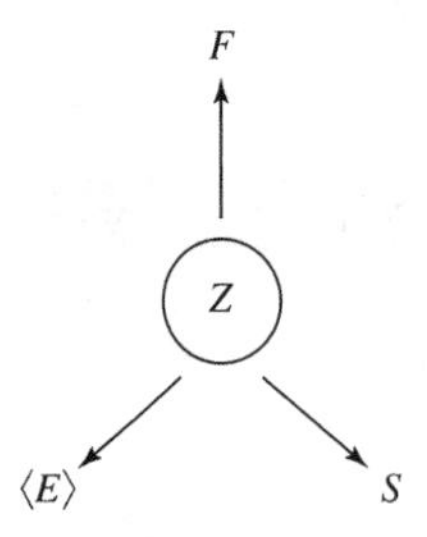

FIGURE 4.3 Illustration of thermodynamic quantities derivable from the partition-function energy, including the Helmholtz free energy, entropy, and average energy. The heat capacity is also derivable from the average energy.

Almost all calculations in statistical mechanics begin with evaluating this factor. The partition function of a statistical ensemble of N particles at temperature T with possible energies E_i is

$$Z = \sum_{i}^{N} \exp\left(-\frac{E_i}{k_B T}\right). \tag{4.16}$$

The partition function is a dimensionless number analogous to the number of microstates Ω. The probability of being in a state with energy E_i is given by the Boltzmann factor divided by Z, or

$$P_i = \frac{\exp\left(-\frac{E_i}{k_B T}\right)}{Z}. \tag{4.17}$$

Requiring that the sum of probabilities is equal to one, $\sum_i P_i = 1$, we may verify

$$\frac{1}{Z}\sum_{i=1}^{N} \exp\left(-\frac{E_i}{k_B T}\right) = \frac{Z}{Z} = 1. \tag{4.18}$$

In practice, it may be difficult to calculate the sum in Equation 4.16 for a given system. However, once this has been accomplished, other thermodynamic quantities can be obtained in a straightforward manner.

4.4.3 Average Energy

The average energy of N particles with possible energies E_i and corresponding probabilities P_i is

$$\langle E \rangle = \sum_{i=1}^{N} E_i P_i. \tag{4.19}$$

As an illustration consider a two-state system with $E_1 = 3$ aJ and $E_2 = 1$ aJ, with $P_1 = 0.25$ and $P_2 = 0.75$. The average energy is obtained from Equation 4.19 as

$$\langle E \rangle = 0.25 \cdot 3 \text{ aJ} + 0.75 \cdot 1 \text{ aJ} = 1.5 \text{ aJ}. \tag{4.20}$$

A typical covalent bond has an energy of 0.6 aJ $= 0.6 \cdot 10^{-18}$ J. A common mistake is to calculate the average energy by summing the energies of each state and dividing by the total number of states. Note that $\langle E \rangle \neq (E_1 + E_2)/2$ unless $P_1 = P_2 = 1/2$.

We can obtain an expression for the average energy $\langle E \rangle$ of a system in thermodynamic equilibrium in terms of the canonical partition function Z. To simplify our calculations, we make use of the common substitution $\beta = 1/k_B T$. The two expressions are often used interchangeably to facilitate derivative operations such as $\partial/\partial\beta$ or to show temperature dependence explicitly. The average energy is thus

$$\langle E \rangle = \sum_{i=1}^{N} E_i \underbrace{\frac{\exp(-\beta E_i)}{Z}}_{P_i} = \frac{1}{Z} \sum_{i=1}^{N} E_i \exp(-\beta E_i). \tag{4.21}$$

The last sum can be expressed as $-\partial Z/\partial\beta$ or

$$\langle E \rangle = \frac{1}{Z}\left(-\frac{\partial}{\partial \beta}\right) \sum_{i=1}^{N} \exp(-\beta E_i) = -\frac{1}{Z}\frac{\partial Z}{\partial \beta} \tag{4.22}$$

so that

$$\langle E \rangle = -\frac{\partial \ln Z}{\partial \beta}. \tag{4.23}$$

Because $\partial/\partial\beta = -k_B T^2 \partial/\partial T$, we can also write the average energy as

$$\langle E \rangle = k_B T^2 \frac{\partial \ln Z}{\partial T}. \tag{4.24}$$

The total energy is then the number of particles times the average energy

$$E = N\langle E \rangle. \tag{4.25}$$

4.4.4 Entropy and Free Energy

Now that we can calculate the total energy of a system provided we have the partition function, we would like to know how much of this energy can be used to drive processes essential to life. The key quantity here is called the Helmholtz free energy F defined by

$$F = E - TS \tag{4.26}$$

or, in terms of the average energy,

$$F = N\langle E\rangle - TS. \tag{4.27}$$

The free energy is the amount of energy available to do useful work. This quantity clearly decreases as the entropy S and temperature T increase. If the temperature dependence of the free energy is known, we can obtain the entropy from Equation 4.26 by differentiation with respect to temperature

$$S = -\left(\frac{\partial F}{\partial T}\right)_{N,V}, \tag{4.28}$$

provided that the volume and number of particles are fixed as denoted by the subscripts N and V. It turns out that the free energy of a system in thermodynamic equilibrium with fixed N and V can be expressed in terms of the canonical partition function Z

$$F = -Nk_BT \ln Z. \tag{4.29}$$

The entropy S can then be obtained by differentiating this expression with respect to T:

$$S = Nk_B(\ln Z + \beta\langle E\rangle) \tag{4.30}$$

where we have made use of Equation 4.24. Also, the entropy is zero at $T = 0$ K, which is a statement of the third law of thermodynamics.

4.4.5 Heat Capacity

The temperature of a system will increase as energy is added. The heat capacity C is numerically equivalent to the energy required to increase the system temperature by 1 K. Written as

$$C = \frac{\partial E}{\partial T}, \tag{4.31}$$

we see that the units of C are J/K. An equivalent formula is obtained by differentiating the average energy with respect to β

$$C = -\frac{N}{k_BT^2}\frac{\partial}{\partial\beta}\langle E\rangle. \tag{4.32}$$

The heat capacity is related to the energy fluctuations of a system in thermodynamic equilibrium

$$C = \frac{\delta E^2}{k_B T^2} \tag{4.33}$$

where the variance in the energy is defined as

$$\delta E^2 = \langle E^2 \rangle - \langle E \rangle^2. \tag{4.34}$$

4.5 Two-State Systems

Two-state systems are characterized by configurations with two distinct energy levels. Examples include enzyme- and ion-channel kinetics. Suppose a system has only two allowed states Σ_1 and Σ_2 with respective energies E_1 and E_2. The probability ratio of being in each state is given by the ratio of Boltzmann factors

$$\frac{P_2}{P_1} = \frac{\exp\left(-\frac{E_2}{k_B T}\right)}{\exp\left(-\frac{E_1}{k_B T}\right)} = \exp\left(-\frac{\Delta E}{k_B T}\right) \tag{4.35}$$

where $\Delta E = E_2 - E_1$. Conservation of probability gives $P_2 + P_1 = 1$ so that

$$P_1 = \frac{1}{1 + \exp\left(-\frac{\Delta E}{k_B T}\right)} \quad \text{and} \quad P_2 = \frac{1}{1 + \exp\left(\frac{\Delta E}{k_B T}\right)}. \tag{4.36}$$

At low temperatures, $\Delta E \gg k_B T$ and all of the particles are in the low energy state so that $P_1 = 1$ and $P_2 = 0$. At high temperatures, $\Delta E \ll k_B T$ where it is equally likely to be in either state $P_1 = P_2 = 1/2$.

The average energy is given by

$$\langle E \rangle = \sum_i E_i P_i = \frac{E_1}{1 + \exp\left(\frac{-\Delta E}{k_B T}\right)} + \frac{E_2}{1 + \exp\left(\frac{\Delta E}{k_B T}\right)}, \tag{4.37}$$

so we see that $\langle E \rangle = (E_1 + E_2)/2$ at high temperatures.

4.6 Continuous Energy Distribution

The partition function corresponding to a system with a continuous energy distribution is obtained by integrating the Boltzmann factor over all internal degrees of freedom. If the energy is only a function of the parameter r, then

$$Z = \int_{\text{all } r} \exp\left(-\frac{E(r)}{k_B T}\right) dr. \tag{4.38}$$

The probability that the system is between r_1 and r_2 is then

$$P(r_1 \le r \le r_2) = \frac{1}{Z}\int_{r_1}^{r_2} \exp\left(-\frac{E(r)}{k_B T}\right) dr. \tag{4.39}$$

The average energy $\langle E \rangle$ of a system with a continuous energy spectrum in thermodynamic equilibrium is similarly obtained:

$$\langle E \rangle = \int_{\text{all } r} E(r) \exp\left(-\frac{E(r)}{k_B T}\right) dr. \tag{4.40}$$

4.7 Composite Systems

Consider a system of two noninteracting particles, each with possible states $1i$ and $2i$. The possible energy levels of the composite system are $E_{1i} + E_{2i}$ so that the Boltzmann factors can be written as a product:

$$Z_{\text{total}} = \sum_{1i}\sum_{2i} \exp\left(-\frac{E_{1i} + E_{2i}}{k_B T}\right) = \sum_{1i}\sum_{2i} \exp\left(-\frac{E_{1i}}{k_B T}\right) \exp\left(-\frac{E_{2i}}{k_B T}\right). \tag{4.41}$$

The partition function thus factors as

$$Z_{\text{total}} = \sum_{1i} \exp\left(-\frac{E_{1i}}{k_B T}\right) \sum_{2i} \exp\left(-\frac{E_{2i}}{k_B T}\right), \tag{4.42}$$

and we can write the product

$$Z_{\text{total}} = Z_1 Z_2. \tag{4.43}$$

As an example, the partition function of an ideal gas molecule in a given state can be separated into translational, rotational, and vibrational components:

$$Z_1 = Z_{\text{trans}} Z_{\text{vib}} Z_{\text{rot}}. \tag{4.44}$$

For an ideal gas containing N indistinguishable molecules,

$$Z_{\text{total}} = \frac{1}{N!} Z_1^N. \tag{4.45}$$

If a system consists of N distinguishable particles, then

$$Z_{\text{total}} = Z_1^N. \tag{4.46}$$

4.7.1 DNA Stretching

Single-molecule techniques have recently been developed using optical tweezers, enabling the manipulation of individual molecules as illustrated in Figure 4.4. A dielectric bead is attached to a molecule that is then unfolded by manipulating the bead with laser tweezers. Before stretching, the molecule assumes one of many possible configurations. The number of possible configurations is reduced as the molecule is stretched, resulting in a decrease in entropy.

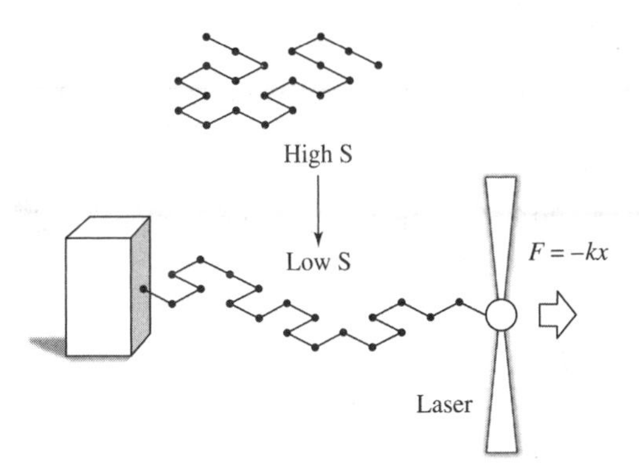

FIGURE 4.4 Stretching a single molecule with laser tweezers. Entropy is reduced as the molecule is unfolded. For small displacements the molecular restoring force obeys Hooke's law $F = -kx$ where k is the effective spring constant.

Detailed experimental studies have been performed measuring the force required to stretch individual polymer chains. For small extensions x, the potential energy function has the form

$$U(x) = -f \cdot x + \text{const}. \tag{4.47}$$

We seek to model the displacement of a one-dimensional chain by considering each individual link of length L as a two-state variable $\sigma_i = \pm 1$ oriented to the right $(+1)$ or to the left (–1). The total displacement is given by

$$x = L\sum_{i=1}^{N}\sigma_i. \tag{4.48}$$

The partition function is then

$$Z = \exp\left(\beta fL\sum_{i=1}^{N}\sigma_i\right), \tag{4.49}$$

which can be written as a product

$$Z = \exp(\beta fL\sigma_1)\exp(\beta fL\sigma_2)\ldots\exp(\beta fL\sigma_N). \tag{4.50}$$

Summing over all possible orientations of each link gives

$$Z = \sum_{\sigma_1=\pm1}\exp(\beta fL\sigma_1)\sum_{\sigma_2=\pm1}\exp(\beta fL\sigma_2)\ldots\sum_{\sigma_N=\pm1}\exp(\beta fL\sigma_N). \tag{4.51}$$

The product can be factored into N identical terms:

$$Z = \left(\sum_{\sigma=\pm1}\exp(\beta fL\sigma)\right)^N = (\exp(\beta fL) + \exp(-\beta fL))^N = (2\cosh\beta fL)^N. \tag{4.52}$$

Equipped with a compact form of the partition function, we can now calculate all of our thermodynamic quantities such as the average energy

$$\langle E\rangle = -\frac{\partial \ln Z}{\partial\beta} = -N\frac{\partial}{\partial\beta}\ln(2\cosh\beta fL) \tag{4.53}$$

that reduces to

$$\langle E\rangle = -N\frac{fL2\sinh\beta fL}{2\cosh\beta fL} = -NfL\tanh\beta fL. \tag{4.54}$$

The entropy, free energy, and heat capacity can then be calculated according to Equations 4.28, 4.29, and 4.31, respectively.

4.8 Casimir Contributions to the Free Energy

4.8.1 Lipid Bilayer Tubes

Dean and Horgan (2005) have estimated the Casimir portion of the free energy at a temperature T in a lipid bilayer tube of thickness δ such as shown in Figure 4.5 to be

$$F_C(L, R) = -\frac{k_B T \kappa_C L}{R} + O\left(\frac{\delta}{R^2}\right) \tag{4.55}$$

where the tube length and radius are L and R, respectively, with $\kappa_C \approx 0.3-0.5$. Because F_C is negative, the Casimir force acts to stabilize the tube against bending forces by providing a contracting radial force. The Casimir force alone is insufficient to stabilize the tube against bending moments, however.

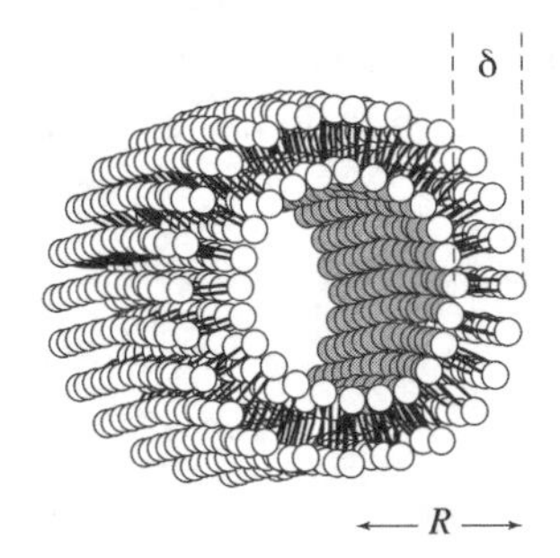

FIGURE 4.5 Lipid bilayer tube of thickness δ and radius R.

4.8.2 Rouleaux

Erythrocytes (red blood cells) are observed to form aggregates resembling stacks of coins under various conditions. These cell stacks, such as the one depicted in Figure 4.6, are known as rouleaux. The stacks are slightly off center because erythrocytes have a biconvex shape being thinner at the center ~0.3 μm compared to the outer edge thickness ~1 μm, with a cell radius between 6 μm and 8 μm.

Rouleaux have been classified as being either "true" or "false" depending on the conditions that these structures are observed to form. True rouleaux form in the presence

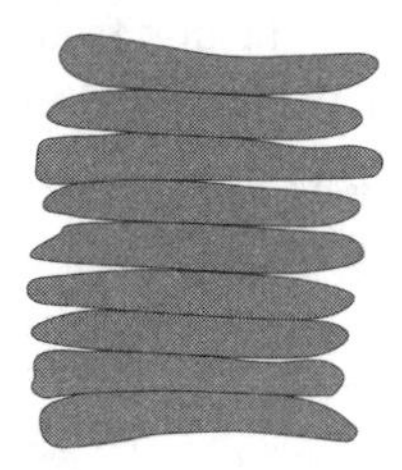

FIGURE 4.6 Schematic of self-aggregating stacks of red blood cells known as rouleaux.

of higher concentrations of certain proteins that minimize the average distance between the blood cells. False rouleaux can form in blood smears and suspensions with low ionic concentrations in the absence of these proteins.

Because erythrocytes are negatively charged, a short-range attractive Van der Waal-type force is required to overcome the net electrostatic repulsion in order for the rouleaux to form. Bradonjie et al. (2009) have proposed a Casimir force mechanism that provides an attractive force to balance Coulomb repulsion at short distances. The interaction energy is given by

$$u(d) = \frac{\sigma^2\Lambda}{2\varepsilon_2}\left\{e^{-d/\Lambda} - \left(\frac{\pi^2\hbar v\sqrt{\varepsilon_0\varepsilon_2}}{360\sigma^2\Lambda}\right)\frac{1}{d^3}\right\} \tag{4.56}$$

with $v = c[(\varepsilon_1 - \varepsilon_2)/(\varepsilon_1 + \varepsilon_2)]^2$ where c is the speed of light and σ is the Steffan–Boltzmann constant. The first term is the screened Coulomb repulsion, and the second term is the Casimir attraction between red blood cells modeled as dielectric plates with permitivity ε_1 separated a distance d in a plasma medium with permitivity ε_2. The Debye screening length Λ is given by

$$\Lambda = \sqrt{\frac{\varepsilon_2 k_B T}{e^2 \sum_i z_i{}^2 C_{i0}}}. \tag{4.57}$$

A dimensionless parameter proportional to the ratio of the electrostatic repulsion to the Casimir attraction at a distance of one Debye screening length is obtained

$$a = \left(\frac{\pi^2\hbar v\sqrt{\varepsilon_0\varepsilon_2}}{360\sigma^2\Lambda^4}\right). \tag{4.58}$$

Including the v factor, the relative strength of the two forces is then

$$a = \left(\frac{\pi^2\hbar c\sqrt{\varepsilon_0\varepsilon_2}}{360\sigma^2\Lambda^4}\right)\left(\frac{\varepsilon_1 - \varepsilon_2}{\varepsilon_1 + \varepsilon_2}\right)^2. \tag{4.59}$$

The Casimir force is a quantum electrodynamical (QED) effect that results from virtual quantum field fluctuations. This force is similar to the Van der Waal attraction that results from fluctuating true dipole moments. The full extent of QED interactions in biology is unknown. Casimir interactions are difficult to calculate requiring renormalization techniques even for the simplest geometries. However, consideration of quantum field effects may forward our understanding of a wide range of biophysical phenomena.

4.9 Protein Folding and Unfolding

The goal of protein-folding simulations is to determine the final folded form, or native state, of a linear polymer chain of amino acid monomers. The three-dimensional structure of a protein is determined by its particular amino acid sequence and its environment. Lipid molecules flank membrane-bound proteins. In aqueous environments, ionic concentrations may affect the protein's conformation. Temperature and the presence of chaperone molecules can also influence the protein conformation and folding rate.

The native state of a protein is the conformation that minimizes its free energy. A protein must be properly folded in order to function. Misfolded proteins suffer a loss of functionality that can result in diseases such as Alzheimer's, Mad Cow, Huntington's, and Parkinson's.

4.9.1 Protein Unfolding

Protein unfolding can occur with the addition of heat Q. The unfolding of a protein is accompanied by an increase in entropy because the unfolded protein can exist in many different conformations, while the folded form is more limited in the number of positions it can assume. This case is slightly different from that of the stretched polymer chain where the entropy of the chain was reduced as it was mechanically extended. If each amino acid can assume M positions, there are a total of

$$\Omega = N^M \tag{4.60}$$

different conformations in a chain of N amino acids corresponding to entropy of

$$S = k_B M \ln N. \tag{4.61}$$

The hydrophobic effect refers to the increased entropy of water molecules as the protein folds. In the native state, nonpolar groups are located in the center of the protein. When the protein becomes unfolded, water becomes ordered around polar groups with a corresponding drop in water entropy. At higher temperatures, water is less ordered around polar groups so that there is a smaller reduction in water entropy upon unfolding.

4.9.2 Levinthal's Paradox

To appreciate the large number of possible folding configurations, consider an idealized chain of 300 amino acids that can each assume three conformations. The total number of ways the protein can fold is $3^{300} \sim 10^{143}$. Levinthal noticed that the time it would take a protein to fold if it randomly sampled all of its possible conformations would be greater than the age of the universe. It turns out, however, that

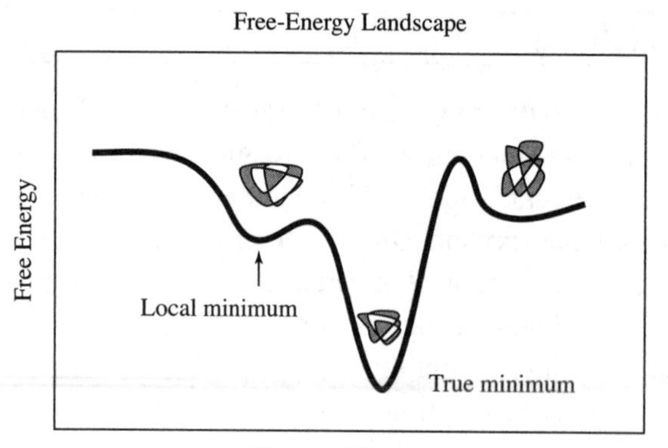

FIGURE 4.7 Schematic of the variation of free energy with an internal degree of freedom. The native state corresponds to the true minimum with metastable states indicated by the local minima. Various conformations are represented at several minima.

single-domain proteins fold on the order of microseconds to a few milliseconds. The resolution to Levinthal's paradox is that the protein folds through a series of metastable states instead of randomly sampling the entire configurational landscape. These metastable states are states with local minima on the energy landscape as illustrated in Figure 4.7.

4.9.3 Energy Landscape

The total protein conformational energy is given by the sum of all energy contributions from bonds, bond angles, torsion energy, Lennard–Jones potential, and electrostatic energy

$$E_{\text{total}} = \underbrace{\sum_k \alpha(x_k - x_{k0})^2}_{\text{bonds}} + \underbrace{\sum_\ell \beta(\theta_\ell - \theta_{\ell 0})^2}_{\text{bond angles}} + \underbrace{\sum_m \gamma(1 - \cos n_m(\phi_m - \phi_{m0}))^2}_{\text{torsion energy}}$$
$$+ \underbrace{\sum_{i,j} \varepsilon\left[\left(\frac{r_{ij0}}{r_{ij}}\right)^{12} - 2\left(\frac{r_{ij0}}{r_{ij}}\right)^{6}\right]}_{\text{Lennard–Jones potential}} + \underbrace{\sum_{i',j'} \delta\frac{q_{i'}q_{j'}}{r_{i'j'}}}_{\text{electrostatic energy}} + \cdots \tag{4.62}$$

with parameters $(\alpha, \beta, \gamma, \varepsilon, \delta)$ proportional to the strength of each interaction. The first three terms in this expression are similar to a spring potential $U_s = kx^2/2$ with a Hooke's law-type restoring force $F = -kx$. The last two terms are the Lennard–Jones potential, resulting from the sum of the Pauli repulsion and the Van der Waal interaction, and the electrostatic interaction between amino acids.

The probability of assuming a given conformation with energy E_C is proportional to the Boltzmann factor

$$P_C \propto \exp\left(-\frac{E_C}{k_B T}\right). \tag{4.63}$$

The partition function is the sum of Boltzmann factors of each conformation:

$$Z_{\text{total}} = \sum_C \exp\left(-\frac{E_C}{k_B T}\right) \tag{4.64}$$

so that the probability of a given conformation is

$$P_C = \frac{1}{Z_{\text{total}}} \exp\left(-\frac{E_C}{k_B T}\right). \tag{4.65}$$

Instead of summing over conformations, we may also sum over the possible energy values E to obtain the partition function

$$Z_{\text{total}} = \sum_E g(E) \exp\left(-\frac{E}{k_B T}\right) \tag{4.66}$$

where the density of states $g(E)$ is the number of states with energy E. Note that it would only be possible to express the total partition function as a product such as

$$Z_{\text{total}} = Z_{\text{bonds}} \times Z_{\text{angles}} \times Z_{\text{tortion}} \times Z_{\text{LJ}} \times Z_{\text{charges}} \times \ldots \tag{4.67}$$

if each energy term Equation 4.62 was independent. Simplifying assumptions can be made, however, and models have been implemented using separable partition functions. Below we consider two models of protein folding geared for computer simulation.

4.9.4 Folding on a Lattice

A simplification for computational folding treats each individual amino acid as single site on a 2-D planar or a 3-D cubic lattice. Each lattice protein is assigned a series of interaction terms like those in Equation 4.62. The simplest model proposed by Dill treats each protein site as being either hydrophilic (H) or hydrophobic (P). In this H–P model, hydrophobic sites are assigned a mutually attractive potential, while hydrophilic sites have repulsive interactions with each other. Figure 4.8 illustrates a protein with amino acid sequence

{HHHPPHHHHHHPPHHH}.

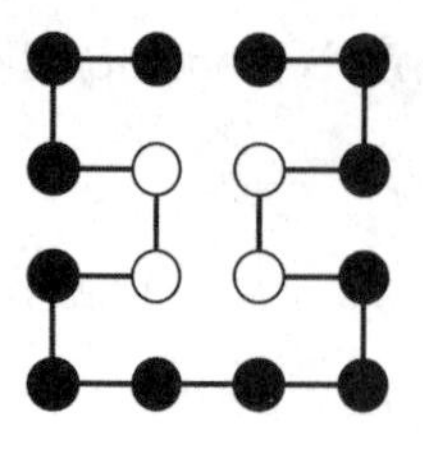

FIGURE 4.8 Simple H–P model of protein folding on a 2-D square lattice with hydrophobic (P) core surrounded by hydrophilic (H) amino acid residues in the native state. (Adapted from Sneppen, K., and E. Zocchi. 2005. Cambridge University Press, U.K.)

In the native state shown in this figure, the hydrophobic sites (open circles) are located at the center of the protein. These are surrounded by the hydrophilic sites (closed circles).

4.9.5 Monte Carlo Methods

Monte Carlo methods applied to protein folding involve randomly perturbing a degree of freedom in the polypeptide chain and then recalculating the conformational energy E_{new} according to Equation 4.62. Criteria are then applied to accept or reject a given permutation based on the energy difference between the new and old conformations $\Delta E = E_{new} - E_{old}$. The metropolis criterion accepts a change in degree of freedom with probability

$$P(\text{accept}) = \begin{cases} \exp\left(-\dfrac{\Delta E}{T}\right) & \Delta E > 0 \\ 1 & \Delta E < 0 \end{cases} \tag{4.68}$$

where T is a control parameter unrelated to the actual temperature. The metropolis algorithm is commonly applied with an "annealing schedule" where T is gradually reduced with an increasing number of updates. The annealing schedule results in a larger rejection probability for a given ΔE early in the simulation to avoid becoming trapped in a local minimum—or metastable configuration. Later in the simulation, a smaller value of T reduces the probability of the protein "hopping out" of the true minimum only to end up in a metastable state. The above steps are repeated many times to increase the likelihood of convergence to the native state.

4.9.6 Folding@home

The Folding@home project was launched on October 1, 2000, by the Panda Group, headed by Vijay Panda with the Department of Chemistry at Stanford University, California. Protein-folding and molecular dynamics simulations are performed using personal computers via a client program that is downloaded from the project site at http://folding.stanford.edu. Distribution of computing over thousands of processors enables the solution of folding problems that would otherwise be impossible.

EXERCISES

Exercise 4.1. Consider a system of $N = 17$ particles and $m = 100$ possible states $\{n_1, n_2, n_3, \ldots, n_{100}\}$ where $\sum_i n_i = N$. Calculate the maximum and minimum possible values of the entropy.

Exercise 4.2. Consider a protein with 50 amino acids that can each assume four configurations. Calculate the maximum number of ways the protein can fold.

Exercise 4.3. Proteins embedded in a planar lipid bilayer have two possible electric dipole **p** orientations corresponding to interaction energies $\pm p_z E$ in an electric field **E** oriented normal to the plane of the bilayer (Figure 4.9). Calculate the partition function Z and the thermodynamic quantities $\langle E \rangle$, F, C, and S for a system with N proteins. Calculate the probability for each dipole orientation.

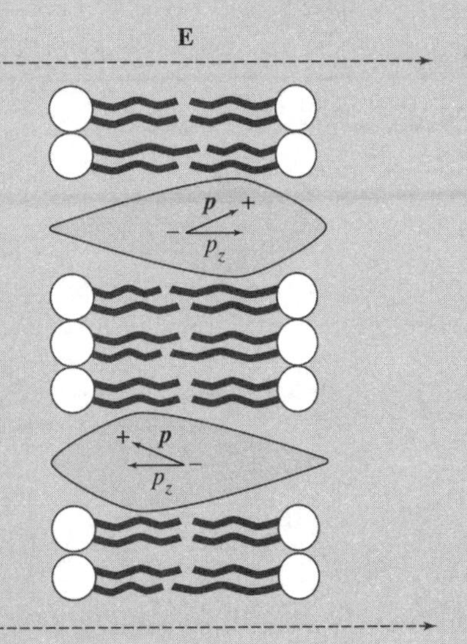

FIGURE 4.9 Proteins with dipole moment ***p*** imbedded in a planar membrane with surface normal parallel to a *z*-directed electric field. The proteins may assume two possible orientations as shown above.

Exercise 4.4. The partition function of the DNA zipper model is given by

$$Z = g^N \frac{z^{N+1} - 1}{z - 1}$$

where $z = \exp(\varepsilon/T - \ln g)$. N is the number of base pairs in the double-stranded DNA molecule, and g is the number of configurations of an unbounded base pair. Calculate the free energy, average energy, and heat capacity in this model (Figure 4.10).

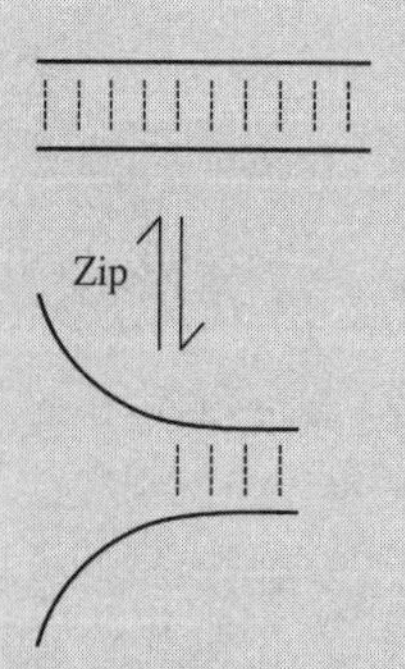

FIGURE 4.10 DNA zipper model.

Exercise 4.5. Calculate the entropy, heat capacity, and free energy in the DNA stretching model in Section 4.7.1.

Exercise 4.6. Calculate the probability that a particle has energy $E > E_1$ for a system with continuous energy distribution according to

$$P(E > E_1) = \frac{1}{Z}\int_{E_1}^{\infty} \exp\left(-\frac{E}{k_B T}\right) dE$$

where

$$Z = \int_{O}^{\infty} \exp\left(-\frac{E}{k_B T}\right) dE.$$

Calculate the probability that a particle with energy E_1 can surmount an energy barrier $\Delta E = E_2 - E_1$.

5 Open Systems and Chemical Thermodynamics

5.1 Enthalpy, Gibbs Free Energy, and Chemical Potential

Many biological processes occur in isobaric conditions with constant pressure P where the total volume V and number of particles N may vary. The thermodynamic quantity H, known as the enthalpy, includes the internal energy E of a system as well as the energy associated with the volume displaced at constant pressure

$$H = E + PV \tag{5.1}$$

where PV has units of Joules. Schroeder gives a humorous illustration of the energy required to spontaneously create a bunny rabbit from the vacuum given by the bunny's internal energy plus the work done in displacing a volume V of atmosphere at pressure P. While constant volume processes minimize the Helmholtz free energy F, spontaneous isobaric processes minimize a quantity known as the Gibbs free energy G given by

$$G = F + PV. \tag{5.2}$$

In terms of the enthalpy, the Gibbs free energy becomes

$$G = H - TS. \tag{5.3}$$

From the thermodynamic identity derived in Appendix 5,

$$dG = -SdT + VdP + \mu dN. \tag{5.4}$$

The change in Gibbs free energy when only the number of particles varies is

$$\Delta G = \mu \Delta N \tag{5.5}$$

under isobaric and isothermal conditions. We may then define the chemical potential as the partial derivative of the Gibb's free energy with respect to particle number at constant pressure and temperature

$$\mu = \left(\frac{\partial G}{\partial N}\right)_{P,T}, \tag{5.6}$$

treating N as a continuous variable. Since N is dimensionless, it is clearly seen that the units of μ are Joules. The chemical potential is numerically equal to the energy required to add one particle to the system. For multiple particle species i,

$$G = \sum_i \mu_i N_i \tag{5.7}$$

and the chemical potential of the ith species is

$$\mu_i = \left(\frac{\partial G}{\partial N_i}\right)_{P,T}. \tag{5.8}$$

For a single species with chemical potential μ the Gibb's free energy is

$$G = \mu N. \tag{5.9}$$

From this relation we can obtain an expression for the chemical potential of a pure ideal gas

$$\frac{\partial \mu}{\partial P} = \frac{1}{N}\frac{\partial G}{\partial P}. \tag{5.10}$$

From the thermodynamic relation expressed by Equation 5.4 $(\partial G/\partial P)_{T,N} = V$ and the ideal gas law $PV = Nk_BT$, we obtain

$$\frac{\partial \mu}{\partial P} = \frac{k_B T}{P}. \tag{5.11}$$

Separating variables and integrating both sides of this equation:

$$\int_{\mu_0}^{\mu} d\mu = k_B T \int_{P_0}^{P} \frac{dP}{P} \tag{5.12}$$

gives the chemical potential per particle

$$\mu = \mu^0 + k_B T \ln \frac{P}{P^0} \tag{5.13}$$

where μ^0 is the chemical potential at $P^0 = 1$ atm $= 1.013 \times 10^5$ Pa. This equation is often expressed in molar form using the same symbols for chemical potential

$$\mu = \mu^0 + RT \ln \frac{P}{P^0} \tag{5.14}$$

where $R = N_A k_B$ is the gas constant and μ is now the chemical potential per mole of substance. For a gas mixture

$$\mu_i = \mu_i^0 + RT \ln \frac{P_i}{P^0} \tag{5.15}$$

where P_i and μ_i are now the partial pressure and the chemical potential, respectively, of the ith component. The chemical potential of a substance is also defined as

$$\mu_i = \mu_i^0 + RT \ln a_i \tag{5.16}$$

where μ_i^0 is the chemical potential in the standard state. The chemical activity is $a_i = \gamma_i[c_i]$ where γ_i is the coefficient of activity and c_i is the concentration of the ith substance.

The chemical potential is also related to the change in entropy with the particle number from the thermodynamic relation

$$dE = TdS - PdV + \sum_i \mu_i dN_i \tag{5.17}$$

where we obtain

$$\mu_i = -T\left(\frac{\partial S}{\partial N_i}\right)_{E,V,N_{j\neq i}}. \tag{5.18}$$

5.2 Chemical Reactions

In this section we first review simple reaction kinetics and then make the connection to thermodynamics considering temperature-dependent rate constants through the Boltzmann factor.

5.2.1 First-Order Reactions

The first-order reaction $A \rightarrow B + C$ is described by the differential equation

$$\frac{d[A]}{dt} = -k[A] \tag{5.19}$$

where the depletion rate is proportional to the concentration. The rate constant k has units of inverse time. Separating variables

$$\frac{d[A]}{[A]} = -kdt \tag{5.20}$$

and integrating both sides

$$\int \frac{d[A]}{[A]} = -k\int dt, \tag{5.21}$$

we obtain

$$\ln[A] = -kt + c \tag{5.22}$$

with integration constant c. Exponentiating both sides of this equation,

$$[A(t)] = [A]_0\, e^{-kt} \tag{5.23}$$

where $e^c = [A]_0$ is the initial concentration at $t = 0$. Equation 5.23 can be expressed as

$$[A(t)] = [A]_0\, e^{-t/\tau} \tag{5.24}$$

with the time constant $\tau = 1/k$.

5.2.2 Second-Order Reactions

The second-order reaction $A + B \rightarrow$ products is described by the differential equations

$$\frac{d[A]}{dt} = -k[A][B] \tag{5.25}$$

and

$$\frac{d[B]}{dt} = -k[A][B]. \tag{5.26}$$

Since the right-hand sides of these equations are equal, we can represent this system by a single differential equation

$$\frac{dx}{dt} = -kx^2. \tag{5.27}$$

Separating variables

$$\frac{dx}{x^2} = -kdt \tag{5.28}$$

and integrating both sides gives

$$-\frac{1}{x} = -kt + c \tag{5.29}$$

with integration constant c. Solving for x,

$$x = \frac{1}{kt - c}. \tag{5.30}$$

At $t = 0$, the initial concentrations $x_0 = -1/c$ or

$$x(t) = \frac{x_0}{x_0 kt + 1}. \tag{5.31}$$

In terms of our reactants we have

$$[A(t)] = \frac{[A]_0}{[A]_0 kt + 1} \quad \text{and} \quad [B(t)] = \frac{[B]_0}{[B]_0 kt + 1}. \tag{5.32}$$

5.3 Activation Energy and Rate Constants

The connection to thermodynamics comes about when we consider temperature-dependent rate constants. We expect chemical reactions to proceed more slowly at lower temperatures and to cease at absolute zero. The rate constant is proportional to a Boltzmann factor

$$k = A\exp\left(-\frac{E_A}{RT}\right) \tag{5.33}$$

where E_A is the activation energy. Taking the logarithm of both sides of this equation

$$\ln k = \ln A - \frac{E_A}{RT}, \tag{5.34}$$

and then taking the derivative with respect to $1/T$, gives the Arrhenius equation

$$\frac{\partial \ln k}{\partial(1/T)} = -\frac{E_A}{R} \tag{5.35}$$

so that E_A/R can be determined experimentally by the negative slope of $\ln k$ versus $1/T$. Equation 5.35 can also be written

$$\frac{\partial \ln k}{\partial T} = \frac{E_A}{RT^2}. \tag{5.36}$$

Arrhenius-like behavior may be found in a diverse range of biophysical phenomena. Figure 5.1 shows a plot of ln k versus $1/T$ with biphasic behavior in the growth rate k of a psychrotropic bacterium due to the inhibition of protein degradation below 290 K. Kinks in Arrhenius plots can result from temperature-dependent activation energies as well as conformational changes in reaction components that occur at specific temperatures.

General Chemical Reaction

Considering the general chemical reaction

$$\upsilon_A A + \upsilon_B B \rightleftharpoons \upsilon_C A + \upsilon_D D \tag{5.37}$$

the change in Gibbs free energy under isothermal and isobaric conditions is

$$\Delta G = \sum_i \mu_i \Delta N_i \tag{5.38}$$

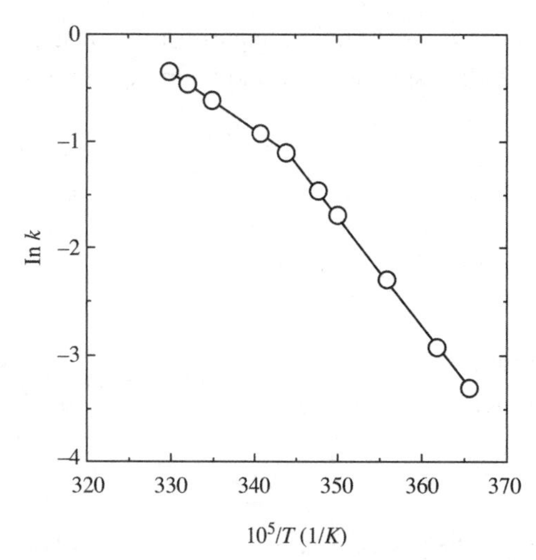

FIGURE 5.1 Arrhenius plot of *Pseudomonas fluorescens* showing biphasic behavior. (Adapted from Guillou and Guespin-Michel. 1996. Evidence for two domains of growth temperature for the psychotropic bacterium *Pseudomonas fluorescens* MFO. *Applied Environmental Microbiology* 62: 3319–3324.)

or

$$\Delta G = \upsilon_C\mu_C + \upsilon_D\mu_D - \upsilon_A\mu_A - \upsilon_B\mu_B. \tag{5.39}$$

In terms of the chemical potential $\mu_i = \mu_i^{\,0} + RT\ln a_i$, the change in Gibb's free energy is

$$\Delta G = \Delta G^0 + RT(\upsilon_C\ln a_C + \upsilon_D\ln a_D - \upsilon_A\ln a_A - \upsilon_B\ln a_B) \tag{5.40}$$

where a_i is the dimensionless chemical activities and the molar standard free energy is

$$\Delta G^0 = \upsilon_C\mu_C^0 + \upsilon_D\mu_D^0 - \upsilon_A\mu_A^0 - \upsilon_B\mu_B^0. \tag{5.41}$$

Combining logarithms we obtain van't Hoff's equation

$$\Delta G = \Delta G^0 + RT\ln\frac{a_C^{\upsilon_C}a_D^{\upsilon_D}}{a_A^{\upsilon_A}a_B^{\upsilon_B}}. \tag{5.42}$$

The change in Gibb's free energy is zero in equilibrium. Setting $\Delta G = 0$, we obtain the equilibrium constant

$$K_p = \left(\frac{a_C^{\upsilon_C}a_D^{\upsilon_D}}{a_A^{\upsilon_A}a_B^{\upsilon_B}}\right)_0 = \exp\left(-\frac{\Delta G^0}{RT}\right). \tag{5.43}$$

For weak concentrations, we can replace $a_i \rightarrow [c_i]$. The equilibrium constant may be written as a product of exponentials using the thermodynamic relation $\Delta G^0 = \Delta H^0 - T\Delta S^0$, thus

$$K_p = \exp\left(-\frac{\Delta H^0}{RT}\right) \times \exp\left(\frac{\Delta S^0}{R}\right) \tag{5.44}$$

where

$$\exp\left(-\frac{\Delta H^0}{RT}\right) = \frac{\text{product Boltzmann factor}}{\text{reactant Boltzmann factor}} \tag{5.45}$$

and

$$\exp\left(-\frac{\Delta S^0}{R}\right) = \frac{\text{\# of accessible product states}}{\text{\# of accessible reactant states}}. \tag{5.46}$$

The first exponential factor represents the barrier potential or activation energy, while the second factor is dependent on the density of states and is independent of temperature.

5.3.1 Detailed Balance

For a reaction to proceed in the presence of an energy barrier such as that shown in Figure 5.2, the reactant molecules must acquire sufficient energy to surmount the barrier. The forward rate constant for a process proceeding from left to right is proportional to the Boltzmann factor

$$k_+ = A\exp\left(\frac{-E_+}{k_B T}\right). \tag{5.47}$$

The rate constant corresponding to the reverse process is given by

$$k_- = A\exp\left(\frac{-E_-}{k_B T}\right). \tag{5.48}$$

At equilibrium,

$$k_+[A] = k_-[B] \tag{5.49}$$

so that

$$\frac{k_+}{k_-} = \frac{[B]}{[A]}. \tag{5.50}$$

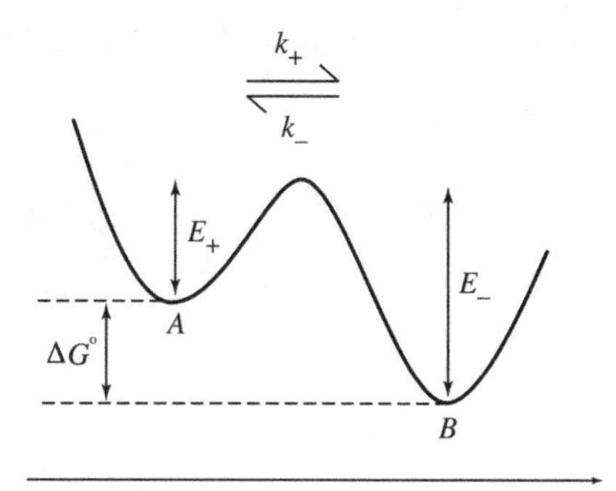

FIGURE 5.2 The ratio of forward to reverse chemical reaction rates is proportional to a Boltzmann factor.

According to the principle of detailed balance, the ratio of the forward k_+ to reverse k_- rate constants is given by the factor

$$\frac{k_+}{k_-} = \exp\left(\frac{\Delta G^0}{k_B T}\right) \tag{5.51}$$

where $\Delta G^0 = E_- - E_+$.

5.3.2 Nitrogen Fixation

Nitrogen fixation involves the conversion of atmospheric nitrogen into ammonia

$$N_2 + 3H_2 \rightleftharpoons 2NH_3. \tag{5.52}$$

This process is essential for the synthesis of nucleic acids and amino acids. Biological nitrogen fixation is carried out by diazotrophs that include various prokaryotes. Nitrogen-fixing bacteria thrive in soil and in legume root nodules. In equilibrium $dG = 0$ so that the sum

$$\sum_i \mu_i dN_i = 0 \tag{5.53}$$

has three terms

$$\mu_{N_2} \underbrace{dN_{N_2}}_{-1} + \mu_{H_2} \underbrace{dN_{H_2}}_{-3} + \mu_{NH_3} \underbrace{dN_{NH_3}}_{+2} = 0 \tag{5.54}$$

so that

$$\mu_{N_2} + 3\mu_{H_2} = 2\mu_{NH_3}. \tag{5.55}$$

The molar standard free energy can then be expressed as the logarithm of partial pressures

$$\Delta G^0 = RT \ln\left(\frac{P_{N_2} P_{H_2}{}^3}{P_{NH_3}{}^2 P_0^2}\right) \tag{5.56}$$

with dimensionless equilibrium constant

$$K_{eq} = \frac{P_{NH_3}{}^2 P_0^2}{P_{N_2} P_{H_2}^3} = \exp\left(-\frac{\Delta G^0}{RT}\right) \tag{5.57}$$

where $P_0 = 1$ atm.

5.4 Enzymatic Reactions

Figure 5.3 shows the reduction in activation energy of a reaction due to an enzymatic catalyst. The reduced activation energy increases the reaction rate of a given biochemical reaction. Enzyme reactions occur in two stages:

$$\underbrace{E + S \rightleftharpoons ES}_{\text{substrate binding}} \xrightarrow{\text{catalytic step}} E + P. \tag{5.58}$$

The substrate binds to the enzyme forming the enzyme–substrate complex during the first step. The product is produced and released during the catalytic second step according to

$$E + S \underset{k_-}{\overset{k_+}{\rightleftharpoons}} ES \xrightarrow{K} E + P. \tag{5.59}$$

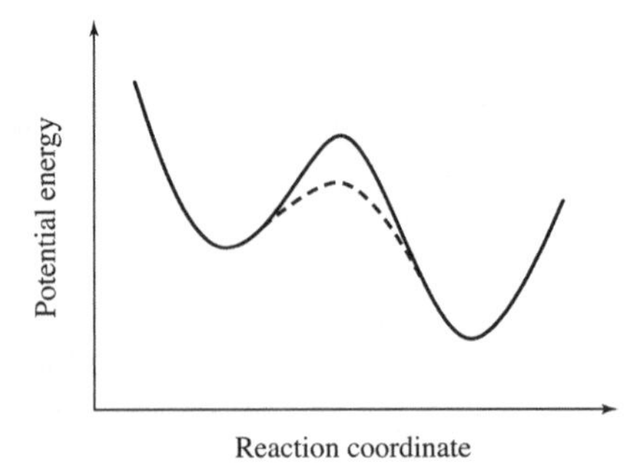

FIGURE 5.3 Reduction of activation energy by an enzyme catalyst. The solid curve represents the reaction without the catalyst. The dashed curve shows a reduction in activation energy due to the catalyst.

The forward rate constants k_+ and K correspond to the formation of the enzyme–substrate complex and the product, respectively. The rate constant k_- corresponds to the disassociation of the ES complex before the reaction is catalyzed.

The time rate of change of the ES concentration is given by

$$\frac{d[ES]}{dt} = k_+[E][S] - k_-[ES] - K[ES]. \tag{5.60}$$

The time rate of change of the product is proportional to the ES concentration

$$\frac{d[P]}{dt} = K[ES]. \tag{5.61}$$

In quasi steady state, we assume that the ES concentration changes much more slowly than the concentration of the product

$$\frac{d[ES]}{dt} \ll \frac{d[P]}{dt} \tag{5.62}$$

so that

$$0 = k_+[E][S] - k_-[ES] - K[ES]. \tag{5.63}$$

Further, if we assume that the total enzyme concentration including the fraction bound to the substrate is a constant $[E]_0$:

$$[E]_0 = [E] + [ES], \tag{5.64}$$

we can then obtain a formula for the product formation rate if we substitute Equation 5.64 into Equation 5.63:

$$0 = k_+([E]_0 - [ES])[S] - k_-[ES] - K[ES], \tag{5.65}$$

giving

$$[ES] = \frac{k_+[E]_0[S]}{k_- + K + k_+[S]}. \tag{5.66}$$

The rate of product formation is then

$$\frac{d[P]}{dt} = K\frac{[E]_0[S]}{K_M + [S]} \tag{5.67}$$

where $K_M = (k_- + K)/k_+$ is the Michaelis constant and $K[E]_0$ is the maximum rate of product formation as $[S]$ is increased.

Calcium Buffer

Ca^{2+}-binding proteins include *troponin C* in skeletal muscle as well as calmodulin found in all eucaryotic cells where it mediates numerous processes regulated by Ca^{2+}. The reaction forming the calcium-buffer complex Ca^{2+} is

$$B + Ca^{2+} \underset{k_-}{\overset{k_+}{\rightleftarrows}} B \cdot Ca \tag{5.68}$$

where k_+ and k_- are the respective forward and reverse rate constants and B is the binding protein. The time rates of change of the calcium and buffer concentrations are given by the differential equations

$$\frac{d}{dt}[Ca^{2+}] = k_-[B \cdot Ca] - k_+[B][Ca^{2+}] \tag{5.69}$$

$$\frac{d}{dt}[B] = k_-[B \cdot Ca] - k_+[B][Ca^{2+}] \tag{5.70}$$

where the depletion rates are reduced by the dissociation of the B · Ca complex. The total buffer concentration is constant and is given by the sum

$$[B]_0 = [B] + [B \cdot Ca]. \tag{5.71}$$

The depletion rates are zero in equilibrium so that our complex concentration is

$$[B \cdot Ca] = \frac{k_+}{k_-}[B][Ca^{2+}]. \tag{5.72}$$

Substituting (5.71),

$$[B \cdot Ca] = \frac{k_+}{k_-}([B]_0 - [B \cdot Ca])\ [Ca^{2+}] \tag{5.73}$$

gives the equilibrium concentration of the calcium-buffer complex

$$[B \cdot Ca] = \frac{[B]_0[Ca^{2+}]}{\frac{k_-}{k_+} + [Ca^{2+}]}. \tag{5.74}$$

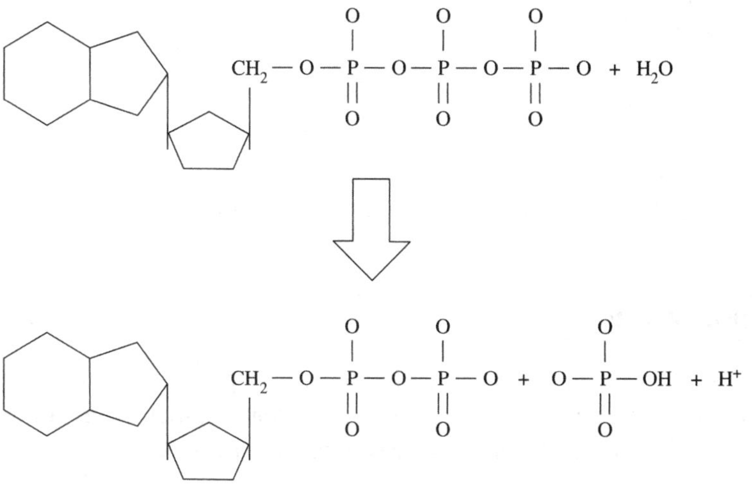

FIGURE 5.4 Adenosine triphosphate (ATP) molecule + water forming adenosine diphosphate (ADP) + inorganic phosphate (P_i).

5.5 ATP Hydrolysis and Synthesis

ATP hydrolysis is given by the overall reaction

$$\text{ATP} + \text{H}_2\text{O} \rightleftharpoons \text{ADP} + \text{P}_\text{i} \tag{5.75}$$

where ADP is adenosine diphosphate and P_i is inorganic phosphate. This reaction is shown schematically in Figure 5.4. The associated change in Gibb's free energy

$$\Delta G = \Delta G^0 + RT \ln\left(\frac{[\text{ADP}][\text{P}_\text{i}]}{[\text{ATP}]}\right) \tag{5.76}$$

with $\Delta G^0 = -7.3$ kcal/mol. Note that H_2O is not included in the rate constant because it is a solvent. ATP synthesis is described by the reverse reaction

$$\text{ADP} + \text{P}_\text{i} \rightleftharpoons \text{ATP} + \text{H}_2\text{O} \tag{5.77}$$

with $\Delta G^0 = +7.3$ kcal/mol.

Figure 5.5 shows the change in Gibbs free energy ΔG as a function of the position of the terminal phosphate of an ATP molecule. The phosphate groups are represented as small circles in this figure. The terminal phosphate is in a metastable state where a small amount of energy must be expended to liberate a larger quantity of stored energy. The energy liberated for each ATP molecule hydrolyzed is 0.06 aJ or 0.01 eV.

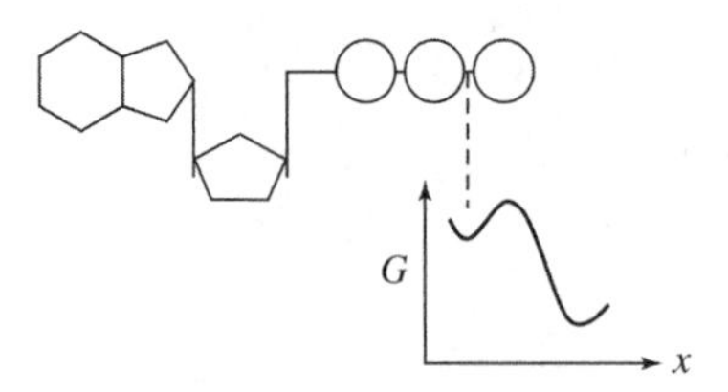

FIGURE 5.5 Change in Gibbs free energy as a function of the position of the terminal phosphate of an ATP molecule.

5.6 Entropy of Mixing

Mixing dissimilar substances will result in a change in entropy as each substance expands to fill the mixing container. If the substances are identical, then there will be no change in entropy. For two different substances with particle numbers N_1 and N_2 where

$$N = N_1 + N_2, \tag{5.78}$$

the change in entropy upon mixing is given by

$$\Delta S = k_B \ln \Omega \tag{5.79}$$

with the density of states

$$\Omega = \frac{N!}{N_1!N_2!}. \tag{5.80}$$

Following the development similar to that of Appendix 4 using Sterling's formula in Appendix 1, we therefore have

$$\Delta S = -k_B\left(N_1 \ln \frac{N_1}{N} + N_2 \ln \frac{N_2}{N}\right). \tag{5.81}$$

The change is Gibbs free energy at constant enthalpy is given by $\Delta G = -T\Delta S$ or

$$\Delta G = k_B T\left(N_1 \ln \frac{N_1}{N} + N_2 \ln \frac{N_2}{N}\right). \tag{5.82}$$

5.7 The Grand Canonical Ensemble

In general, biological cells and organisms are open systems and can exchange particles with their environment. We can modify our thermodynamic analysis of the previous

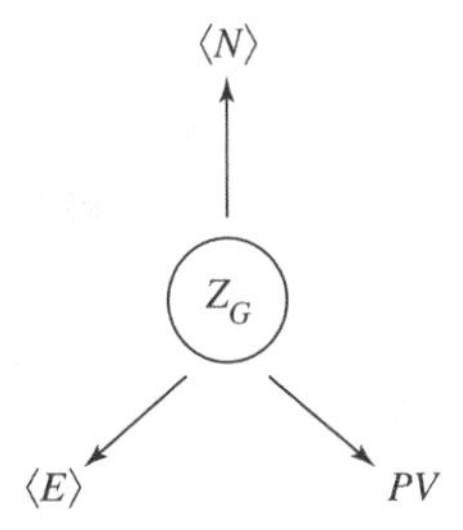

FIGURE 5.6 Illustration of several thermodynamic quantities derivable from the grand canonical partition function including average particle number, average energy, and the pressure volume product. The heat capacity and entropy may also be obtained from the grand canonical partition function.

chapter considering diffusive systems in contact with a reservoir at temperature T. The Gibbs factor is written similar to the Boltzmann factor in the previous chapter:

$$P_{ij} \sim \exp\left(-\frac{E_{i,j} - \mu_j N_j}{k_B T}\right) \tag{5.83}$$

where μ_j is the chemical potential of the jth species. The grand canonical partition function Z_G for an ensemble of states with variable number of particles N_j with possible energies $E_{i,j}$ in thermodynamic equilibrium is given by a sum of Gibbs factors:

$$Z_G = \sum_{i,j} \exp\left(-\frac{E_{i,j} - \mu_j N_j}{k_B T}\right). \tag{5.84}$$

Like the canonical partition function in Chapter 4, key thermodynamical quantities may be calculated from the grand canonical partition function as illustrated in Figure 5.6. The probability that a system has a specific energy and particle number is obtained by normalizing the Gibbs factor by Z_G:

$$P_{i,j} = \frac{\exp\left(-\frac{E_{i,j} - \mu_j N_j}{k_B T}\right)}{Z_G}. \tag{5.85}$$

The average energy $\langle E \rangle$ of a system in thermodynamic equilibrium is given by

$$\langle E \rangle = \sum_{i,j} P_{i,j} E_{i,j}. \tag{5.86}$$

Substituting Equation 5.84 for $P_{i,j}$,

$$\langle E \rangle = \frac{1}{Z_G} \sum_{i,j} E_{i,j} \exp\left(-\frac{E_{i,j} - \mu_j N_j}{k_B T}\right), \tag{5.87}$$

which can be written as a derivative of the partition function

$$\langle E \rangle = -\frac{\partial \ln Z_G}{\partial \beta} = k_B T^2 \frac{\partial \ln Z_G}{\partial T}. \tag{5.88}$$

A similar analysis gives the average number of particles:

$$\langle N \rangle = \frac{1}{Z_G} \sum_{i,j} N_j \exp\left(-\frac{E_{i,j} - \mu_j N_j}{k_B T}\right). \tag{5.89}$$

For a single species with chemical potential μ,

$$\langle N \rangle = \frac{\partial \ln Z_G}{\partial \gamma} \tag{5.90}$$

where the quantity $\gamma = \mu / k_B T$ is related to the fugacity $z = \exp \gamma$.

5.8 Hemoglobin

Hemoglobin is an iron-containing molecule that transports oxygen in the blood. Hemoglobin contains four myoglobin molecules each with one iron atom that serves to bind oxygen molecules. The binding energy of O_2 with myoglobin is $E_{O_2} = -0.7$ eV. Taking the energy of the unbound state to be zero, the grand canonical partition function is

$$Z_G = \exp\left(\frac{0}{k_B T}\right) + \exp\left(-\frac{E_{O_2} - \mu_{O_2}}{k_B T}\right) = 1 + \exp\left(-\frac{E_{O_2} - \mu_{O_2}}{k_B T}\right). \tag{5.91}$$

The occupation probabilities are then

$$P(\text{occupied}) = \frac{\exp\left(-\frac{E_{O_2} - \mu_{O_2}}{k_B T}\right)}{1 + \exp\left(-\frac{E_{O_2} - \mu_{O_2}}{k_B T}\right)} \tag{5.92}$$

and

$$P(\text{unoccupied}) = \frac{1}{1 + \exp\left(-\frac{E_{O_2} - \mu_{O_2}}{k_B T}\right)}. \tag{5.93}$$

This gives P(occupied) = 0.98 and P(occupied) = 0.02 with $\mu_{O_2} = -0.6$ eV at body temperature T = 310 K. A two-site model of hemoglobin takes into account the increased tendency to bind an oxygen molecule as the other binding sites become occupied. The occupation energies are $E_{O_2}^{(0)} = 0$ (one possible state), $E_{O_2}^{(1)} = -0.55$ eV (two possible states), and $E_{O_2}^{(2)} = -1.3$ eV (one possible state). The grand canonical partition function is then

$$Z_G = 1 + 2\exp\left(-\frac{E_{O_2}^{(1)} - \mu_{O_2}}{k_B T}\right) + \exp\left(-\frac{E_{O_2}^{(2)} - \mu_{O_2}}{k_B T}\right), \tag{5.94}$$

from which other thermodynamic quantities may be calculated in Exercise 5.5.

EXERCISES

Exercise 5.1 Calculate the change in entropy and Gibbs free energy per milliliter combining solutions of red and white blood cells with respective concentrations of 10^7 and 10^4 cells/ml at body temperature T = 310 K.

Exercise 5.2 Consider the two-state reaction

$$A \underset{k_-}{\overset{k_+}{\rightleftarrows}} B$$

and write down the differential equations describing the time evolution of the concentrations $[A(t)]$ and $[B(t)]$. Solve the corresponding differential equations with initial concentrations $[A]_0$ and $[B]_0$. Identify a characteristic time constant of the reaction.

Exercise 5.3 From the thermodynamic relations

$$G = \mu N = E + PV - TS$$

and

$$TS = \langle E\rangle - \mu\langle N\rangle + k_B T \ln Z_G$$

show that

$$\langle P\rangle = k_B T\left(\frac{\partial \ln Z_G}{\partial V}\right)_{T,\mu}.$$

Exercise 5.4 Calculate the average energy in myoglobin at body temperature (310 K).

Exercise 5.5 Calculate the occupation probabilities and the average energy of the two-site hemoglobin model at 310 K.

Exercise 5.6 Carbon monoxide more easily binds to hemoglobin with $E_{CO} = -0.85$ eV compared to the oxygen-binding energy $E_{O_2} = -0.7$ eV. In an environment with an oxygen to carbon monoxide ratio of 100:1 we have $\mu_{CO} = -0.72$ eV with $\mu_{O_2} = -0.6$ eV. Calculate the binding probability of both CO and O_2 in myoglobin and comment on the significance of your finding.

Exercise 5.7 Derive the Henderson–Hasselbalch equation

$$\text{pH} = \text{pK}_a + \log x$$

for the dissociation of the weak acid HA

$$\text{HA} \overset{\text{K}_a}{\rightleftarrows} \text{H}^+ + \text{A}^-$$

with equilibrium constant K_a

$$\text{where } x = \frac{[\text{A}^-]}{[\text{HA}]}, \text{pH} = -\log[\text{H}^+], \text{ and } \text{pK}_a = -\log \text{K}_a$$

Comment of the significance of the inflection point $\text{pH} = \text{pK}_a$ on a plot of pH versus pK_a.

6 Diffusion and Transport

6.1 Maxwell–Boltzmann Statistics

The mobility of small molecules and the motility of microorganisms are governed in large part by diffusion. We will first study the random motion of molecules to better understand the diffusion processes that play important roles in biology. In a gas, the distribution of molecular speeds is proportional to a Boltzmann factor where the probability that the ith molecule has energy E_i is proportional to

$$\exp\left(-\frac{E_i}{k_B T}\right) = \exp\left(-\frac{mv^2}{2k_B T}\right) \tag{6.1}$$

where the energy $E_i = mv^2/2$. Gas molecules obeying Maxwell–Boltzmann (M–B) statistics have a normalized distribution of velocities described by

$$n(v)dv = 4\pi\left(\frac{m}{2\pi k_B T}\right)^{3/2} v^2 \exp\left(-\frac{mv^2}{2k_B T}\right)dv. \tag{6.2}$$

To calculate the average speed, we evaluate the integral

$$\langle v \rangle = \int_0^\infty vn(v)dv = \sqrt{\frac{8k_B T}{\pi m}}. \tag{6.3}$$

To calculate the average of v^2 we evaluate

$$\langle v^2 \rangle = \int_0^\infty v^2 n(v)dv = \frac{3k_B T}{m}. \tag{6.4}$$

Techniques for integrating Equations 6.3 and 6.4 are reviewed in Appendix 1. The root mean square (rms) speed is given by

$$v_{\text{rms}} = \sqrt{\langle v^2 \rangle} = \sqrt{\frac{3k_B T}{m}}. \tag{6.5}$$

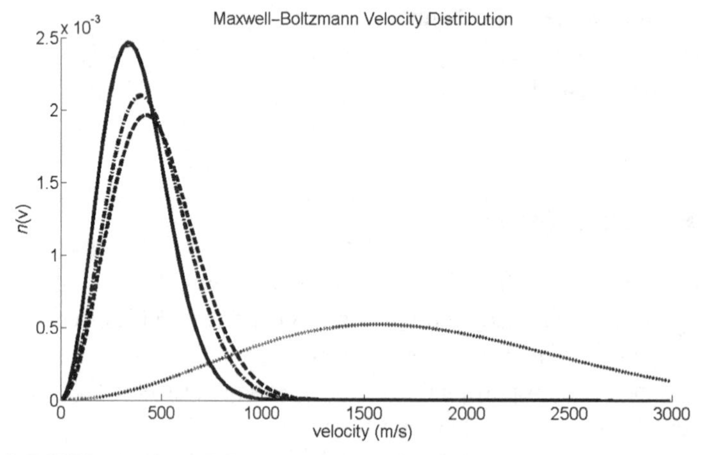

FIGURE 6.1 Maxwell–Boltzmann velocity distributions peaked from left to right: carbon dioxide, oxygen, nitrogen, and hydrogen at 300 K. The area under each curve is equal to one. Hydrogen, the lightest gas, has a most probable speed near 1500 m/s; the molecular speeds of the heavier molecules peaked near 500 m/s.

The most probable speed is obtained by setting $dn/dv = 0$ giving $v_p = \sqrt{2k_BT/m}$. The M–B velocity distribution is plotted using the MATLAB script below for several gases at 300 K in Figure 6.1.

```
%define molecular masses and constants

m_O2=5.31*10^-26 ;
m_H2=3.35*10^-27;
m_N2=4.65*10^-26;
m_CO2=7.32*10^-26;

k_B=1.38e-23;

T=300;

%define the velocity range and distributions

v=10:20:3000;

n_O2=4*pi*(m_O2/(2*pi*k_B*T))^(3/2)*(v.^2).*exp(-m_O2*v.^2/
(2*k_B*T));

n_H2=4*pi*(m_H2/(2*pi*k_B*T))^(3/2)*(v.^2).*exp(-m_H2*v.^2/
(2*k_B*T));

n_N2=4*pi*(m_N2/(2*pi*k_B*T))^(3/2)*(v.^2).*exp(-m_N2*v.^2/
(2*k_B*T));

n_CO2=4*pi*(m_CO2/(2*pi*k_B*T))^(3/2)*(v.^2).*exp(-m_CO2*v.^2/
(2*k_B*T));

%plot the distributions on the same graph

hold on

plot(v,n_O2,‘-.k’,‘LineWidth’, 4);

plot(v,n_H2,‘:k’,‘LineWidth’, 4);

plot(v,n_N2,‘—k’,‘LineWidth’, 4);

plot(v,n_CO2,‘k’,‘LineWidth’,4);

hold off

set(gca, ‘FontSize’, 15)
xlabel(‘velocity (m/s)’,‘FontSize’,15)
ylabel(‘n(v)’,‘FontSize’,15)
title(‘Maxwell-Boltzmann Velocity Distribution’,‘FontSize’,15)
```

6.2 Brownian Motion

In 1828 Robert Brown noticed the erratic motion of pollen grains suspended in water when observed under a microscope. Brown initially attributed this incessant jiggling of the micron-sized grains to some unknown biological process involving the pollen. Subsequent experiments substituting black soot and powdered stone from the sphinx revealed similar behavior. These observations, and the fact that the pollen grains never seemed to run out of energy when deprived of nutrients, led Brown to conclude that the motion was not related to life.

Albert Einstein published papers in 1905 describing the photoelectric effect, special relativity, and Brownian motion. In explaining Brownian motion, Einstein realized that molecules with very high velocity $v \gg v_{\text{rms}}$ corresponding to the "tail region" of the M–B distribution occasionally strike a solute particle resulting in its seemingly erratic motion when viewed under the microscope.

As an analogy, imagine walking straight through a crowed mall, oblivious to the counterflow of human traffic. Occasionally you might brush against shoppers walking in the opposite direction, resulting in very minor sideways displacements. Every great once in a while, however, you unfortunately stumble into an irate professional boxer who knocks you with a healthy blow, and you execute a quite a large displacement. In the case of Brownian motion, most of the collisions do not impart a sufficient impulse to result in a noticeable displacement of the solute particle. It is only the rare, highly energetic molecular collisions that are responsible for the particles' incessant dance.

For a particle undergoing a 1-D random walk $x(t)$, the room mean square displacement is

$$R_{\text{rms}} = \left\langle x(t)^2 \right\rangle^{1/2}. \tag{6.6}$$

The time-dependent displacement has the form

$$R_{\text{rms}} = \sqrt{2Dt} \tag{6.7}$$

where D is the diffusion constant in m²/s. In 2-D and 3-D we have $R_{\text{rms}} = \sqrt{4Dt}$ and $\sqrt{6Dt}$, respectively. Figure 6.2 shows a Brownian motion simulation with a single particle executing a random walk in two dimensions. This simulation is conducted using the following MATLAB script:

```
n_steps=5000;

x=zeros(n_steps,1);

y=zeros(n_steps,1);

for i=1:n_steps-1

          x(i+1)=x(i)+randn;
          y(i+1)=y(i)+randn;

end

plot(x,y,'k');
```

Here randn is normally distributed with mean 0 and standard deviation 1.

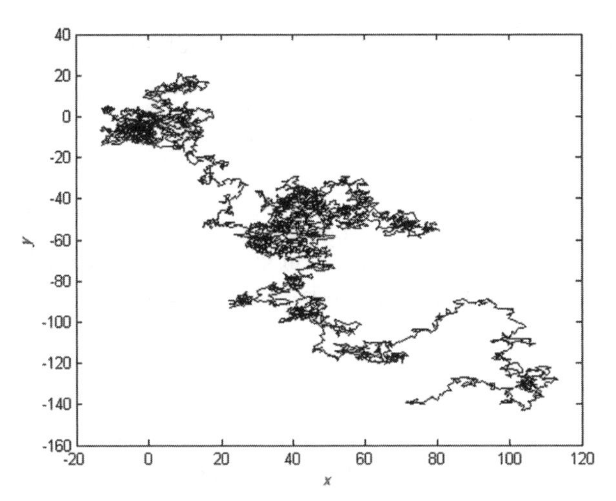

FIGURE 6.2 Path of a particle undergoing Brownian motion in two dimensions.

Temperature Dependence of Diffusion Constant

The diffusion constant of a body with characteristic length a is proportional to the thermal energy k_BT and inversely proportional to the viscosity η according to the Einstein relation

$$D = \frac{k_BT}{6\pi a\eta}. \tag{6.8}$$

The drag coefficient γ_{drag} is related to the viscosity

$$\gamma_{\text{drag}} = 6\pi a\eta \tag{6.9}$$

so that the Einstein relation between the diffusion constant and the drag coefficient is given by

$$D = \frac{k_BT}{\gamma_{\text{drag}}}. \tag{6.10}$$

A particle moving though a viscous medium at low speeds will experience a drag force proportional to its speed v according to

$$F_{\text{drag}} = v\gamma_{\text{drag}}. \tag{6.11}$$

The diffusion constant D is of order 10^{-5} cm^2/s for small biological molecules in aqueous solution. The diffusion constant of protein molecules is roughly an order of magnitude smaller because of their larger molecular weight. If diffusion is governed by processes with activation energy E_a, then D is proportional to the Boltzmann factor

$$D = D_0 \exp\left(-\frac{E_a}{k_B T}\right). \tag{6.12}$$

This could correspond to diffusion of particles across an energy barrier. In such systems, the probability for a particle to surmount the energy barrier per second is

$$P \approx f_{\text{vib}} \exp\left(-\frac{E_a}{k_B T}\right) \tag{6.13}$$

where f_{vib} is a characteristic atomic or molecular vibration frequency. In the Eyring rate theory, $f_{\text{vib}} = k_B T/h$ or $0.63 \cdot 10^{13}$ Hz at room temperature, where h is Planck's constant.

6.3 Fick's Laws of Diffusion

Fick's laws were developed in 1855 by Adolf Fick to describe diffusion in fluids; however, these laws are also applicable to diffusion in gases and solids. These laws are mathematically similar to the heat equation and Schrödinger's equation and can be used to model transport processes in lipids, living cells, and neurons. Essential to cognitive function is the diffusion of neurotransmitter molecules across the synapse between nerve cells as discussed in Chapter 10.

6.3.1 Fick's First Law

If an ink drop is introduced into a glass of water, the ink molecules will diffuse from regions of higher to lower concentration C as shown in Figure 6.3.

The diffusive flux J is proportional to the negative concentration gradient times the diffusion constant according to Fick's first law

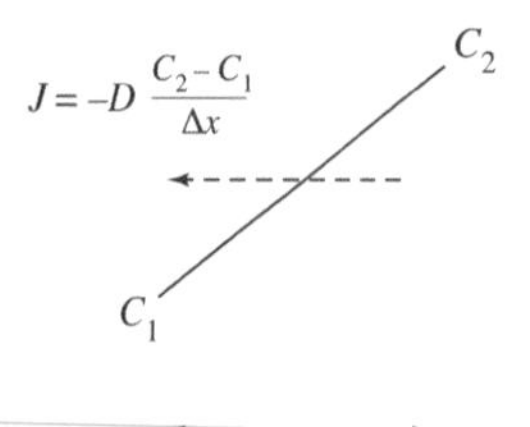

FIGURE 6.3 Flux vector pointing opposite to the concentration gradient according to Fick's first law.

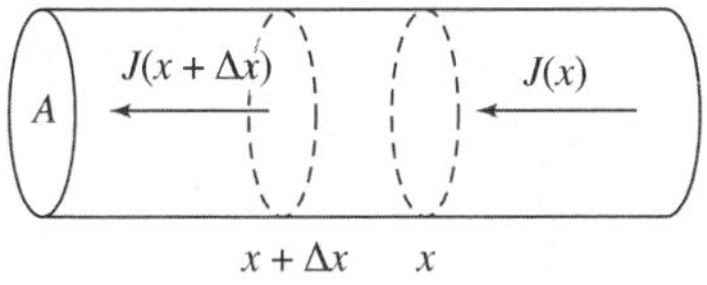

A

$J(x + \Delta x)$ $J(x)$

$x + \Delta x$ x

FIGURE 6.4 Spatial variation of particle flux J that is proportional to the time rate of change of concentration as specified by Fick's second law.

$$J = -D\frac{\partial C}{\partial x}. \tag{6.14}$$

Note that there is no diffusive flux if the concentration is everywhere spatially uniform. In three dimensions, we write Fick's first law as

$$\mathbf{J} = -D\nabla C. \tag{6.15}$$

Conservation of mass requires that the spatial variation of the diffusive flux is equal to the time rate of change of the concentration. Consider the cylindrical tube in Figure 6.4 containing a substance with concentration $C(x)$. If the flux $J(x + \Delta x) = J(x)$, then the concentration does not change with time. If the flux leaving the volume element is greater than flux entering the element, then the concentration decreases with time and vice versa. Thus, the flux gradient is minus the time rate of change of concentration or,

$$\frac{\partial J}{\partial x} = -\frac{\partial C}{\partial t}. \tag{6.16}$$

6.3.2 Fick's Second Law

Taking the derivative of Fick's first law with respect to x gives

$$\frac{\partial J}{\partial x} = -D\frac{\partial^2 C}{\partial x^2}. \tag{6.17}$$

Substituting Equation 6.16 into this equation gives Fick's second law

$$\frac{\partial C}{\partial t} = D\frac{\partial^2 C}{\partial x^2}. \tag{6.18}$$

For concentration varying in three spatial dimensions we write

$$\frac{\partial C}{\partial t} = D\nabla^2 C. \tag{6.19}$$

This equation may be solved by separation of variables. The simplest example with one spatial dimension is a solution maintained at concentration $C(0, t) = C_0$. The concentration along the x-axis is then given by $C(x, t) = C_0 \text{erfc}\left(x/\sqrt{4Dt}\right)$ where erfc (z) is the complementary error function erfc (z) = 1 − erf (z).

6.3.3 Quantum Diffusion

Fick's second law has the same mathematical form as the heat equation encountered in Chapter 3 as well as Schrödinger's equation for a free particle in the absence of external potentials where $V(x) = 0$:

$$i\hbar \frac{\partial \Psi}{\partial t} = \frac{-\hbar^2}{2m} \nabla^2 \Psi \tag{6.20}$$

where $\hbar$ is Planck's constant $\div 2\pi$ and $\Psi^*\Psi = |\Psi(x, t)|^2$ is the probability of locating the electron at (x, t) as discussed in Appendix 7. An electron at a particular location in empty space will evidently wander off on its own accord. This is quantum diffusion. The quantum flux, or probability current density, is

$$J = \frac{\hbar}{2mi}(\Psi^* \nabla \Psi - \Psi \nabla \Psi^*), \tag{6.21}$$

similar to Fick's first law where $\Psi^*\Psi$ is analogous to the concentration C. The tendency to diffuse would appear to be a fundamental aspect of nature. Diffusion will tend to increase the entropy of a system. That is consistent with the second law of thermodynamics. In fact the universe as a whole is expanding with the "concentration" of galaxies decreasing with time. We return to the subject of quantum mechanics considering the light absorption of biomolecules and the vibrational spectra of atoms in Chapter 8.

6.3.4 Time-Independent Concentrations

Returning to Fick's second law, when the concentration is not changing with time, we have $dC/dt = 0$, so that C obeys Laplace's equation $\nabla^2 C = 0$. The steady state solution to Fick's equation in spherical coordinates is thus

$$\frac{1}{r^2}\frac{\partial}{\partial r}\left(r^2 \frac{\partial C}{\partial r}\right) = 0. \tag{6.22}$$

Multiplying by r^2 and integrating gives

$$r^2 \frac{\partial C}{\partial r} = c_1. \tag{6.23}$$

Dividing by r^2 and integrating a second time gives

$$C(r) = -\frac{c_1}{r} + c_2. \tag{6.24}$$

The integration constants c_1 and c_2 are obtained from the initial conditions $C(R) = C_0$ and $C(\infty) = 0$, respectively, in turn giving $c_1 = -C_0R$ and $c_2 = 0$ so that

$$C(r) = C_0\frac{R}{r}. \tag{6.25}$$

The flux is then obtained from the concentration gradient

$$\mathbf{J} = -D\frac{\partial C}{\partial r}\hat{r} = DC_0\frac{R}{r^2}\hat{r} \tag{6.26}$$

where $\hat{r}$ is a unit vector pointing in the radial direction.

6.3.5 Fick's Law for Growing Bacterial Cultures

Fick's second law can be modified to model the spatial distribution of bacteria reproducing at a rate proportion to their concentration λC:

$$\frac{\partial C}{\partial t} = D\nabla^2 C + \lambda C \tag{6.27}$$

where λ is the reproduction rate constant. The diffusion constant of a bacterium swimming with a speed v over straight runs with average duration τ is

$$D = \frac{v^2\tau}{3(1-\alpha)} \tag{6.28}$$

where the average value $\alpha = \langle\cos\theta\rangle$ and θ is the angle between exponentially distributed runs. If the run directions are completely random, then

$$D = v^2\tau/3. \tag{6.29}$$

The following MATLAB M-file script creates an animation of bacterial growth and diffusion beginning from the center of a two-dimensional square plate.

```
time_steps=1000;
n =100; D=0.2; lambda = 0.5;

C=zeros(n);                  %initialize concentration array
grad_C=zeros(n);             %initialize derivative array
```

```
C(n/2,n/2)=1.0;
i= 2:n-1; j=2:n-1;
for step = 1:time_steps
            grad_C(i,j)=C(i,j-1)+C(i,j+1)+C(i-1,j)+C(i+1,j);
            C=(1-4*D)*C+ D*grad_C+ lambda*C;

            pause(.01);
            imagesc(C);
end
```

The final result of this simulation can be viewed without the animation by removing the pause command and placing the imagesc command outside of the time loop.

6.4 Sedimentation of Cell Cultures

A colloidal suspension of particles such as yeast cells will settle out in a gravitational field so that the concentration C increases with depth. The suspended particles are acted on by gravity, the buoyant force, and thermal fluctuations. Archimedes' principle states that the buoyant force is given by the weight of the water displaced by a submerged particle so that the net vertical force in absence of thermal fluctuations is m^*g where

$$m^* = (\rho - \rho_{\text{water}})V. \tag{6.30}$$

The gravitational potential energy is m^*gz so that C is given by a Boltzmann factor

$$C(z) = C_0 \exp\left(-\frac{m^*gz}{k_B T}\right) \tag{6.31}$$

where the concentration of particles is C_0 at the bottom of the container.

(a) Standard cell culture in 1g

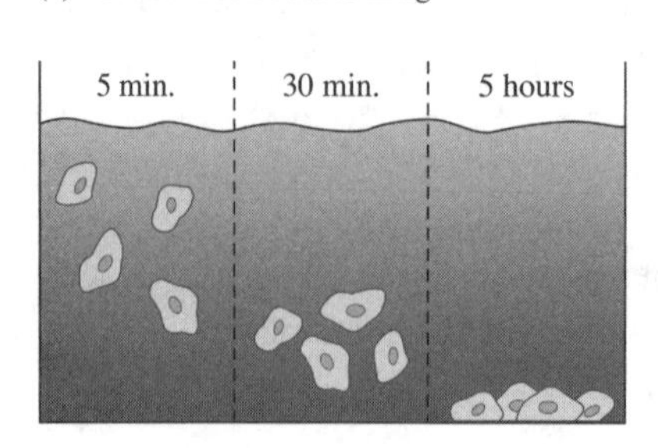

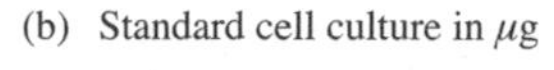

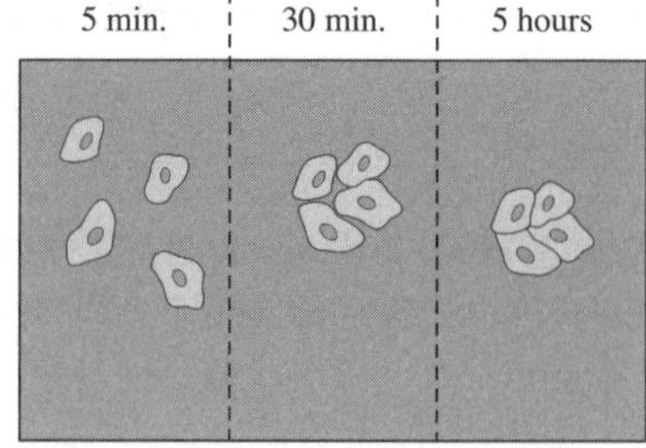

FIGURE 6.5 Settling of a cell culture in (a) a 1-g gravitation field and in (b) a microgravity environment. (Courtesy of NASA Marshall Space Flight Center Collection/Courtesy of nasaimages.org.)

Figure 6.5 shows a schematic of the settling of a cell culture in Earth's gravitational field and in a microgravity environment. Cells can more freely form large three-dimensional aggregates in the absence of settling. Also, there is no buoyant force in microgravity since the buoyant force is equal to the weight of the suspension displaced by the cells according to Archimedes' principle.

6.5 Diffusion in a Centrifuge

Fick's laws can be used to calculate the concentration profile in a centrifuge rotating with angular speed ω. From Newton's second law

$$F = ma = m\frac{v^2}{r} \tag{6.32}$$

and with $v = r\omega$, we have that the centripetal acceleration is supplied by the frictional drag force

$$mr\omega^2 = \gamma_{\text{drag}} v_r \tag{6.33}$$

where v_r is the radical speed. The drift flux is then

$$Cv_r = C\frac{mr\omega^2}{\gamma_{\text{drag}}}. \tag{6.34}$$

Including the drift flux in Fick's first law gives the Nernst equation

$$J = C\frac{mr\omega^2}{\gamma_{\text{drag}}} - D\frac{dC}{dr}. \tag{6.35}$$

Using the Einstein relation describing the temperature dependence of the drag coefficient and factoring gives

$$J = D\left(C\frac{mr\omega^2}{k_B T} - \frac{dC}{dr}\right). \tag{6.36}$$

For equilibrium concentrations, we have $J = 0$ so that

$$C\frac{mr\omega^2}{k_B T} = \frac{dC}{dr}. \tag{6.37}$$

Integrating this expression, we first separate variables

$$\frac{mr\omega^2}{k_B T}dr = \frac{dC}{C} \tag{6.38}$$

so that

$$\int \frac{mr\omega^2}{k_B T}\,dr = \int \frac{dC}{C}, \tag{6.39}$$

which gives

$$\ln C = \frac{mr^2\omega^2}{2k_B T} + \text{const.} \tag{6.40}$$

Exponentiating this expression with $C(0) = C_0$ gives the radial concentration

$$C(r) = C_0 \exp\left(\frac{mr^2\omega^2}{2k_B T}\right). \tag{6.41}$$

The NASA Bioreactor

The NASA bioreactor rotates cell cultures thus simulating microgravity conditions that allow the formation of three-dimensional cell aggregates. The bioreactor is slowly rotated to minimize turbulence and to allow circulation of nutrients. Slow rotation

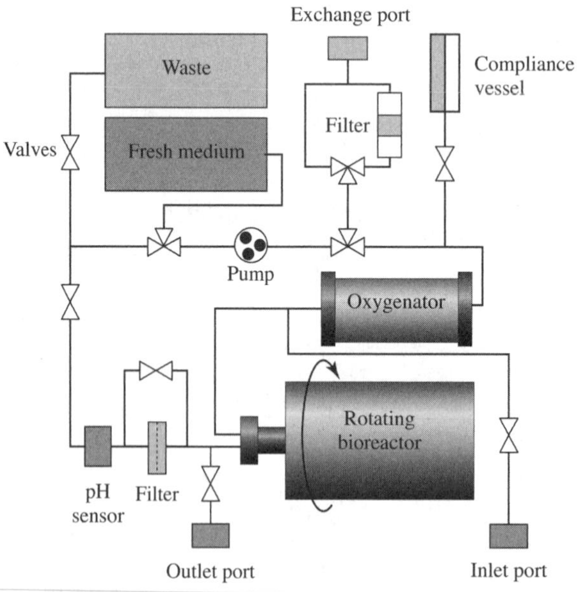

FIGURE 6.6 Schematic of the NASA bioreactor. (Courtesy of NASA Marshall Space Flight Center Collection/Courtesy of nasaimages.org.)

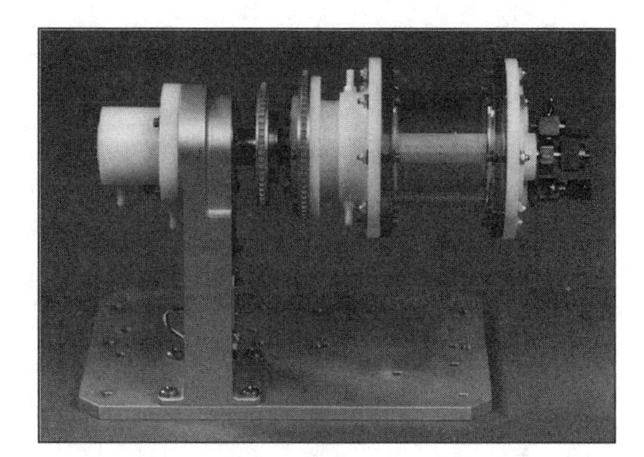

FIGURE 6.7 Rotating sample cell portion of the NASA bioreactor. (Courtesy of NASA Marshall Space Flight Center Collection/Courtesy of nasaimages.org.)

reduces shear forces that would prevent aggregation under magnetic stirring. This technique is used to simulate tumor growths as well as the growth of tissues such as heart, bone marrow, and muscle. A schematic of the bioreactor is shown in Figure 6.6, depicting the main elements including the rotating sample cell, oxygenation chamber, and the nutrient and filtration systems.

The rotating portion of the bioreactor is shown in Figure 6.7. The flow pattern is established by periodically squeezing the plastic tubing thereby pumping the fluid by peristaltic action to avoid contamination by mechanical contact with the cell medium.

6.6 Diffusion in an Electric Field

Particles with charge q will experience a net force in an electric field of magnitude E:

$$m\frac{dv}{dt} = qE - \gamma_{\text{drag}} v. \tag{6.42}$$

The terminal drift speed

$$v_t = \frac{qE}{\gamma_{\text{drag}}} \tag{6.43}$$

is attained when the acceleration is zero. The electrophoretic flux, with units of number of particles per second, is $J = Cv_t$. Fick's first law now becomes

$$J = C\frac{qE}{\gamma_{\text{drag}}} - D\frac{\partial C}{\partial x}. \tag{6.44}$$

The Nernst–Planck equation is obtained from the Einstein relation $D = k_B T/\gamma_{\text{drag}}$:

$$J = D\left(-\frac{\partial C}{\partial x} + C\frac{qE}{k_B T}\right). \tag{6.45}$$

In three dimensions, the Nernst–Planck equation is

$$\mathbf{J} = D\left(-\nabla C + C\frac{q\mathbf{E}}{k_B T}\right) \tag{6.46}$$

or

$$\mathbf{J} = -D\nabla C + C\mu\mathbf{E} \tag{6.47}$$

where

$$\mu = \frac{qD}{k_B T} \tag{6.48}$$

is the mobility of charged particles.

6.7 Lateral Diffusion in Membranes

Biological membranes behave as a two-dimensional viscous fluid. Proteins and lipids embedded in the membrane are not rigidly constrained but can diffuse in the plane of the membrane as illustrated in Figure 6.8.

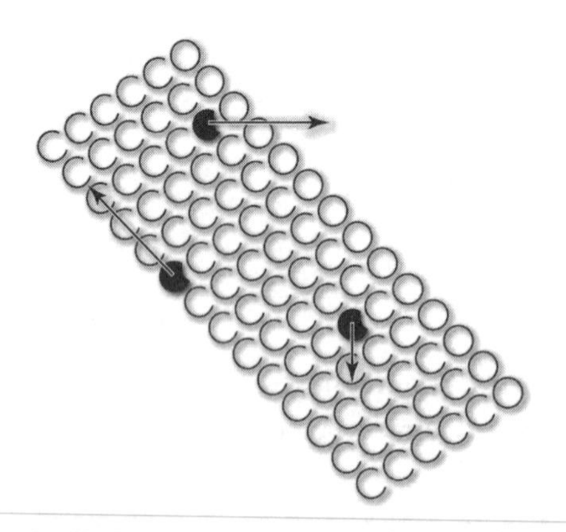

FIGURE 6.8 Lateral diffusion of proteins (dark circles) in the plasma membrane. Lipid molecules are represented by open circles.

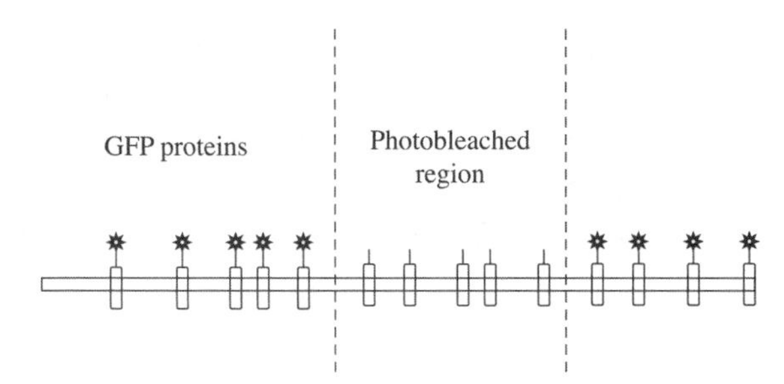

FIGURE 6.9 Illustration of the technique of fluorescence recovery after photobleaching (FRAP).

The diffusion of proteins through the fluid mosaic membrane structure has been studied using fluorescent labels. The technique of fluorescent recovery after photobleaching (FRAP) can be used to determine the membrane diffusion constant by proteins tagged with green fluorescent protein (GFP). A small region with GFP-tagged protein is photobleached as shown in Figure 6.9. The recovery time for GFP proteins to diffuse into the photobleached region is monitored to measure the diffusion constant. Using this technique, Cole et al. measured the diffusion constants of Golgi membrane proteins with lateral diffusion constants ranging between 3 and $5 \cdot 10^{-9}$ cm²/s.

The lateral diffusion constant of a cylinder of radius a imbedded in a membrane of thickness h was calculated by Saffman and Delbrück:

$$D_L = \frac{k_B T}{4\pi\eta_{\mathrm{m}} h}\left(\ln\frac{\eta_{\mathrm{m}} h}{\eta_{\mathrm{w}} a} - 0.5772\right) \tag{6.49}$$

where η_{m} and η_{w} are the viscosities of the membrane and surrounding aqueous medium, respectively. The theoretical value of D_L turns out to be lower than the measured value because of crowding of membrane proteins that can occupy roughly 50% of the plasma membrane. Also, some proteins are attached to cytoskeletal filaments and are further impeded in their diffusion.

6.8 Navier–Stokes Equation

The Navier–Stokes equation describing the flow of a fluid with viscosity η and density ρ can be written as a sum of inertial, pressure, viscous, and driving terms:

$$\underbrace{\rho\left(\frac{\partial \mathbf{v}}{\partial t} + (\mathbf{v}\cdot\nabla)\mathbf{v}\right)}_{\text{inertial terms}} = -\underbrace{\overbrace{\nabla P}^{\text{pressure}} + \overbrace{\eta\nabla^2\mathbf{v}}^{\text{viscous term}}}_{\text{fluid stress}} + \overbrace{F_{\mathrm{ext}}}^{\text{external driving term}} \quad . \tag{6.50}$$

The solution of this partial differential equation gives the spatial and temporal fluid velocity **v**. The Navier–Stokes equation is essentially a generalization of Newton's second law applied to a fluid with motion resulting from driving forces such as gravity, and fluid stresses supplied by pressure differentials and viscosity. The inertial terms can be expressed as a convective derivative $D\mathbf{v}/Dt$ where

$$\frac{D}{Dt} = \frac{\partial}{\partial t} + (\mathbf{v} \cdot \nabla). \tag{6.51}$$

Dividing by the fluid density, Equation 6.50 becomes

$$\frac{\partial \mathbf{v}}{\partial t} + (\mathbf{v} \cdot \nabla)\mathbf{v} = -\frac{\nabla P}{\rho} + \upsilon \nabla^2 \mathbf{v} + f_{\text{ext}} \tag{6.52}$$

where $f_{\text{ext}} = F_{\text{ext}}/\rho$ and the kinematic velocity υ is given by the ratio of the fluid viscosity and its density $\upsilon = \eta/\rho$. Each term in the above equation has units of acceleration.

6.8.1 Reynolds Number

The Reynolds number is a useful measure of the fluid flow proportional to the ratio of inertial to viscous forces

$$\frac{(\mathbf{v} \cdot \nabla)\mathbf{v}}{\upsilon \nabla^2 \mathbf{v}} \sim \frac{v^2/a}{\upsilon v/a^2} = \frac{va}{\upsilon}. \tag{6.53}$$

The dimensionless Reynolds number is given by

$$R = \frac{\rho v a}{\eta}. \tag{6.54}$$

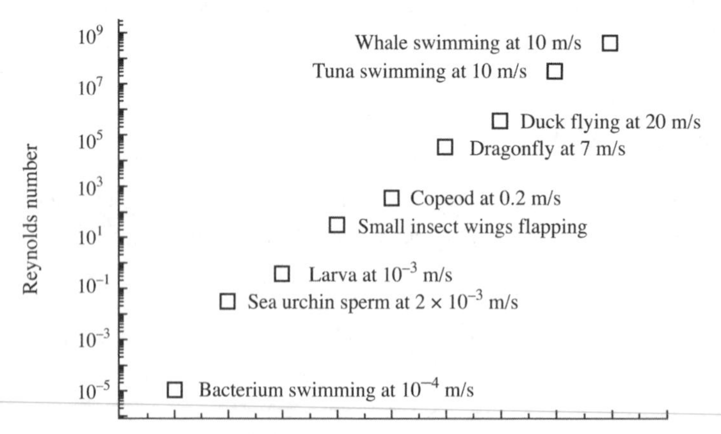

FIGURE 6.10 Reynolds numbers of moving organisms ranging over 14 orders of magnitude. (Data from Vogel. 1994. *Life in Moving Fluids.* Princeton University Press, Princeton, New Jersey.)

This number may be used to characterize the flow around a body of spatial extent a moving through a medium with speed v, or to describe a viscous fluid flowing through a pipe (or blood vessel) of diameter a. Low Reynolds number is a characteristic of laminar flow dominated by viscous forces, while turbulent flow is associated with a high Reynolds number where inertial forces dominate. Figure 6.10 illustrates Reynolds numbers extending over 14 orders of magnitude corresponding to different speeds of various organisms.

■ 6.9 Low Reynolds Number Transport

6.9.1 Purcell's "Life at Low Reynolds Number"

Purcell's seminal *Physics Today* article "Life at Low Reynolds Number" explores the low Reynolds number world of microorganisms where transport is dominated by diffusion. Because viscous forces dominate at low Reynolds number, an organism's locomotion is not assisted by inertia; for example, *E. coli* do not coast as illustrated below.

Swimming using a back and forth-type of reciprocal motion will be ineffective unless accompanied by cyclic conformational changes. However, microorganisms may propel themselves using rotary motion with the biological equivalent of propellers—the flagella.

Purcell posed the question of whether an organism could increase its food supply by stirring its environment. The characteristic transport time for stirring is obtained by dividing the organism's length by the stirring speed $t_s = l/v$. This time is compared to the diffusion time $t_D = l^2/D$. We expect stirring to be effective in supplying the organism with nutrients if the Sherwood number $S = t_D/t_s$ is greater than one, or $lv/D > 1$. It turns out that $S \sim 10^{-2}$ for $(l, D, v) \sim (1\ \mu\text{m}, 10^{-5}\ \text{cm}^2/\text{s}, 10\ \mu\text{m/s})$, so there is no advantage for a bacterium to stir its surroundings at low Reynolds number.

Another possibility is that an organism might increase its nutrient absorption rate by traveling through the surrounding medium. For an organism to increase its food supply even 10%, it must travel 700 μm/s to outrun diffusion. Although microorganisms cannot travel this fast, they can swim to areas with higher nutrient concentrations. The purpose of swimming at low Reynolds number is therefore to find "greener pastures" in which to graze.

6.9.2 Coasting Distance of a Bacterium

As mentioned in the previous section, the distance a bacterium will coast without self-propulsion is miniscule at low Reynolds number. Here we explicitly calculate the coasting distance with the drag force equal to the coating speed times the drag coefficient

$$F_{\text{drag}} = \gamma_{\text{drag}} v. \tag{6.55}$$

From Newton's second law, the drag force is equal to the bacterium's mass times its acceleration

$$m\frac{dv}{dt} = -\gamma_{\text{drag}}\, v. \tag{6.56}$$

Separating variables

$$\frac{dv}{v} = -\frac{\gamma_{\text{drag}}}{m}dt \tag{6.57}$$

and integrating

$$\int \frac{dv}{v} = -\frac{\gamma_{\text{drag}}}{m}\int dt \tag{6.58}$$

gives

$$\ln v = -\frac{\gamma_{\text{drag}}}{m}t + \text{const.} \tag{6.59}$$

Exponentiating both sides and taking $v(0) = v_0$ gives

$$v(t) = v_0 \exp\left(-\frac{\gamma_{\text{drag}}}{m}t\right). \tag{6.60}$$

Integrating the velocity to obtain the distance

$$d = \int_0^\infty v(t) = v_0 \int_0^\infty \exp\left(-\frac{\gamma_{\text{drag}}}{m}t\right)dt, \tag{6.61}$$

we finally obtain

$$d = \frac{v_0 m}{\gamma_{\text{drag}}}. \tag{6.62}$$

For a typical bacterium this length is $\sim 10^{-11}$ m for $v_0 \sim 10^{-5}$ m/s.

6.10 Active and Passive Membrane Transport

The sodium–potassium pump Na^+/K^+-ATPase located in the cellular plasma membrane establishes ionic concentration gradients resulting in a resting potential difference across the membrane. The pump shown schematically in Figure 6.11 maintains a higher sodium concentration outside the cell with a higher potassium concentration inside the cell.

Initially, the pump binds three sodium ions inside the cell along with one ATP molecule. The pump changes its conformation, accompanied by decreased sodium affinity, through ATP hydrolysis. The sodium ions are then released outside of the membrane. Two potassium ions are then bound outside the cell, causing a release of the phosphate. The released phosphate induces a conformational change, and the potassium ions flow inside the cell. The cycle then repeats itself. Potassium ions can exit the cell through leak channels that are always open. Sodium can come back into the cell, moving down its concentration gradient while bringing in larger molecules like glucose. The sodium–glucose transporter binds two sodium ions and a glucose molecule. Conformation change of this pump upon binding then causes glucose and sodium to enter the cell.

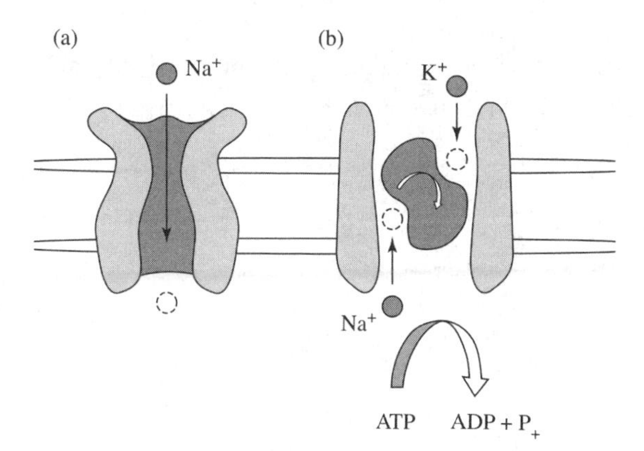

FIGURE 6.11 Schematic of (a) passive ion channel and (b) active transport by the sodium–potassium pump.

EXERCISES

Exercise 6.1 Show that the concentration of molecules described by

$$C(x,t) = \frac{A}{t^{1/2}} \exp\left(-\frac{x^2}{4\mu t}\right)$$

satisfies the diffusion equation

$$D\frac{\partial^2 C}{\partial x^2} = \frac{\partial C}{\partial t}.$$

Calculate the diffusion constant D in terms of the parameter μ.

Exercise 6.2 Calculate the average, the root mean square, and the most probable speed of gas molecules obeying Maxwell–Boltzmann statistics with velocity distribution

$$n(v)dv = 4\pi\left(\frac{m}{2\pi k_B T}\right)^{3/2} v^2 \exp\left(-\frac{mv^2}{2k_B T}\right).$$

Show that the Maxwell–Boltzmann velocity distribution is normalized. Refer to Appendix 1 to evaluate the required integrals.

Exercise 6.3 Show that, for a spherical absorber of radius R in an infinite medium with $C(R) = 0$ and $C(\infty) = C_0$, the time-independent concentration is given by

$$C(r) = C_0\left(1 - \frac{R}{r}\right),$$

and the magnitude of the flux is

$$J = -DC_0\frac{R}{r^2}.$$

Exercise 6.4 Fick's equation for time-independent concentrations in cylindrical coordinates is

$$\frac{1}{r}\frac{\partial}{\partial r}\left(r\frac{\partial C}{\partial r}\right) = 0.$$

Show that, for a cylindrical emitter of radius R in an infinite medium with $C(R) = C_0$ and $C(\infty) = 0$, the radial flux magnitude is given by

$$J = DC_0\frac{R}{r}.$$

Exercise 6.5 Develop an analytical expression for the drift speed of a protein with mass M and charge q imbedded in a plasma membrane with an electric field $\mathbf{E}$ parallel to the membrane at temperature T.

Exercise 6.6 Perform a 2-D Brownian motion simulation. Calculate the average distance a particle travels as a function of the total number of steps taken N. Simulate 1000 particles beginning random walks at the origin. Plot the average distance versus the number of steps the particles travel on a log–log plot. Interpret your results in terms of the average distance versus total time traveled, the where the total time is proportional to N.

Exercise 6.7 Repeat the previous exercise with the particles executing 3-D random walks.

Exercise 6.8 Perform a 3-D simulation of bacterial growth and diffusion using Equation 6.27 with $C(x, y, z)$.

7 Fluids

7.1 Laminar and Turbulent Fluid Flow

The flow of fluids, including both gases and liquids essential for life processes, can be characterized as either laminar or turbulent. Laminar flow proceeds with the smooth motion of adjacent fluid layers sliding continuously past each other without breaking into whirlpools or vortices. This type of fluid dynamics, also known as streamline flow, can be visualized by lines of constant fluid speed called streamlines. Fluid speeds are higher in regions where the density of streamlines is greater as shown in Figure 7.1. Laminar streamlines never cross because a fluid cannot have two velocity values at a single point. Steady state fluid flow is a special case of laminar flow where fluid velocity is everywhere time independent.

Turbulent flow occurs with a breakup of adjacent fluid layers resulting in whirlpools and vortices as shown in Figure 7.2. The fluid velocity at a given point in a turbulent region changes rapidly with time. A flow may transition from laminar to turbulent as the fluid speed is increased. Turbulence also results from the rapid motion of bodies through a fluid.

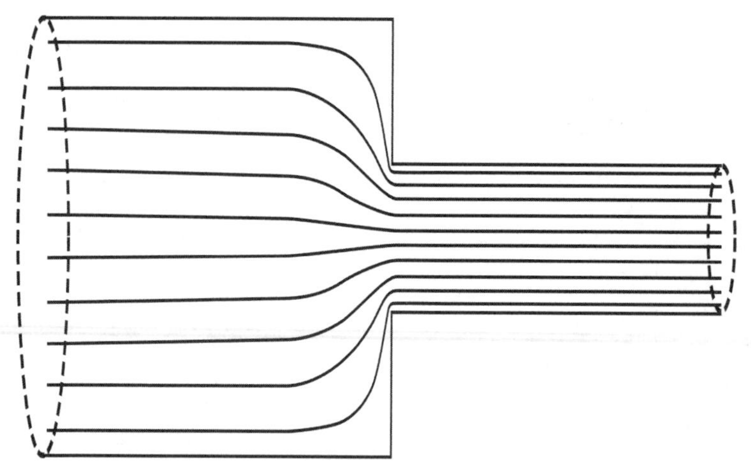

FIGURE 7.1 Streamline flow though a pipe with a greater density of streamlines in the constricted region corresponding to a higher fluid speed with reduced pressure.

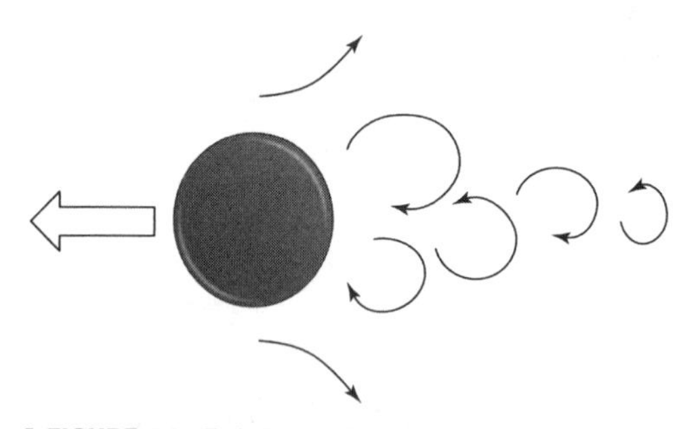

FIGURE 7.2 Turbulent wake behind an object traveling through a viscous medium.

7.2 Bernoulli's Equation

The Navier–Stokes equation may be written for fluid flow in a gravitation field

$$\frac{\partial \mathbf{v}}{\partial t} + (\mathbf{v} \cdot \nabla)\,\mathbf{v} = -\frac{\nabla P}{\rho} + \upsilon \nabla^2 \mathbf{v} + f_{\text{body}} \tag{7.1}$$

where $\upsilon = \eta/\rho$ and $f_{\text{body}} = -\mathbf{g}$. We may derive Bernoulli's equation neglecting viscous drag and acceleration of the fluid. Discarding $\upsilon\nabla^2\mathbf{v}$ and $\partial\mathbf{v}/\partial t$ gives

$$(\mathbf{v} \cdot \Delta)\,\mathbf{v} = -\frac{\nabla P}{\rho} - \mathbf{g}. \tag{7.2}$$

Integrating both sides of this equation

$$\rho\int(\mathbf{v}\cdot\nabla)\mathbf{v}\cdot\mathbf{dy} = -\int\nabla P\cdot\mathbf{dy} - \rho\int\mathbf{g}\cdot\mathbf{dy} \tag{7.3}$$

where $\mathbf{g} = g\mathbf{j}$ and $\mathbf{dy} = dy\mathbf{j}$ gives

$$\rho\int_{v_1}^{v_2} v\,dv = -\int_{P_1}^{P_2} dP - \rho g\int_{y_1}^{y_2} dy. \tag{7.4}$$

Thus we have Bernoulli's equation

$$P_1 + \frac{1}{2}\rho v_1^2 + \rho g y_1 = P_2 + \frac{1}{2}\rho v_2^2 + \rho g y_2 \tag{7.5}$$

so that the quantity

$$P + \frac{1}{2}\rho v^2 + \rho g y = \text{const.} \tag{7.6}$$

between any two points along the streamline flow. Bernoulli's equation can be regarded as a statement of conservation of energy density where each term in Equation 7.6 has dimensions of energy/volume.

7.3 Equation of Continuity

The steady state flow rate through a given pipe or blood vessel is

$$\frac{dv}{dt} = \text{const.} \tag{7.7}$$

with units of volume per time. Consider a plug of fluid entering the pipe in Figure 7.3 with a cross-sectional area A_1.

The plug of fluid travels a distance Δx_1 in a time Δt. The fluid exits the pipe with cross-section A_2 traveling a distance Δx_2 in the same time interval. Since the volume of fluid entering the pipe must be equal the volume leaving the pipe in the same time Δt, we have

$$\frac{1}{\Delta t}(A_1\Delta x_1) = \frac{1}{\Delta t}(A_2\Delta x_2). \tag{7.8}$$

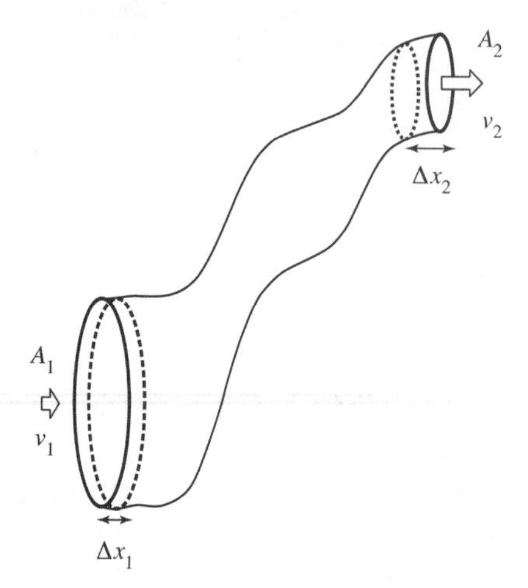

FIGURE 7.3 Fluid flow inside a pipe with varying cross section.

Taking the limit as $\Delta t \to 0$ gives

$$A_1 \frac{dx_1}{dt} = A_2 \frac{dx_2}{dt}, \tag{7.9}$$

and we obtain the continuity equation

$$A_1 v_1 = A_2 v_2, \tag{7.10}$$

or $Av =$ const. Fluid will flow faster through a constriction in a pipe or a blood vessel since $A\downarrow \Rightarrow v\uparrow$ because the product of area and fluid speed is a constant. Placing one's thumb partially over the opening of a garden hose results in a higher-velocity stream. The increase in fluid speed may lead to a transition from laminar to turbulent flow. A familiar example would be the flow of water from a kitchen faucet. The fluid speed increases as the water accelerates downward due to gravity. The column of water begins to constrict as the fluid speed increases because of the continuity relation. Finally, the stream breaks into a white water turbulent flow before it reaches the bottom of the sink.

7.4 Venturi Effect

The Venturi effect describes the change in pressure when fluid flows through a pipe with changing cross-sectional area. For simplicity we consider the constant height Bernoulli equation $y_1 = y_2$ with with no change in gravitational potential:

$$P + \frac{1}{2}\rho v^2 = \text{const.} \tag{7.11}$$

Comparing two points along the streamline flow, we can calculate the change in pressure:

$$P_1 - P_2 = \frac{1}{2}\rho\left(v_2^2 - v_1^2\right). \tag{7.12}$$

From the continuity equation

$$P_1 - P_2 = \frac{1}{2}\rho v_1^2\left[\left(\frac{A_1}{A_2}\right)^2 - 1\right] \tag{7.13}$$

so that a constriction in a blood vessel results in a *drop* in pressure where $A\downarrow \Rightarrow v\uparrow \Rightarrow P\downarrow$, counter to what one might expect. An example of the Venturi effect is given by airflow across a wing in Figure 7.4(a). In the laminar regime, fluid across the upper surface of the airfoil travels a longer distance compared to the lower surface. The fluid speed is therefore greater on the upper surface so that the pressure is lower than the bottom surface. The lifting force is then given by the pressure difference times the wing area $(P_2 - P_1)A$. Momentum transfer of the fluid also contributes to the lift as illustrated in Figure 7.4(b).

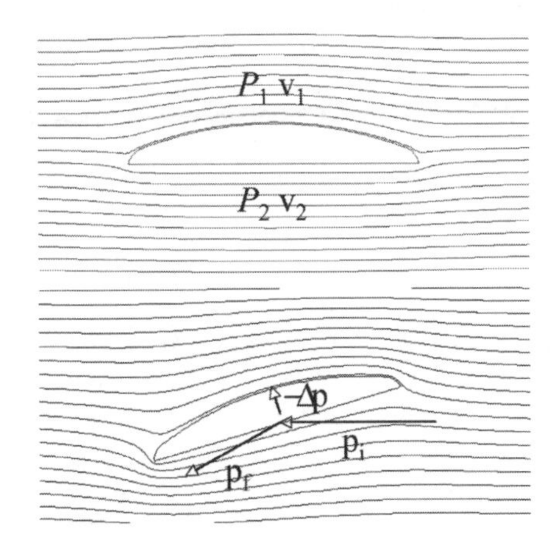

FIGURE 7.4 Streamline flow around a wing airfoil with varying angles of attack. (a) Faster fluid flow across the upper surface of the wing results in a lower pressure than at the bottom surface contributing to the lift provided by the wing. (b) Change in momentum of a fluid element by the wing contributing to the lift.

7.5 Fluid Dynamics of Circulatory Systems

7.5.1 Viscous Flow

For viscous flow through a blood vessel, frictional forces with the walls and internal friction in the fluid cause it to flow faster at the center of the blood vessel compared to the walls, as shown in Figure 7.5.

Considering the blood flow in concentric layers, a shear force between adjacent fluid layers is given by

$$F = \eta A_{\parallel} \frac{dv}{dr} \tag{7.14}$$

where η is the fluid viscosity and the cylindrical area surrounding a bundle of streamlines is

$$A_{\parallel} = 2\pi rL. \tag{7.15}$$

From the pressure differential

$$\delta P = \frac{F}{\pi r^2} \tag{7.16}$$

and the area in Equation 7.15, the fluid speed as a function of the radial coordinate is obtained by integrating Equation 7.14

$$v(r) = \int_r^R \frac{\delta P r}{2\eta L} dr, \tag{7.17}$$

which gives

$$v(r) = \frac{\delta P \left(R^2 - r^2\right)}{4\eta L}. \tag{7.18}$$

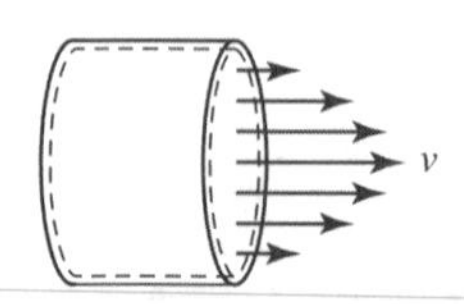

FIGURE 7.5 Fluid velocity profile in a blood vessel. Blood flows faster at the center of the vessel than near the walls of the vessel.

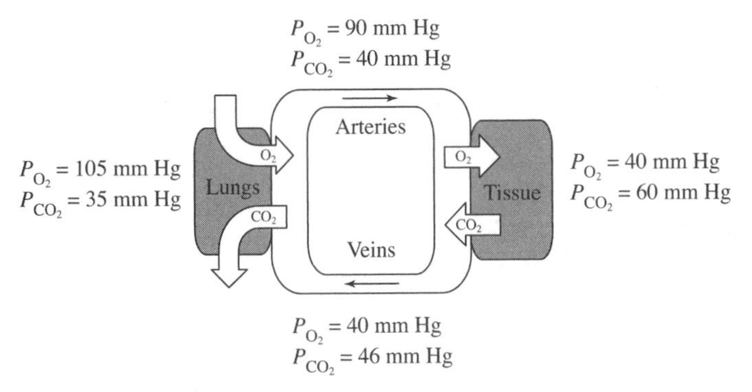

FIGURE 7.6 Schematic of gas exchange facilitated by the circulatory system. (Adapted from Tuszynski. 2003. *Introduction to Molecular Biophysics*. CRC series in pure and applied physics. CRC Press, 351.)

The flow rate in volume/sec is obtained by integration over the surface

$$\frac{dV}{dt} = 2\pi\int_0^R v(r)r\,dr = 2\pi\int_0^R \frac{\delta P\,(R^2 - r^2)}{4\eta L} r\,dr = \frac{\pi\delta P}{2\eta L}\left(\frac{R^2r^2}{2} - \frac{r^4}{4}\right)_0^R = \frac{\pi\delta PR^4}{8\eta L}, \tag{7.19}$$

which is Poiseuille's law. Because of this R^4 dependence, a small change in the area of the blood vessel will result in a large change in the volume flow rate.

7.5.2 Vessel Constriction and Aneurysm

Blood flows through a network of vessels in the human circulatory system. A simplified schematic of gas exchange by the circulatory system is shown in Figure 7.6.

Arteries deliver blood to a vast network of capillaries that provide nourishment to tissues throughout the body. Veins return blood from the capillaries back to the heart and to the lungs. The constriction of a blood vessel due to vulnerable plaque in Figure 7.7(a) can result in a transition from laminar to turbulent flow. The fluid speed is higher in the constricted region so that the pressure inside the vessel is lower than the external pressure. The vessel can momentarily collapse as a result of this pressure drop. The fluid quickly pushes the vessel open again, and the cycle repeats. Distinctive sounds of both fluid turbulence and vascular flutter can be heard with a stethoscope.

An aneurysm can form in a blood vessel where a portion of the vessel is enlarged with a balloon-like bulge as depicted in Figure 7.7(b). The blood pressure is higher in the bulge compared to the normal vessel pressure because of the Venturi effect. Higher pressure in the aneurysm tends to expand it even farther until the blood vessel can eventually burst.

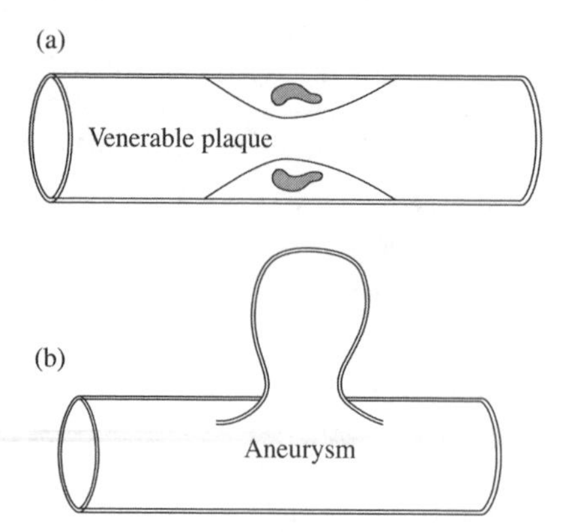

FIGURE 7.7 Blood vessel (a) constricted by vulnerable plaque and (b) expanded by an aneurysm.

7.5.3 Variation of Blood Pressure with Depth

In the static case, where the fluid speed is zero, the Navier–Stokes equation has vanishing terms

$$\underbrace{\frac{\partial \mathbf{v}}{\partial t} + (\mathbf{v} \cdot \nabla)\mathbf{v}}_{0} = -\frac{\nabla P}{\rho} + \underbrace{\upsilon \nabla^2 \mathbf{v}}_{0} + f_{\text{body}} \tag{7.20}$$

where $f_{\text{body}} = -\mathbf{g}$ in a gravitational field. Integrating the remaining terms,

$$\int_0^h \rho g \mathbf{j} \cdot \mathbf{j}\, dy = \int_{P_0}^{P} \nabla P \cdot \mathbf{j}\, dy \tag{7.21}$$

gives us the familiar result that the gauge pressure, or variation of pressure as a function of depth, is $\rho g h = P - P_0$. The absolute pressure is

$$P = P_0 + \rho g h \tag{7.22}$$

where P_0 is the atmospheric pressure at the surface of the fluid.

7.5.4 Microgravity Effects

Because of gravity, blood pressure in the brain can range between 60 mm Hg and 80 mm Hg compared to 200 mm Hg in the feet. The circulatory system must perform work to overcome gravitational potential energy in supplying blood to upper portions of the body. In microgravity environments encountered during human spaceflight, the body does not reduce its workload when the effective gravitational potential is removed.

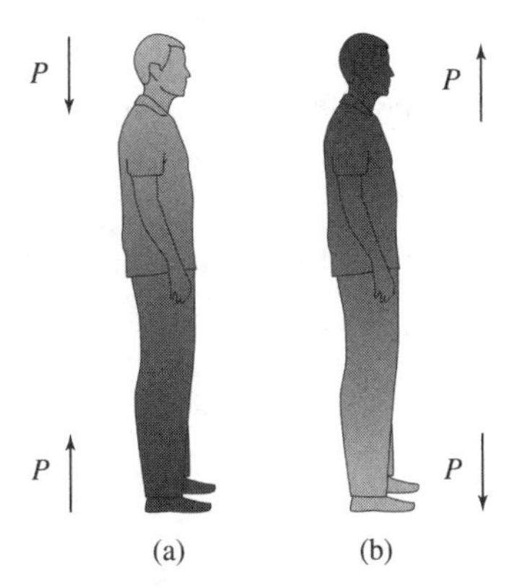

FIGURE 7.8 Blood pressure profile of an astronaut (a) on Earth in a downward gravitational field and (b) in orbit in a micro-gravity environment. The astronaut experiences a reverse blood pressure gradient in micro-g because the circulatory system is accustomed to doing work overcoming gravitational potential energy in supplying blood to the upper body.

Overcompensation by the circulatory system in microgravity thus causes an inversion in the blood pressure differential normally experienced on Earth, with an increase in upper-body blood pressure as illustrated in Figure 7.8. This overcompensation attenuates with longer-duration missions so that astronauts experience a lower than normal blood pressure in the upper extremities upon returning to Earth.

7.6 Capillary Action

Einstein's first published paper on the capillary effect helped establish the existence of molecules at the turn of the last century. Capillary action results from surface tension due to intermolecular forces acting at the boundary of a fluid as shown in Figure 7.9(a).

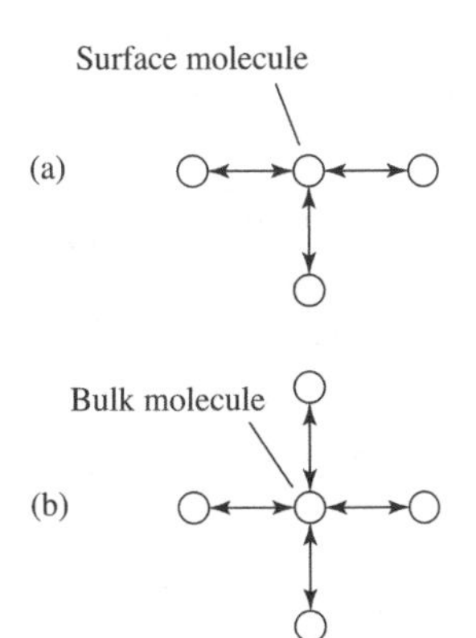

FIGURE 7.9 (a) Surface tension resulting from electrostatic attraction of the nearest neighboring molecules resulting in a net downward force. (b) Zero net force acting on a molecule in the bulk suspension.

Surface molecules experience a net downward force in the absence of neighboring molecules outside of the fluid. Bulk fluid molecules, however, experience a zero net average force because of canceling neighboring interactions with molecules on all sides similar to Figure 7.9(b).

Specially adapted insects can take advantage of this effect to stride across the surface of still water. Surface tension is especially important for plants in absorbing water and nutrients in a gravitational field. The maximum height h a liquid of density ρ will rise in a capillary of radius a is given by

$$h = \frac{2\gamma\cos\theta}{\rho g a} \tag{7.23}$$

where γ is the surface tension of the liquid and θ is the angle that the liquid makes with respect to the surface normal of the capillary wall as depicted in Figure 7.10. Capillary action can be observed by dipping a porous material such as a paper towel or a celery stick into a colored liquid. The liquid will rise to some maximum height above the liquid's surface.

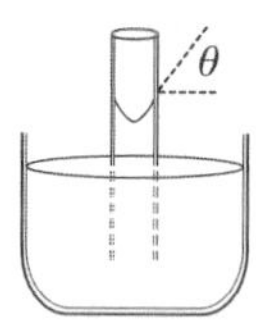

FIGURE 7.10 Capillary effect resulting from surface tension causing the fluid to rise inside a capillary tube. The fluid makes an angle θ with respect to the walls of the capillary tube.

EXERCISES

Exercise 7.1 Caffeine is a vassal constrictor. Derive an expression for the change in average blood pressure in a capillary that increases in area δA due to caffeine withdrawal. Make use of the relation Av = const.

Exercise 7.2 Fluid pressures in the human cranial cavity and spinal column range between 100 and 200 mm Hg above atmospheric pressure. A spinal-tap procedure is conducted by inserting a hollow tube into the spinal cavity to measure the fluid pressure. To what range of heights will the fluid rise if the density of the spinal fluid is the same as that of water: $\rho_w = 10^3$ kg/m^3?

8 Bioenergetics and Molecular Motors

8.1 Kinesins, Dyneins, and Microtubule Dynamics

Microtubules play a significant role in intracellular locomotion. These proteins consist of α- and β-tubulin dimers linked to form cylindrical filaments. The ends of each microtubule consist of either an α- or a β-tubulin subunit. The α terminus is denoted as the minus end, while the β terminus is denoted as the plus end. Organelles and other intracellular cargo are dependent on microtubules to travel from one location to another within the cell.

Kinesins are often associated with microtubules within the cell. Fluorescent staining of kinesin and microtubules have shown the two to exist in close proximity. The kinesin proteins facilitate membrane transport, mitosis and meiosis, mRNA and protein transport, ciliary and flagellar genesis, signal transduction, and microtubule dynamics. Unlike their functional counterpart, dynein, kinesin tends to travel toward

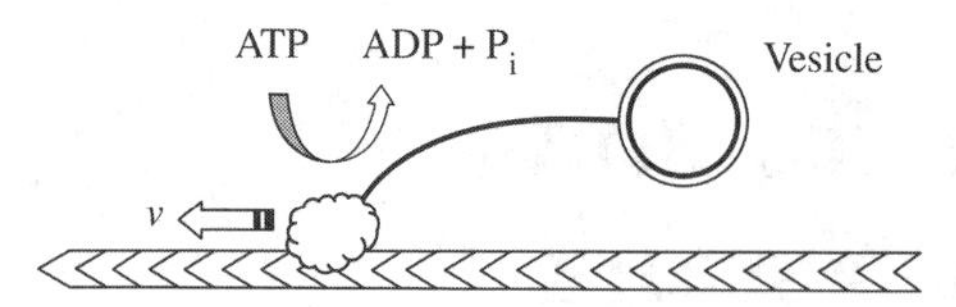

FIGURE 8.1 Transport of a vesicle along a microtubule by a molecular motor with an average speed v.

the plus end of microtubules. It is a molecular dimer consisting of two polypeptides with heavy and light chains. The globular heads of each kinesin protein contain a microtubule-binding site and an ATP-binding site. ATP hydrolysis provides the chemical energy necessary for mechanical movement along microtubules as shown in Figure 8.1. Upon the release of ADP from the ATP-binding site, a conformational change in the protein occurs and results in the displacement of kinesin along the microtubule.

There are two proposed mechanisms for kinesin movement along microtubules. The more popular "hand-over-hand" mechanism suggests that the two globular heads of kinesin alternate lead positions, one stepping past the other in an alternating fashion. The "inchworm" theory suggests that one kinesin head maintains the lead position and inches forward toward the plus end of a microtubule, leaving the second globular head to follow.

Both kinesin and myosin originated from a similar evolutionary ancestor and thus contain a similar core structure. This similarity in structure leads to a similarity in function. Myosin motors, aside from facilitating muscle contraction, play a role in cytokinesis, cell movement, membrane transport, cell architecture, and signal transduction. Both kinesin and myosin are able to convert the chemical energy of ATP hydrolysis into the mechanical energy of motion.

Dynein, like kinesin, also acts as a motor protein and facilitates the movement of cilia. Unlike kinesin, dynein travels toward the minus end of microtubules. These motor proteins fall into two classes: cytoplasmic and axonemal dyneins. Cytoplasmic dyneins are involved in transporting cargo within the cell, such as golgi vesicles and neuronal vesicles along axonal microtubules. These microtubules are the tracks along which synaptic vesicles travel from the cell body of a neuron to the axon terminus. Axonemal dyneins facilitate the beating of cilia and the movement of flagella. Dynein attaches to one of the microtubules within cilia and walks along a neighboring microtubule. The globular end of dynein attached to the second microtubule contains an ATP-binding site in addition to a microtubule-binding site. When ATP is cleaved, the dynein protein has sufficient energy to travel toward the minus end. This walking motion induces a sliding action and causes relative motion among microtubule doublets within each cilium. The wave-like bending of each cilium ultimately stems from the coordinated waves of dynein activity.

8.2 Brownian Motors

Brownian motion is the random movement of particles in a fluid under the influence of thermal energy and the random collisions of molecules. Many biological motors use Brownian motion coupled with ATP hydrolysis to power their motors. During brief moments when thermal motion is unidirectional, these motors are able to harness this energy for use. Many biological motors have a locking mechanism, or ratchet, whereby only favorable collisions are selected.

8.2.1 Feynman Ratchet

Richard Feynman demonstrated a conceptual Brownian motor that could operate in the presence of a temperature gradient. The gear in Figure 8.2 (usually with asymmetric teeth) can rotate easily in the clockwise direction but is prevented from rotating in the counterclockwise direction by the pawl.

The gear is connected to a paddle wheel by a rod. Both the gear and paddle wheel undergo collisions with molecules executing Brownian motion. It turns out that it is not possible to perform useful work by extracting thermal energy with the entire mechanism immersed in a single reservoir. The reason is that molecular collisions with the gear and pawl mechanism will occasionally reset any advancement made by collisions with the paddle wheel. The ratchet will therefore not operate if both the paddle wheel and the gear are at the same temperature in accord with the second law of thermodynamics. The device can be made to do useful work, such as lifting a weight, if the gear

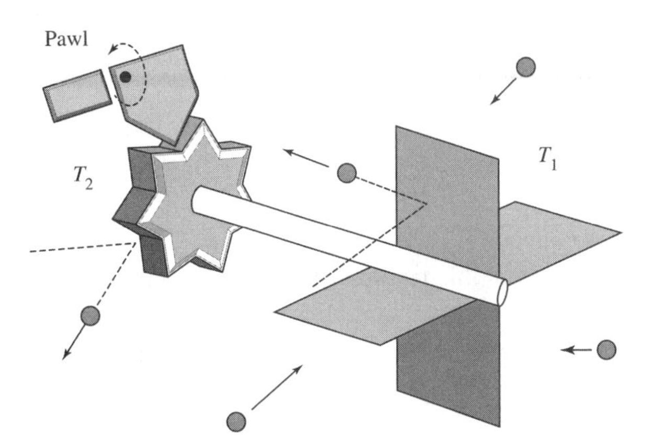

FIGURE 8.2 Feynman's Brownian ratchet consisting of a paddle wheel, gear, and pawl. The device can be made to perform useful work if $T_1 > T_2$. If the temperatures are equal, then collisions with the gear and pawl mechanism will offset any forward progress made by the ratchet.

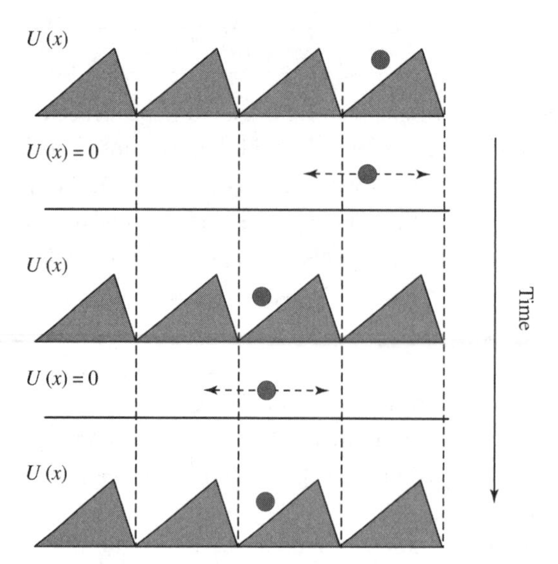

FIGURE 8.3 Time-varying ratchet potential resulting in an average velocity to the left. The particle has probability P_1 (stepping into the left well), P_2 (remaining in the same well), and P_3 (stepping into the right well) with $P_1 + P_2 + P_3 = 1$.

and pawl are at a lower temperature than the paddle wheel. The ratio of counterclockwise to clockwise rotation rates is given by the ratio of Boltzmann factors

$$\frac{\text{cw rate}}{\text{ccw rate}} = \frac{\exp\left(\dfrac{-E}{k_B T_1}\right)}{\exp\left(\dfrac{-E}{k_B T_2}\right)} \tag{8.1}$$

where E is the energy required to raise the pawl mechanism, T_2 and T_1 are the temperatures of the gear and paddle wheel, respectively.

Another type of Brownian ratchet illustrated in Figure 8.3 consists of a sawtooth potential switching on and off with frequency f.

A particle executing Brownian motion in this potential will have a nonzero average velocity determined by the asymmetry of the potential. Space–time diagrams of both biased and unbiased Brownian motion are shown in Figure 8.4. The average velocity of a particle is zero without the ratcheting potential as shown in Figure 8.4(a). The particle has a net velocity to the left in the presence of the ratcheting potential in Figure 8.4(b). During one period of the ratchet potential, the particle has a probability of stepping to the left, to the right, or remaining in the same location.

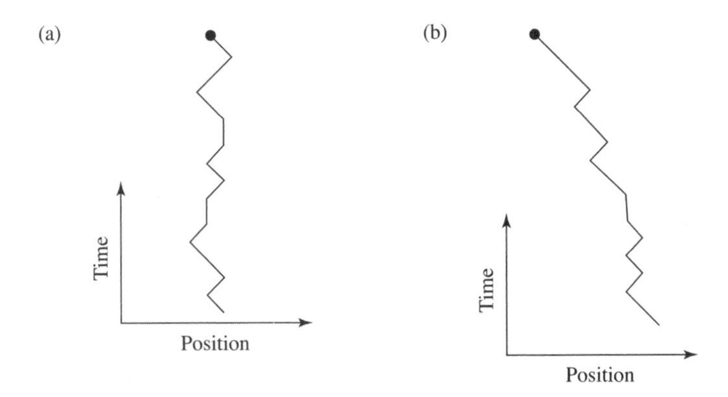

FIGURE 8.4 (a) Space–time diagram of a random walk in one dimension. (b) Biased random walk with the particle making net progress to the left in a time-varying ratchet potential.

8.2.2 Kinesin Brownian Dynamics

Kinesin molecules are capable of directed locomotion along microtubules with only one molecular head. Normally, these molecules transport intracellular vesicles along microtubules toward a specific destination by a controlled walking motion with two heads. The ability to travel along microtubules with only one head supports the classification of kinesin molecules as Brownian motors.

Molecular motors tend to move unidirectionally along protein polymers. For example, kinesin motors carry intracellular cargo unidirectionally along microtubules, while myosin motors tend to travel unidirectionally along actin. ATP hydrolysis lowers the energy barrier necessary for kinesin to travel along microtubules. With this energy barrier sufficiently lowered, kinesin is able to move forward or backward along a microtubule in a Brownian fashion. However, the release of ADP and inorganic phosphate from the site of hydrolysis locks the kinesin molecule in place and prevents any backward movement. Thus, only forward Brownian motion is permitted, allowing kinesin molecules with a single head to move in a single direction.

8.2.3 Myosin Brownian Dynamics

The second example of a Brownian motor is the myosin motor protein, which travels along actin filaments in muscle tissue. Myosin heads have been discovered to not only move forward but also backward along actin fibers. The net movement is unidirectional; however, this is a result of specific mechanisms that prevent counterproductive Brownian motion. Thus, biological proteins can move in a seemingly deterministic and controlled manner in the presence random Brownian motion of molecules.

8.3 ATP Synthesis in Mitochondria

8.3.1 Electron Transport Chain

The majority of ATP synthesis in eukaryotes takes place on the inner membrane of mitochondria. The bacterial equivalent of the inner membrane is the bacterial plasma membrane. This is due to the absence of mitochondria within the bacterial cytoplasm. This fact also supports the endosymbiotic theory, postulated by Mereschkowski in 1905, that mitochondria and chloroplasts originated from prokaryotic organisms living within eukaryotic organisms.

The electron transport chain is a series of protein complexes lining the inner mitochondrial membrane that is responsible for creating the proton gradient that drives ATP synthesis. Nicotinamide adenine dinucleotide (NADH) molecules from glycolysis and the citric acid cycle donate high-energy electrons to these protein enzymes. The first complex to capture two electrons is NADH dehydrogenase (Complex I). For each electron transferred to NADH dehydrogenase, one proton, or hydrogen ion, is pumped from the matrix into the intermembrane space. The electrons are then shuttled to cytochrome b-c_1 (Complex III) via a mobile carrier called ubiquinone (coenzyme Q). Succinate dehydrogenase (Complex II) is not involved in pumping protons into the intermembrane space; however, it does shuttle additional electrons to ubiquinone, which can later be used to pump hydrogen ions into the intermembrane space via Complex III.

Cytochrome b-c_1 then transfers the electrons to another mobile carrier known as cytochrome c. In this process, cytochrome b-c_1 pumps one proton into the intermembrane space for each electron transferred to cytochrome c. The mobile carrier then transfers the electrons to the last protein complex in the electron transport chain, cytochrome oxidase (Complex IV). Complex IV captures four electrons from cytochrome c and shuttles them to molecular oxygen, the final electron acceptor. During this process, four hydrogen ions are transported into the intermembrane space and a molecule of oxygen combines with four additional hydrogen ions to form two molecules of water.

The proton gradient is produced and maintained by the movement of electrons through the protein complexes discussed above. The potential energy from this gradient ultimately drives the production of ATP via ATP synthase. Oxygen is necessary in the electron transport chain for accepting the electrons shuttled through the protein complexes. Hence, oxygen is commonly known as the final electron acceptor in the electron transport chain. Furthermore, without oxygen, the supply of NAD^+ cannot be replenished, and glucose cannot be oxidized to pyruvate in glycolysis. (Recall that in glycolysis, NAD^+ is reduced to the high-energy electron carrier NADH when glucose is oxidized to pyruvic acid). Thus, during anaerobic conditions (i.e., during intense physical exercise), lactic acid is produced in order to replace the NAD^+ that is reduced to NADH in cellular respiration.

8.3.2 ATP Synthase

ATP synthase is the enzyme that converts adenosine diphosphate (ADP) and inorganic phosphate (P_i) to adenosine triphosphate (ATP) as shown in Figure 8.5. Synthesis depends on proton translocation, which is possible across the proton-tight, inner mitochondrial membrane due to electron transport through the specialized, inner membrane proteins previously discussed. The enzyme that drives ATP synthesis consists of an electrical rotary nanomotor (F_0) that drives a chemical nanomotor (F_1) by elastic power transmission.

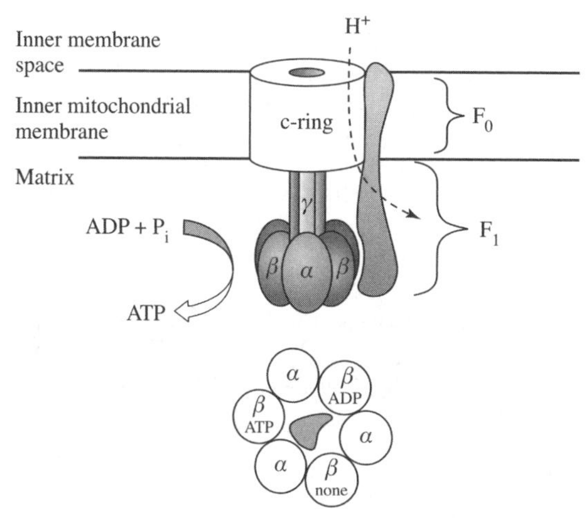

FIGURE 8.5 ATP synthase. (Adapted from Serdyuk et al. 2007. *Methods in Molecular Biophysics: Structure, Dynamics, Function.* Cambridge University Press, Cambridge, U.K.)

In elastic power transmission, an elastic element, characterized by a spring constant κ, stores and transmits elastic deformation energy between a motor and a mechanical device. In the case of ATP synthase, a rotary nanomotor drives the pump that allows protons to flow into the matrix.

The F_0 subunit of ATP synthase is the membrane-intrinsic portion that is responsible for the transport of hydrogen ions from the intermembrane space into the mitochondrial matrix. The F_1 subunit of ATP synthase is the soluble portion of the enzyme, which is in contact with the mitochondrial matrix. It is responsible for the actual conversion of ADP and P_i to ATP, which is coupled to the transport of protons from the intermembrane space to the mitochondrial matrix. The action of these two motors in concert with each other via elastic power transmission produce ATP with a high kinetic efficiency. The presence of a proton gradient, with more hydrogen ions in the intermembrane space than the matrix, creates the ion motive force that pushes protons through the ATP synthase complex. The F_0 portion of ATP synthase rotates with each proton that enters the matrix from the intermembrane space and the translocation of three protons into the matrix produces one molecule of ATP.

8.3.3 Torque Generation in ATP Synthase

Electrostatic forces are believed to generate torque in the F_0 portion of the ATP synthase that is coupled to the F_1 portion by the γ subunit. Protons are translated from the inner membrane space to the matrix through a-subunit half channels that are offset to allow the protons to enter and exit the rotating F_0 rotor as depicted in Figure 8.6.

A potential difference exists between the half channels equal to the inner membrane potential difference because each half channel is separately in contact with opposite sides of the membrane. This results in a transverse dipole moment with tangential electric field components shown in Figure 8.7, that supply torque to the protonated c-ring occupation sites.

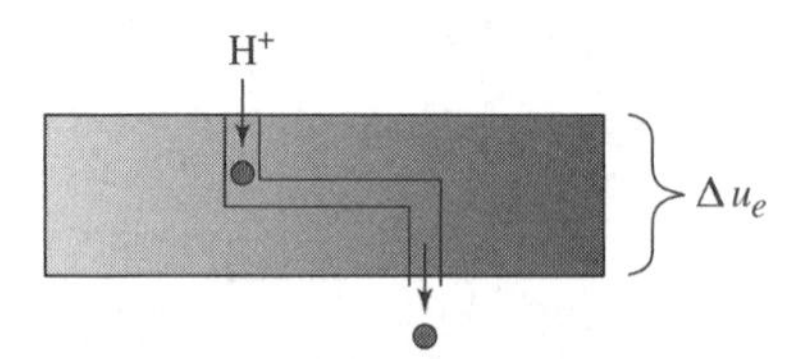

FIGURE 8.6 Half channels in the ATP synthase a-subunit at different potentials resulting in a torque-generating transverse dipole moment in Miller's model. Protons entering the upper half of the channel bind to the cAsp-61 sites and complete one revolution in the rotating c-ring before exiting the bottom half of the channel.

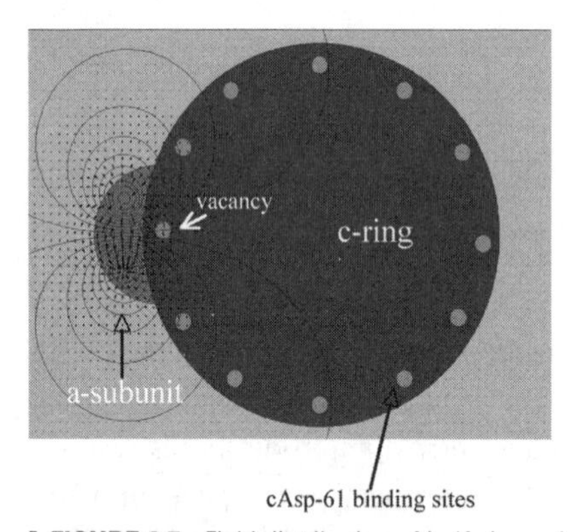

FIGURE 8.7 Field distribution of half channels at different potentials in the a-subunit of ATP synthase. The resulting transverse dipole moment generates torque in the protonated c-ring. (As described by Miller et al. 2008. Physical mechanisms of biological molecular motors. *Physica B: Condensed Matter* 404: 503–506.) The electronegative vacancy between the half channels aligns with the positive half channel to minimize the electrostatic energy thus becoming occupied, whereupon the next binding site becomes vacant, and so on.

8.4 Photosynthesis in Chloroplasts

In cellular respiration, glucose and oxygen are used to produce ATP, water, and carbon dioxide. Photosynthesis is the reverse process by which energy from sunlight, water, and carbon dioxide is used to produce oxygen and glucose. This can be summarized by the following equation:

$$6CO_2 + 6H_2O + \text{Energy (photons)} \rightarrow C_6H_{12}O_6 + 6O_2. \quad (8.2)$$

Water is supplied through xylem vessels, while CO_2 and O_2 enter and exit the plant, respectively, through stomata pores.

In the light-dependent reactions of photosynthesis, ATP is produced using the energy from photons ($E = hf$). These photons excite electrons that pass through an electron transport chain analogous to its mitochondrial counterpart. The proteins involved in the photosynthetic electron transport chain represented in Figure 8.8

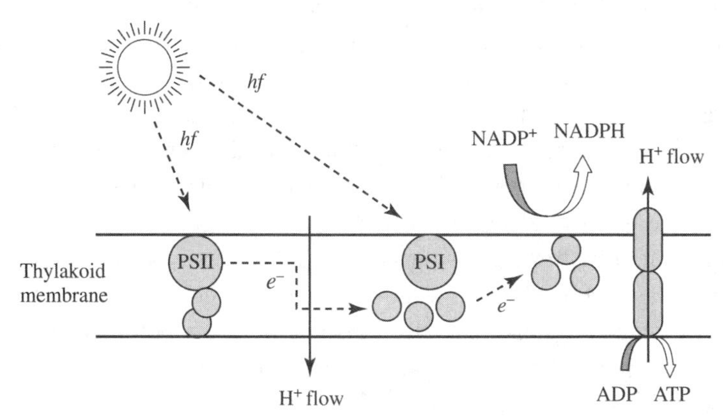

FIGURE 8.8 Light-activated electron transport chain. H^+ flow into chloroplasts drives ATP synthesis. (Adapted from Plaxco K. W. and M. Gross. 2006. *Astrobiology: A Brief Introduction*, The John Hopkins University Press, Baltimore, Maryland.)

include photosystem II, cytochrome b_6f, photosystem I, ferredoxin NADP reductase (FNR), and ATP synthase. Associated with these protein complexes are the mobile carrier molecules: plastoquinone Qb, plastocyanin, and ferredoxin.

Table 8.1 Type of Chlorophyll, Type of Plant, and Wavelengths Absorbed

Type of Chlorophyll	Occurrence	Peak Absorption
Chlorophyll a	Higher plants	428 nm & 660 nm
Chlorophyll b	Higher plants	452 nm & 642 nm
Chlorophyll d	Red algae/Cyanobacteria	447 nm & 688 nm
Bacteriochlorophyll a	Purple bacteria	356 nm & 771 nm
Bacteriochlorophyll b	Purple bacteria	372 nm & 791 nm

Photosynthesis in plants and algae occur in organelles known as chloroplasts. Cyanobacteria and certain bacteria such as green sulfur and heliobacteria are also capable of photosynthesis but do not contain chloroplasts. The protein complexes in the electron transport chain of plants are embedded within the tightly folded membranes of thylakoids, which increase the surface area exposed to sunlight.

8.4.1 Photosystem II

Light-harvesting, chlorophyll-binding proteins surround the Photosystem II reaction center. When a chlorophyll molecule comes in contact with a photon, resonance energy is created and transferred to other chlorophyll molecules surrounding the reaction center. This vibrational energy is ultimately transferred to the P680 reaction center, which consists of a pair of chlorophyll molecules capable of donating an electron to the electron transport chain. The electron is transferred to a pheophytin molecule and plastoquinone Qa before eventually being captured by plastoquinone Qb. When plastoquinone Qb receives two electrons, it transports the electrons to the next component in the electron transport chain; cytochrome b6f. Plastocyanin then transfers the electrons from cytochrome b6f to Photosystem I. Similar to the electron transport chain in mitochondria, the transfer of electrons from one protein complex to the next is coupled with the pumping of hydrogen ions into the thylakoid lumen, producing a proton gradient that is necessary for ATP synthesis. Electrons derived from the splitting of water molecules eventually replace the electrons lost from the P680 reaction center. For every two molecules of water split, one molecule of oxygen is produced and four hydrogen ions are released into the lumen.

8.4.2 Photosystem I

Photosystem I functions similarly to Photosystem II in that a photon creates resonance energy among chlorophyll molecules, which eventually causes the release of an electron into the electron transport chain. The electron is transferred to the FNR complex via the mobile carrier ferredoxin. It is at the FNR complex that NADPH is created from $NADP^+$, a hydrogen ion, and two electrons transferred from Photosystem I. The ATP and NADPH produced during the light-dependent reactions are ultimately used in the Calvin cycle or light-independent reactions to produce glucose.

8.5 Light Absorption in Biomolecules

Chlorophyll, hemoglobin, and light-absorbing pigments have absorbencies at specific frequencies because of characteristic electronic configurations along the length of these molecules. In these molecules, electrons flow freely along conjugated paths consisting of alternating single and double bonds as illustrated in Figure 8.9.

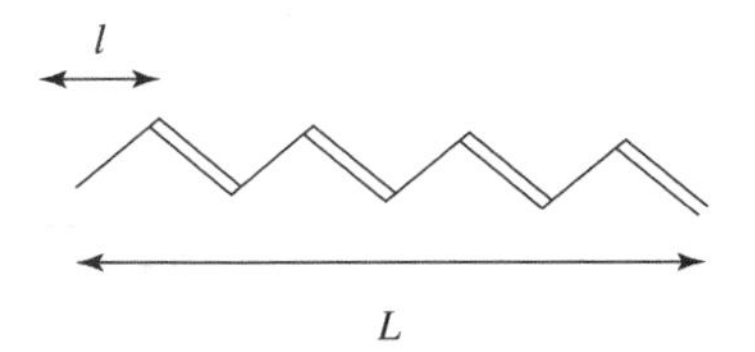

FIGURE 8.9 Schematic of light-absorbing molecule with alternating single and double bonds. The number of atoms is $p = 9$.

The quantum mechanical behavior of electrons confined to biomolecules can be modeled as an electron in a one-dimension well of length L equal to the length of the biomolecules. The quantized energy levels are given by

$$E_n = n^2 \frac{\hbar^2 \pi^2}{2m_e L^2} \qquad n = 1, 2, 3, \ldots. \tag{8.3}$$

A photon with energy $\Delta E = hf$ will be absorbed during the transition $n_i \rightarrow n_f$

$$\Delta E = (n_f^2 - n_i^2) \frac{\hbar^2 \pi^2}{2m_e L^2}. \tag{8.4}$$

Here, $n_i = p/2$ is the highest occupied molecular orbital (HOMO), and $n_f = p/2 + 1$ is the lowest unoccupied molecular orbital (LUMO) where p is the number of atoms in the conjugated path of the molecule. The absorbed energy is then

$$hf = (p + 1) \frac{\hbar^2 \pi^2}{2m_e L^2}. \tag{8.5}$$

Using the relation $f\lambda = c$,

$$\frac{c}{\lambda} = (p + 1) \frac{\hbar \pi}{4m_e L^2} \tag{8.6}$$

so that the wavelength of absorbed light is given by

$$\lambda = \frac{4m_e c L^2}{(p + 1)\hbar \pi}. \tag{8.7}$$

Substituting $L = (p - 1)\ell$ where ℓ is the bond length along the molecule we have

$$\lambda = \frac{8\ell^2 (p - 1)^2}{\lambda_c (p + 1)} \tag{8.8}$$

where the Compton wavelength of the electron $\lambda_c = h/m_e c = 2.426$ pm. From this equation we see that the absorption wavelength increases with p.

8.6 Vibrational Spectra in Biomolecules

8.6.1 Square Well Potential

The vibrations of individual carbon atoms in biomolecules can result in a fine structure in their absorption spectra. Small peaks are observed superimposed on the much broader absorption spectra. Carbon atoms with mass $m_c = 2.0 \times 10^{-26}$ kg can be treated as particles confined to 1-D square wells of width $\delta \sim 2 \times 10^{-11}$ m with energy levels

$$E_n = n^2 \frac{\hbar^2 \pi^2}{2m_e \delta^2} \qquad n = 1, 2, 3, \ldots. \tag{8.9}$$

The wavelength corresponding to the transition $n_i = 1 \rightarrow n_f = 2$ is

$$\lambda = \frac{8m_e c\delta^2}{3h} \sim 10^4 \, \text{nm}, \tag{8.10}$$

which is in the far infrared.

8.6.2 Harmonic Oscillator Potential

The vibrational spectra of atoms and molecules can also be modeled as particles confined to a potential well with linear restoring force described by

$$F(x) = -kx \tag{8.11}$$

corresponding to a potential

$$U(x) = \frac{1}{2}kx^2. \tag{8.12}$$

The solution to the one-dimensional Schrödinger equation with this potential gives the energy spectrum

$$E_n = \left(n + \frac{1}{2}\right)\hbar\omega \qquad n = 0, 1, 2, \ldots. \tag{8.13}$$

The partition function is given by the sum of Boltzmann factors

$$Z = \sum_{n=0}^{\infty} \exp\left[-\left(n + \frac{1}{2}\right)\frac{\hbar\omega}{k_B T}\right]. \tag{8.14}$$

Factoring the term proportion to 1/2

$$Z = \exp\left(-\frac{\hbar\omega}{2k_BT}\right)\sum_{n=0}^{\infty}\exp\left[-n\frac{\hbar\omega}{k_BT}\right] \tag{8.15}$$

and using the identity

$$\sum_{n=0}^{\infty} x^n = \frac{1}{1-x} \tag{8.16}$$

gives the partition function

$$Z = \frac{\exp\left(-\frac{\hbar\omega}{2k_BT}\right)}{1-\exp\left(-\frac{\hbar\omega}{k_BT}\right)}. \tag{8.17}$$

The vibrational contribution to the free energy N molecules is then

$$F = -Nk_BT\ln Z = N\left(\frac{h\omega}{2} + k_BT\ln\left(1-\exp\left(-\frac{\hbar\omega}{K_BT}\right)\right)\right). \tag{8.18}$$

We may also obtain the average vibrational energy as

$$\langle E\rangle = -\frac{\partial \ln Z}{\partial\beta} = -\frac{\partial}{\partial\beta}\ln\left[\frac{\exp\left(-\frac{1}{2}\beta\hbar\omega\right)}{1-\exp(-\beta\hbar\omega)}\right]. \tag{8.19}$$

Expanding the logarithm

$$\langle E\rangle = -\frac{\partial}{\partial\beta}\left[-\frac{1}{2}\beta\hbar\omega - \ln(1-\exp(-\beta\hbar\omega))\right] \tag{8.20}$$

and then differentiating gives

$$\langle E\rangle = \frac{1}{2}\hbar\omega + \frac{\hbar\omega\exp(-\beta\hbar\omega)}{1-\exp(-\beta\hbar\omega)}. \tag{8.21}$$

The vibrational contribution to the heat capacity is then given by

$$C = \frac{\partial}{\partial T} N\langle E \rangle \tag{8.22}$$

or

$$C = N\frac{(\hbar\omega)^2[-\exp(-\beta\hbar\omega)(1-\exp(-\beta\hbar\omega)) - \exp(-2\beta\hbar\omega)]}{[1-\exp(-\beta\hbar\omega)]^2}\left(\frac{-1}{k_B T^2}\right). \tag{8.23}$$

Simplifying this expression gives

$$C = N\frac{(\hbar\omega)^2\exp(-\beta\hbar\omega)}{[1-\exp(-\beta\hbar\omega)]^2}\left[\frac{1}{k_B T^2}\right]. \tag{8.24}$$

Dividing top and bottom by $(-\beta\hbar\omega)$ we have

$$C = Nk_B\left(\frac{\hbar\omega}{k_B T}\right)^2\frac{\exp(\beta\hbar\omega)}{[\exp(\beta\hbar\omega)-1]^2}. \tag{8.25}$$

Note that the heat capacity has the same units as the Boltzmann constant as does the entropy S that is calculated according to

$$S = -\frac{\partial F}{\partial T} = -Nk_B\left(\ln\left(1-\exp\left(-\frac{\hbar\omega}{k_B T}\right)\right) - \frac{\hbar\omega}{k_B T}\frac{\exp\left(-\frac{\hbar\omega}{k_B T}\right)}{\left(1-\exp\left(-\frac{\hbar\omega}{k_B T}\right)\right)}\right). \tag{8.26}$$

Simplifying this expression we have

$$S = Nk_B\left(\frac{\hbar\omega}{k_B T}\frac{1}{\left(\exp\left(\frac{\hbar\omega}{k_B T}\right)-1\right)} - \ln\left(1-\exp\left(-\frac{\hbar\omega}{k_B T}\right)\right)\right). \tag{8.27}$$

The following MATLAB code plots S/Nk_B vs. $k_BT/\hbar\omega$ in Figure 8.10:

```
t=0.01:.01:100;

S= 1./(t.*exp(1./t-1))-log(1-exp(-1./t));
semilogx(t,S,'k','LineWidth', 2);

xlabel('k_BT/h\omega','FontSize',20);
ylabel('S/N k_B','FontSize',20)

set(gca, 'FontSize', 20)
title('Entropy of Harmonic Oscillators','FontSize',20)
```

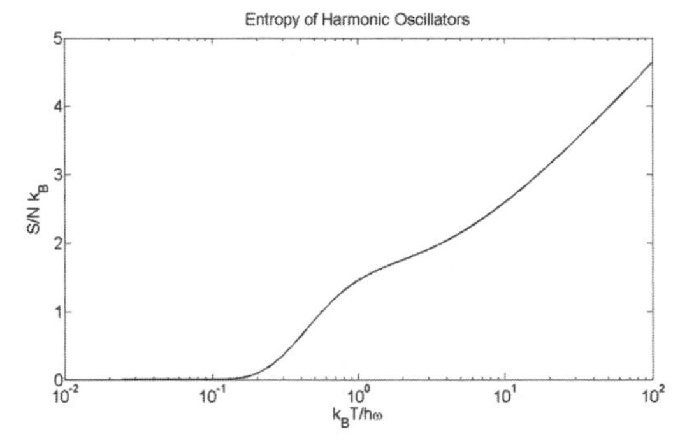

FIGURE 8.10 Plot of entropy vs. dimensionless temperature in the harmonic oscillator vibrational spectra model.

EXERCISES

Exercise 8.1. Estimate the absorption wavelengths by a ring-shaped chlorophyll molecule with effective radius R using the de Broglie relation $\lambda = \dfrac{h}{mv}$ for the electron standing wave with $n\lambda = 2\pi R$. Take the energy of the electron to be $E = \dfrac{1}{2}m_e v^2$ and the energy of the photon $E = hf = hc/\lambda$.

Exercise 8.2. Calculate the vibrational contribution to the free energy, entropy, and heat capacity of the square well potential model. Compare the temperature dependence of these thermodynamic quantities to those obtained in the harmonic oscillator model.

9 Passive Electrical Properties of Living Cells

■ 9.1 Poisson–Boltzmann Equation

Electrostatic interactions govern many biological processes from the establishment of membrane potentials to the operation of molecular motors. In charge-free regions, the electric potential U_e is described by Laplace's equation

$$\nabla^2 U_e = 0. \quad (9.1)$$

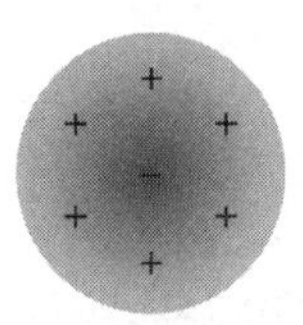

FIGURE 9.1 Negatively charged cell surrounded by a positive counterion cloud.

A counterion cloud will surround charges immersed in ionic solutions. Living cells have an overall negative charge and will therefore attract positive ions distributed with charge density ρ as shown in Figure 9.1.

The electric potential is then described by Poisson's equation in a region with permittivity ε

$$\nabla^2 U_e = -\frac{\rho}{\varepsilon}. \tag{9.2}$$

The distribution of ions thus gives rise to an electric field $\mathbf{E} = -\nabla U_e$. As discussed in Chapter 2, the electrostatic energy associated with a charge q_i in an electrostatic potential U_e is given by $W_e = q_i U_e$. The ionic concentration C_i has temperature dependence described by the Boltzmann factor

$$C_i = C_{i0} \exp\left(-\frac{q_i U_e}{k_B T}\right) \tag{9.3}$$

where C_{i0} is the concentration in field-free regions. The charge of the ith species is $q_i = z_i e$ with valence z_i. The total charge density is given by summing over each species in suspension

$$\rho = \sum_i C_{i0} q_i \exp\left(-\frac{q_i U_e}{k_B T}\right). \tag{9.4}$$

The Poisson–Boltzmann (P–B) equation then becomes

$$\nabla^2 U_e = -\frac{1}{\varepsilon} \sum_i C_{i0} q_i \exp\left(-\frac{q_i U_e}{k_B T}\right). \tag{9.5}$$

The P–B equation may be used to calculate the potential near electrical double layers forming at electrode–electrolyte interfaces where fixed electrode charges attract mobile counterions in solution. This equation must be solved by applying boundary conditions on either U_e or the normal displacement $\mathbf{D} \cdot \mathbf{n}$ equal to the charge density σ at any electrode surfaces

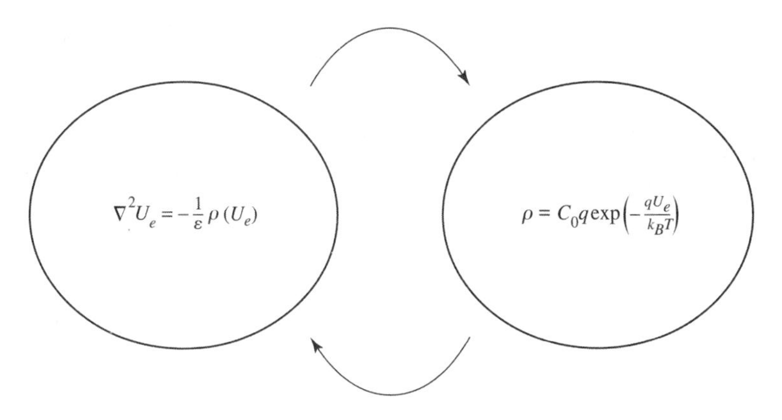

FIGURE 9.2 Procedure for solving the Poisson–Boltzmann equation by iteratively calculating the charge density and the potential.

$$\sigma = \mathbf{D} \cdot \mathbf{n} = -\varepsilon \left(\frac{\partial U_e}{\partial n}\right)_{\text{surf}} \tag{9.6}$$

where **n** is a unit vector normal to the surface. In most situations, the P–B equation will require a numerical solution alternately solving for the potential U_e and the charge density ρ as indicated by Figure 9.2.

9.1.1 One–One Valent Electrolytes

For suspensions of one–one valent electrolytes, we sum Equation 9.4 over opposite charges. Taking $q_1 = q$ and $q_2 = -q$, Equation 9.5 becomes

$$\nabla^2 U_e = -\frac{1}{\varepsilon} C_0 q \left[\exp\left(\frac{qU_e}{k_B T}\right) - \exp\left(-\frac{qU_e}{k_B T}\right)\right]. \tag{9.7}$$

Combining the exponentials using $2\sinh(x) = \exp(x) - \exp(-x)$ gives

$$\nabla^2 U_e = -\frac{1}{\varepsilon} C_0 q \left[2\sinh\left(\frac{qU_e}{k_B T}\right)\right]. \tag{9.8}$$

This formula is applicable to NaCl solutions with ionic charges $q_{Na^+} = +e$ and $q_{Cl^-} = -e$.

9.1.2 Low Field Limit

In the low field limit, with electrostatic field energy much less than the thermal energy $q_i U_e << k_B T$, we can approximate the Boltzmann factor

$$\exp\left(-\frac{q_i U_e}{k_B T}\right) \approx 1 - \frac{q_i U_e}{k_B T}. \tag{9.9}$$

The P–B equation then becomes

$$\nabla^2 U_e = -\frac{1}{\varepsilon}\sum_i C_{i0} q_i \left(1 - \frac{q_i U_e}{k_B T}\right). \tag{9.10}$$

Electroneutrality gives that the net charge of the solution should be zero:

$$\sum_i C_{i0} q_i = 0, \tag{9.11}$$

so the first term in the sum (Equation 9.10) vanishes and

$$\nabla^2 U_e = \left[\frac{1}{\varepsilon k_B T}\sum_i C_{i0} q_i^2\right] U_e, \tag{9.12}$$

which is the Helmholtz equation

$$\nabla^2 U_e = \kappa^2 U_e. \tag{9.13}$$

The quantity κ is the inverse Debye screening length $\Lambda = 1/\kappa$ discussed in Chapter 2. The Helmholtz equation may be solved by the technique of separation of variables in several coordinate systems as shown in Appendix 10.

9.1.3 Spherical Solution

In the low field limit, we consider a solution to the Helmholtz form of the P–B equation in spherical coordinates where the potential is only a function of the radial coordinate

$$\nabla^2 U_e = \frac{1}{r^2}\frac{\partial}{\partial r}\left(r^2 \frac{\partial U_e}{\partial r}\right) = \kappa^2 U_e. \tag{9.14}$$

We assume a solution of the form

$$U_e \sim \frac{\exp(-\kappa r)}{r}. \tag{9.15}$$

Substitution into the P–B equation gives

$$U_e = \frac{q}{4\pi\varepsilon}\frac{\exp(-\kappa r)}{r} \tag{9.16}$$

where

$$\kappa = \sqrt{\frac{q^2 C_0}{\varepsilon k_B T}} = \sqrt{4\pi \ell_B z^2 C_0} \tag{9.17}$$

for a single species and

$$\kappa = \sqrt{\frac{1}{\varepsilon k_B T}\sum_{i=1}^{n} C_{i0} q_i^2} = \sqrt{4\pi \ell_B \sum_{i=1}^{n} C_{i0} z_i^2} \tag{9.18}$$

for n species where $q_i = z_i e$. The quantity ℓ_B is the Bjerrum length discussed in Chapter 2.

9.2 Intrinsic Membrane Potentials

The Boltzmann factor plays a key role in determining the potential difference across biological membranes with differing ionic concentrations on each side of the membrane. We take the potentials outside and inside the membrane as illustrated in Figure 9.3 to be U_{e1} and U_{e2}, respectively.

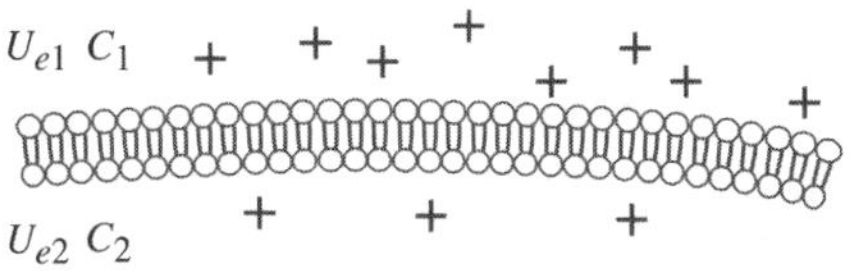

U_{e1} C_1

U_{e2} C_2

FIGURE 9.3 Membrane with ionic concentrations C_1 and C_2 on each side of the membrane. The potential difference $U_{e1}-U_{e2}$ is given by the Nernst equation.

The probabilities for locating an ion P_1 and P_2 outside or inside the membrane are proportional to the Boltzmann factor in each region so that the probability ratio for locating the ions on each side of the membrane is given by

$$\frac{P_1(\text{outside})}{P_2(\text{inside})} = \frac{\exp(-qU_{e1}/k_B T)}{\exp(-qU_{e2}/k_B T)}. \tag{9.19}$$

The probability ratio is equal to the concentration ratio

$$\frac{P_1}{P_2} = \frac{C_1}{C_2} \tag{9.20}$$

where C_1 and C_2 are ionic concentrations inside and outside the membrane, respectively,

$$\frac{C_1}{C_2} = \exp\left(\frac{-q(U_{e1} - U_{e2})}{k_B T}\right). \tag{9.21}$$

Taking the natural logarithm of both sides of this equation gives the well-known Nernst equation

$$U_{e1} - U_{e2} = -\frac{k_B T}{q} \ln\left(\frac{C_1}{C_2}\right) \tag{9.22}$$

where the quantity $k_B T/e = 61$ mV at 310 K.

9.3 Induced Membrane Potentials

9.3.1 Plasma Membrane

Intrinsic membrane potentials resulting from ionic concentration gradients will be modified in external electric fields. An empirical formula was developed by H. P. Schwan describing the frequency-dependent change in membrane potential induced in a spherical cell of radius R given by

$$\Delta U_e = \frac{3}{2}\cos\theta \frac{E_0 R}{1 + i\omega\tau}. \tag{9.23}$$

where E_0 is the field amplitude, $\omega = 2\pi f$ the angular frequency, θ is the polar angle measured from the direction of the incident field, and τ is the membrane charging time constant given by

$$\tau = RC_{\mathrm{m}}\left(\rho_i + \frac{\rho_a}{2}\right) \tag{9.24}$$

where C_{m} is the membrane capacitance and ρ_i and ρ_a are the resistivities inside and outside the cell, respectively. Figure 9.4 shows a plot of the $\mathrm{Re}\Delta U_e$ versus $\mathrm{Im}\Delta U_e$ as a function of frequency for $\tau = 10^{-3}$ s. Equation 9.23 is modified for conductive membranes for $\omega\tau < 1$:

$$\Delta U_e = \frac{3}{2}\cos\theta\frac{E_0 R}{1 + RG_m(\rho_i + \rho_a/2)} \tag{9.25}$$

where G_m is the membrane conductance per unit area. The change in membrane potential for membranes with surface conductance G_s is

$$\Delta U_e = \frac{3}{2}\cos\theta\frac{E_0 R}{1 + \rho_a G_s/R} \tag{9.26}$$

where G_s is in units of Siemens. Equations 9.23–9.26 give the peak membrane potential in different cases. However, a numerical solution is required if we seek to obtain the spatial field distribution near the cell membrane as illustrated in the following section.

The following MATLAB script plots the frequency-dependent induced membrane potential of a spherical cell in an ac electric field shown in Figure 9.4.

```
E0=100; %Volts/meter
R=1.0e-6; %Cell radius in meters
tau=1.0e-3; %time constant in seconds
theta=0; % maximum change in potential at the polar caps

nmax=1000;
w=zeros(nmax,1);
delta_Ue=zeros(nmax,1);
Re_Ue=zeros(nmax,1);
Im_Ue=zeros(nmax,1);

for n=1:nmax

w(n)=n^1.5;  % increase frequency exponentially

delta_Ue(n)=(1+i*w(n)*tau)\1.5*cos(theta)*E0*R;

Re_Ue(n)=real(delta_Ue(n));
Im_Ue(n)=imag(delta_Ue(n));

end

% graph the output
set(gca, 'FontSize', 15)
plot(Re_Ue,Im_Ue,'k','LineWidth', 2)
xlabel('Re U_e (V)','FontSize',15)
ylabel('Im U_e (V)','FontSize',15)
axis equal
title('Induced membrane potential','FontSize',15)
```

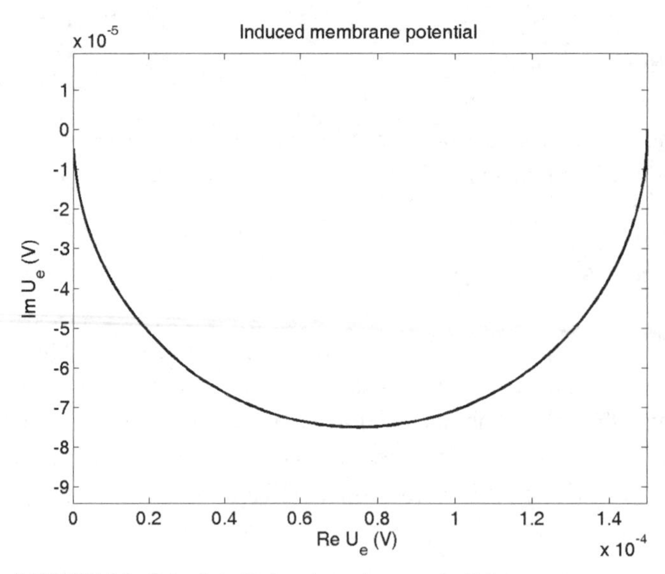

FIGURE 9.4 Cole plot of induced membrane potential.

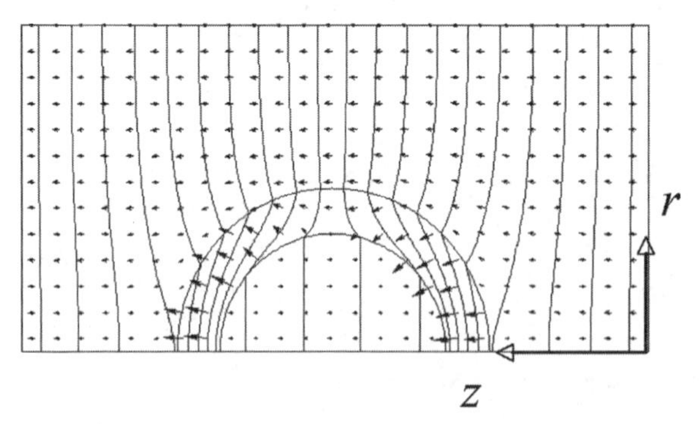

FIGURE 9.5 Electric field and equipotentials near a liposome in an external static electric field pointing along the *z*-axis. (The finite element simulation was performed using Electrostatic Field Analysis in QuickField.)

9.3.2 Liposome in a Static Electric Field

Liposomes are lipid bilayer vesicles that can be produced in the laboratory and used for drug delivery and other applications. Figure 9.5 shows a finite element calculation of a liposome in a static electric field incident from right to left.

The model has rotational symmetry about the z-axis horizontal to the bottom of the graph. The liposome has a diameter of 50 nm and a wall thickness of 7 nm. The equipotential lines are perpendicular to the electric field vectors. The E field is amplified in the membrane and reduced inside the liposome. The relative dielectric constant of liposome membrane is $\varepsilon_r = 10$ compared to the internal and external region with

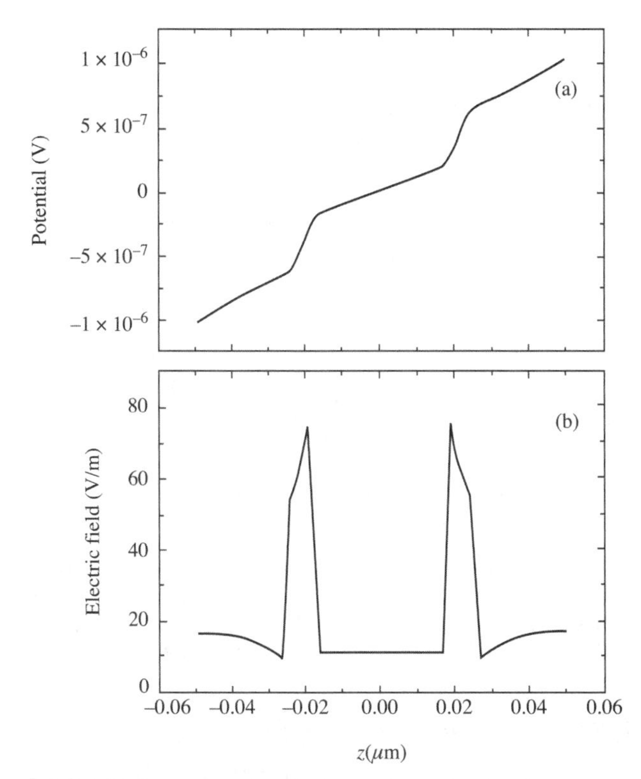

FIGURE 9.6 Contour plots of (a) potential and (b) electric field along the axis of the liposome in Figure 9.5 The electric field is given by the slope of the potential in (a) and is amplified inside the membrane and constant inside the liposome.

$\varepsilon_r = 78$. In Figure 9.6, the potential and electric field are plotted along the z-axis passing through the center of the liposome. The peak electric field in the membrane is nearly 80 V/m.

9.3.3 Spherical Cell in an Alternating Electric Field

A spherical cell in an alternating electric field is modeled in Figure 9.7 showing equipotentials and current flow vectors resulting from (a) 50 Hz and (b) 1.0 MHz excitations. The cell radius is 0.25 μm with a 5-nm radius requiring an extremely fine finite element mesh density near the plasma membrane. The current density is almost perfectly screened from the interior of the cell by the membrane at low frequencies as seen in Figure 9.7(a). At high frequencies, the ac current is coupled through the membrane in Figure 9.7(b). Also, the electric field and current density is uniform inside the cell at all frequencies.

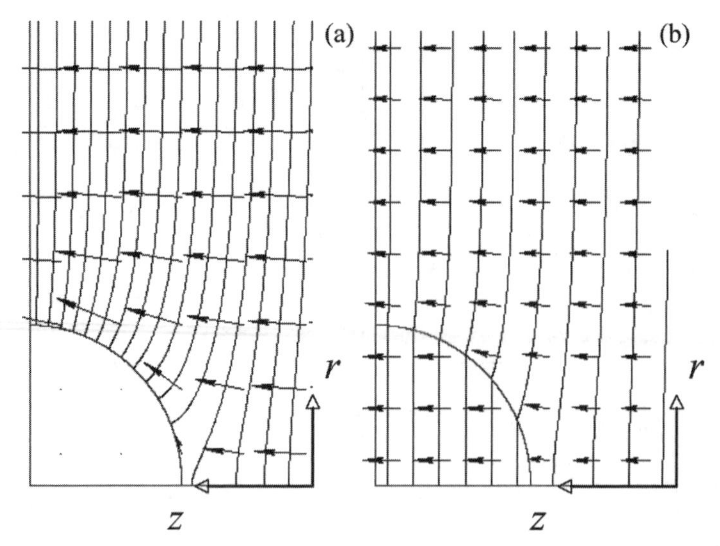

FIGURE 9.7 Finite element calculation of equipotentials and current flow vectors resulting from (a) 50 Hz and (b) 1.0 MHz excitations of a plasma membrane. The current density is nearly zero inside the cell at low frequencies. At high frequencies, the AC current readily penetrates the cell. The current density is uniform inside the cell at all frequencies. (Simulation performed using AC Current Analysis in QuickField.)

9.3.4 Inertial Effects in Field-Induced Counterion Motion

To illustrate how inertial effects contribute to the motion of counterions, we consider the ideal case of ions surrounding a cell in a medium with zero viscosity while neglecting diffusion. An ion of mass m and charge q will experience a force $F = qE$ in an electric field E. From Newton's second law, the acceleration of the ions is $a = \pm qE/m$ for a square wave excitation. Ions accelerated from rest will travel a distance $d = \frac{1}{2}at^2$ in a time $t = \sqrt{2d/a} = \sqrt{2dm/qE}$, neglecting the time it takes to change directions. The ions will periodically transverse this distance at the excitation frequency $f \approx 1/t$ of the square wave. Thus the distance transversed by the ions decreases along with the induced dipole moment ($\mathbf{p}$ = charge $\times$ displacement) and dielectric permittivity $\varepsilon \propto \mathbf{p}$ at higher frequencies.

9.3.5 Induced Potentials in Organelle Membranes

Figure 9.8 shows an equivalent circuit model of a cell membrane surrounding the cytoplasm with the nuclear membrane surrounding the nucleoplasm. The outer cell membrane effectively acts as a high-pass filter so that membrane potentials are developed only across internal membranes at higher frequencies. Isolated organelles extracted from cells may be electrically stimulated, or microelectrodes may be inserted inside the cell to electrically excite organelles thus circumventing the plasma membrane.

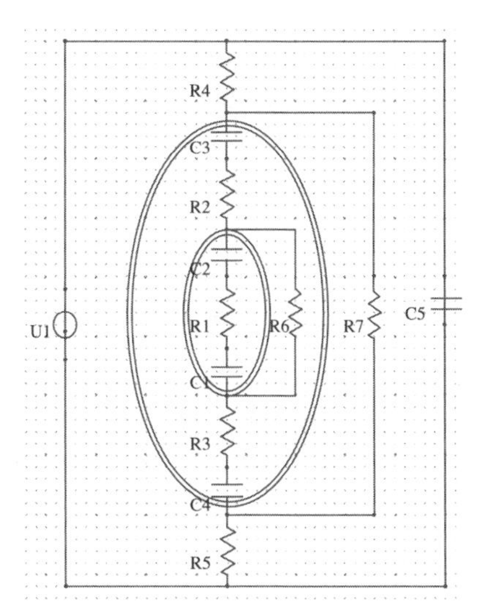

FIGURE 9.8 Equivalent circuit model of a cell membrane surrounding the cytoplasm with the nuclear membrane surrounding the nucleoplasm. (Constructed in QuickField.)

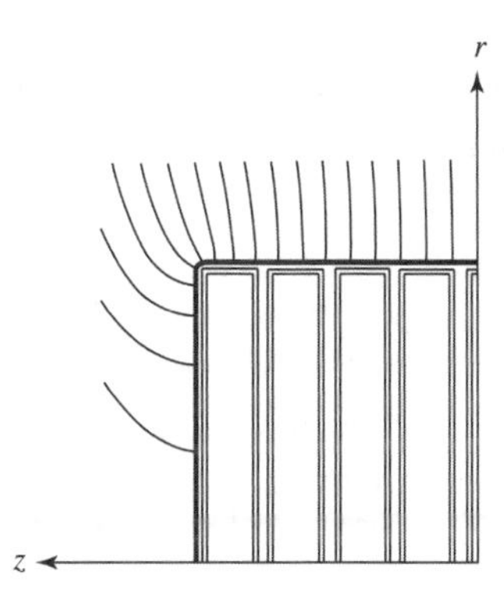

FIGURE 9.9 Finite element calculation of the equipotentials near a segment of a mitochondrion under electrical excitation. Modeled regions include the cytoplasm, outer mitochondrial membrane, inner membrane space, inner membrane, and matrix space. The model has rotational symmetry about the *z*-axis horizontal to the page. (Simulation performed using AC Current Analysis in QuickField.)

9.3.6 Mitochondrion in an Alternating Electric Field

Mitochondria are two-membrane organelles consisting of an outer membrane and a convoluted inner membrane separating the central matrix, inner membrane space, and the outside intracellular region. Figure 9.9 shows a mitochondrion under AC excitation modeled in axial symmetry with rotational symmetry about the z-axis at the bottom of the figure.

Conductivity and permittivity values are taken as $\varepsilon_r = 40$ and $\sigma = 5.0$ S/m outside the mitochondrion and $\varepsilon_r = 7$ and $\sigma = 2.0$ S/m inside the mitochondrial membranes, and $\varepsilon_r = 5.0$ and $\sigma = 0.001$ S/m in the inner membrane space. This figure illustrates a corner segment of the mitochondrion with a convoluted inner membrane. At low frequencies, the induced membrane potential is higher across the outer mitochondrial membrane with the inner membrane potential becoming greater near 0.5 MHz.

9.4 Bioimpedance

9.4.1 Time Harmonic Current Flow

Biological tissue is composed of mostly water, is weakly conducting, and behaves as a lossy dielectric in AC electromagnetic fields. In general, AC currents flowing in tissue under electrical stimulation will consist of a sum of conduction and displacement current densities

$$J = \sigma E + i\omega\varepsilon E. \tag{9.27}$$

Equation 9.27 is sometimes written as $J = \sigma^* E$ where the complex conductivity is $\sigma^* = \sigma' + i\sigma''$ so that $\sigma = \sigma' = \text{Re}\sigma^*$ and $\varepsilon = \text{Im}\sigma^*/\omega$. Other conventions exist involving complex permittivity or complex resistivity. As well, there are slight notational variations in the literature. Comparison with Equation 9.27 can help sort out the meaning of a given notational convention.

The active and reactive components of the current density J are given by

$$J_{\text{active}} = \sigma E \quad J_{\text{reactive}} = \omega\varepsilon E \tag{9.28}$$

so that the conductivity and permittivity can be determined from experimentally measured current components

$$\sigma = \frac{J_{\text{active}}}{E} \quad \varepsilon = \frac{J_{\text{reactive}}}{E\omega}. \tag{9.29}$$

Biological tissue is highly inhomogeneous, with local variations in conductivity and permittivity, so Equation 9.29 will give effective values for a given tissue sample or cell suspension. A useful figure of merit is the loss tangent (or loss factor) defined as

$$\tan\delta = \frac{J_{\text{active}}}{J_{\text{reactive}}} = \frac{\sigma}{\omega\varepsilon}. \tag{9.30}$$

The loss tangent is a measure of the rms or peak energy lost per cycle divided by the energy stored per cycle. The loss angle

$$\delta = \tan^{-1}\frac{\sigma}{\omega\varepsilon} \quad (9.31)$$

will be zero for a simple capacitor with no resistive losses. Given a uniform current density $J = I/A$, the total current is expressed as the admittance (or inverse impedance) times the voltage $I = Y^*V$, or, from Equation 9.27,

$$I = \left(\frac{\sigma A}{d} + i\omega\frac{\varepsilon A}{d}\right)V \quad (9.32)$$

where $E = V/d$ and Y^* is the complex admittance

$$Y^* = G + i\omega C. \quad (9.33)$$

The conductance G and the capacitance C are proportional to the conductivity and permittivity, respectively. For inhomogeneous tissue regions, C and G will each be dependent on the conductivity and permittivity distributions in a complicated way. Equation 9.33 is often expressed as

$$Y^* = G + iB \quad (9.34)$$

where $B = \omega C$ is the susceptance. In terms of impedance, Ohm's law is $V = IZ$, so

$$Z = \left(\frac{\sigma A}{d} + i\omega\frac{\varepsilon A}{d}\right)^{-1} \quad (9.35)$$

with imaginary and real parts

$$Z = R + iX. \quad (9.36)$$

Note that resistance $R = \mathrm{Re}Z$ and reactance $X = \mathrm{Im}Z$ are not proportional to σ and ε. For this reason, admittance is often preferable to impedance data for determining these parameters. Table 9.1 lists some common bioimpedance quantities and their units.

The quantities (ε, σ), (R, X), and (G, B) can be thought of as complementary pairs related by the Kramer–Kronig (K–K) integral transformations tabulated in Appendix 6. For example, the K–K relations enable the calculation of $\sigma(\omega)$ from $\varepsilon(\omega)$ and vice versa.

9.4.2 Dielectric Spectroscopy

The electrical response of living cells to AC electric fields exhibits similar behavior for cells in suspension and tissue. In general, the effective dielectric constant of biological material decreases, while the conductivity increases at higher frequencies.

Table 9.1 Bioimpedance Symbols and Units

Symbol	Quantity	Units
V	Voltage	Volts (V)
I	Current	Amperes (A)
J	Current density	A/m^2
R	Resistance	Ohms Ω
C	Capacitance	Farads (F) or (Coulombs/V)
σ	Conductivity	Siemens/meter (S/m)
ε	Permittivity	F/m or (C/V/m)
Z	Impedance	Ω
X	Reactance	Ω
Y	Admittance	Siemens S = Ω^{-1}
G	Conductance	S
B	Susceptance	S

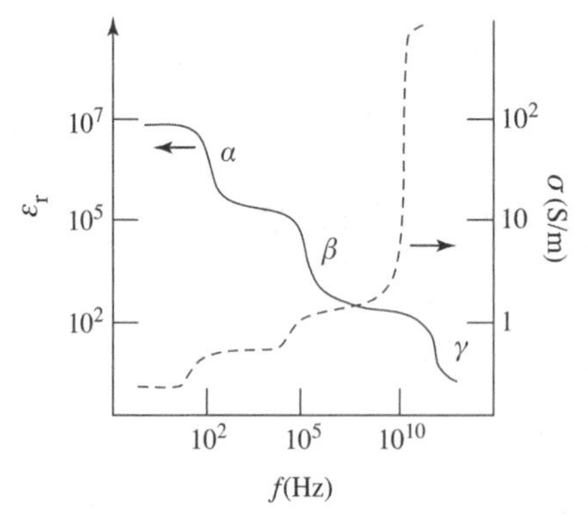

FIGURE 9.10 Sketch of the frequency-dependent permittivity (solid line) and conductivity (dashed line) of living cells in suspension or tissue. (Adapted from Schwan. 1994. Electrical properties of tissues and cell suspensions: Mechanisms and models. *Engineering in Medicine and Biology Society* 6: 70a–71a.) The permittivity and conductivity changes in three major steps corresponding to the α, β, and γ dispersions.

Three rather broad frequency ranges have been identified where different processes affect the dielectric permittivity and conductivity in biological material. These frequency ranges correspond to the α-, β-, and γ-dispersions as shown in Figure 9.10.

α-dispersion: (mHz–kHz) results mostly from the motion of charges, or counterions, next to the cell's plasma membrane in suspension where the relative dielectric permittivity can exceed 10^7. At lower frequencies, counterions travel some fraction of the cell's length in one period of the excitation as shown in Figure 9.11.

The mechanism for α-dispersion is different in tissue than in cell suspensions. The AC response of gap junctions connecting cells contributes to the enormous low frequency dielectric permittivity of tissue. Figure 9.12 illustrates a gap junction between two cells.

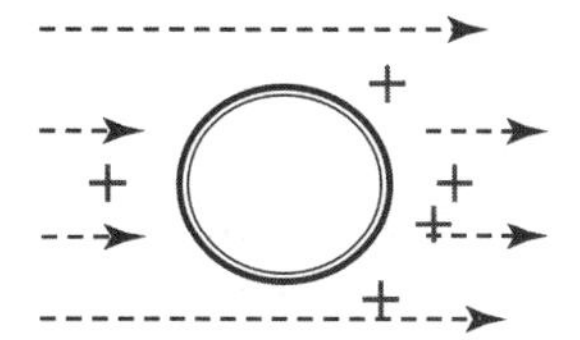

FIGURE 9.11 Redistribution of counterions surrounding a cell in an electric field resulting in an induced dipole moment.

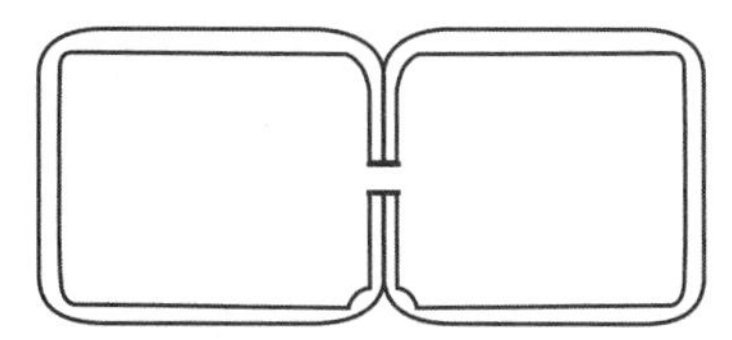

FIGURE 9.12 Schematic of two cells connected by a gap junction.

β-dispersion: (0.1–100 MHz) results from the passive cell membrane capacitance, Maxwell–Wagner effects with organelle membrane contributions, and dispersion due to the cytoplasm. The β-dispersion may be simulated in QuickField AC Current Analysis.

γ-dispersion: (0.1–100 GHz) results mainly from relaxations of molecular dipole moments such as water, salts, and proteins. QuickField models of cells in eternal fields will not include the γ-dispersion.

δ-dispersion: Weaker dispersion near the γ-range due to protein-bound water.

Table 9.2 lists these dispersions as they appear in (1) aqueous solutions of electrolytes, amino acids, proteins, and nucleic acids, (2) vesicles with and without surface charges, and (3) cells with surface charges, organelles, and cytosolic protein.

Table 9.2 Dispersions in Biological Suspensions

Biological Suspension		α	β	γ	δ
Water	+ electrolytes			•	
	+ amino Acids			•	•
	+ proteins		•	•	•
	+ nucleic acids	•	•	•	•
Vesicles	– surface charge		•	•	
	+ surface charge	•	•	•	
Cells	+ surface charge	•	•	•	
	+ organelles		•	•	•
	+ proteins		•	•	•

(Adapted from Schwan. 1994. Electrical properties of tissues and cell suspensions: Mechanisms and models. *Engineering in Medicine and Biology Society* 6: 70a–71a.)

9.4.3 Debye Relaxation Model

The Debye relaxation model describes the AC response of systems such as living cell suspensions and polymers consisting of dipole populations to alternating fields. There is a time delay between the excitation and the dipole response characterized by a single relaxation time τ in the Debye model. The dielectric relaxation is described by a frequency-dependent complex permittivity

$$\varepsilon^*(\omega) = \varepsilon_\infty + \frac{\Delta\varepsilon}{1 + i\omega\tau} \tag{9.37}$$

where $\Delta\varepsilon = \varepsilon_s - \varepsilon_\infty$ with respective static DC and high-frequency permittivity values ε_s and ε_∞. The complex permittivity is often represented on a Cole–Cole plot of ε' versus ε'' where $\varepsilon^* = \varepsilon' - i\varepsilon''$. In terms of the real part of the permittivity and conductivity, we have the relation $\varepsilon^* = \varepsilon + i\sigma/\omega$ where $\varepsilon = \mathrm{Re}\varepsilon^*$ and $\mathrm{Im}\varepsilon^* = \sigma/\omega$. Debye equivalent circuit models may be constructed to simulate the electrical response of cells and membranes to external fields and direct electrode stimulation. The AC and transient response of electrically excited cells and tissue can be modeled in QuickField in many situations. Figure 9.13 shows the Debye single dispersion equivalent circuit model constructed in QuickField consisting of an RC–C parallel combination.

9.4.4 Cole Equation

An empirical formula highly successful in fitting bioimpedance data was developed by Cole in 1941. The frequency-dependent complex impedance is

$$Z = R_\infty + \frac{\Delta R}{1 + (i\omega\tau)^\alpha} \tag{9.38}$$

where $\omega = 2\pi f$ and the Cole exponent α is associated with a distribution of relaxation times. The spread in relaxation times may be due to cellular and molecular interactions, gap junctions, tissue anisotropy, fractal dimension, and size effects. Cole plots formed by plotting Im(Z) versus Re(Z) are semicircles in the complex plane. For multiple dispersions such as observed in the myocardium, the impedance formula can be extended

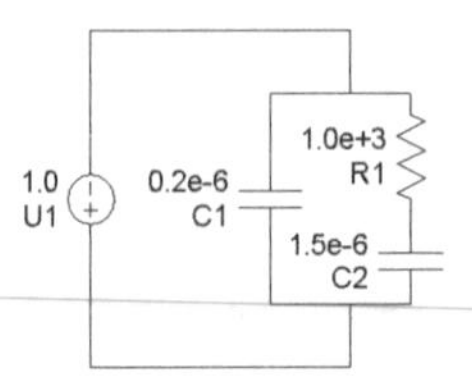

FIGURE 9.13 Debye relaxation circuit model. (Constructed in QuickField.)

$$Z = R_\infty + \frac{\Delta R_1}{1+(i\omega\tau_1)^{\alpha_1}} + \frac{\Delta R_2}{1+(i\omega\tau_2)^{\alpha_2}} + \ldots \tag{9.39}$$

with parameters $\{R_\infty, \Delta R_1, \tau_1, \alpha_1, \Delta R_2, \tau_2, \alpha_2 \ldots\}$ adjusted to fit a given data set. The complex permittivity models the dielectric response of tissue and is given by the Cole–Cole expression

$$\varepsilon^*(\omega) = \varepsilon_\infty + \frac{\Delta\varepsilon}{1+(i\omega\tau)^{1-\alpha}}. \tag{9.40}$$

This expression is similar to the Debye expression for the complex permittivity of a cell suspension with the exponent $1 - \alpha$.

9.4.5 Maxwell–Wagner Effect

The Maxwell–Wagner (M–W) effect describes processes that occur at the interface between different dielectric regions, giving rise to the β-dispersion in living cell suspensions. Spatial variations of dielectric and conducting matter capacitively couple to alternating electric fields. The M–W effect depends on the geometry of the dielectric interface with respect to the incident field direction. The simplest electrical model exhibiting this type of dispersion consists of two lossy dielectrics slabs in series as shown in Figure 9.14(a). Figure 9.14(b) shows a circuit model constructed in QuickField equivalent to two lossy dielectric slabs in series. M–W contributions vanish in dielectrics with interfaces parallel to the external field.

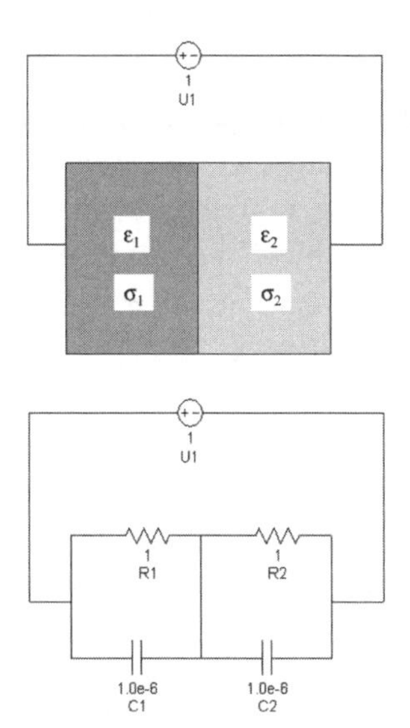

FIGURE 9.14 (a) Two lossy dielectric slabs in series. (b) Circuit model equivalent to two lossy dielectric slabs in series giving rise to Maxwell–Wagner dispersion. (Constructed in QuickField.)

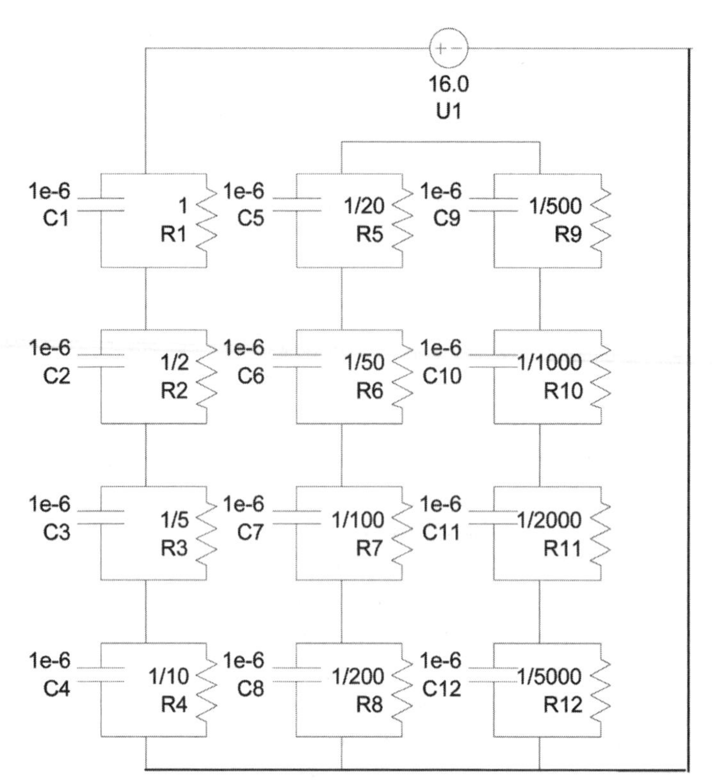

FIGURE 9.15 Tregear's electrical circuit model of skin impedance consisting of 12 resistors and 12 equal capacitors. (Constructed in QuickField.)

9.4.6 Skin Impedance

R. T. Tregear developed a model of skin impedance in 1965. Tregear's electrical circuit model consists of a network of 12 resistors and 12 capacitors where the capacitors C have equal values and the resistances have decreasing values according to $R_i = R, R/2, R/5, R/10, R/100, R/200$, and so on. Figure 9.15 shows Tregear's electrical circuit model of skin impedance with 12 resistors of decreasing value and 12 equal capacitors.

9.4.7 Electrode Polarization

Dielectric spectroscopy measurements involving contacting metal electrodes with conducting solutions are subject to large errors at low frequencies due to the electrode–electrolyte interface. Electrode polarization is generated by an electrical doublelayer that forms at the solution–electrode interface. The total measured impedance includes contributions from the solution and the interface. The measured low-frequency dielectric response contains this polarization component in addition to Maxwell–Wagner and α-dispersion contributions.

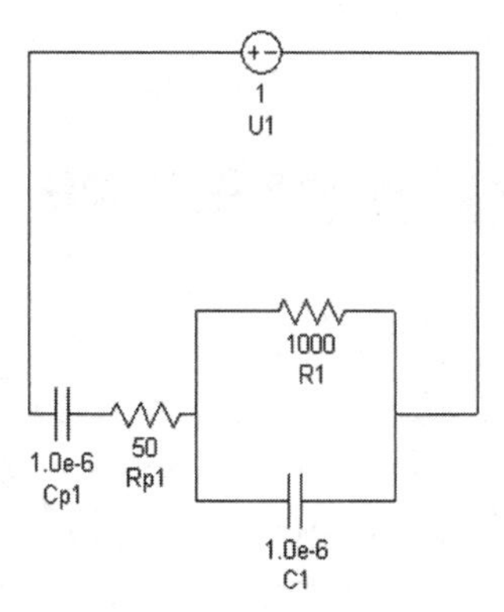

FIGURE 9.16 Equivalent circuit model of electrode polarization due to the electrode–electrolyte interface with polarization capacitance C_p and resistance R_p. (Constructed in QuickField.)

Figure 9.16 illustrates an idealized circuit model of electrode polarization where C_p and R_p are the polarization capacitance and resistance, respectively. The suspension capacitance and resistance are given by C_2 and R_2, respectively. The polarization impedance depends on the type of electrode and electrolyte, as well as the excitation frequency. Platinum black electrodes and plate-separation techniques can be used to eliminate the polarization impedance where measurements are made at different electrode separations and then subtracting the common electrode artifact signal. Noncontacting techniques involving torroidal induction coils with torroidal pickup or SQUID magnetometers can also be used to avoid electrode polarization effects.

The frequency response of the electrode–electrolyte interface has been modeled using a frequency-independent phase element ϕ over a wide range of frequencies with impedance parameters K and β

$$Z = K(i\omega)^{-\beta}$$

$$\phi = -\beta\left(\frac{\pi}{2}\right). \tag{9.41}$$

9.5 Bioimpedance Simulator

Bioimpedance Simulator® is a circuit-based modeling software than can simulate the α- and β-dispersions of various cells and tissues with two-dimensional x–y geometry. The software can model AC impedance of tissue with inhomogeneities, tumors, and ischemic injury. The program is based on the Simulated Program with Integrated Circuit Emphasis (SPICE) developed at the University of California at Berkeley. BioZsim and related files can be downloaded at http://www.cnm.es/~mtrans/BioZsim/.

A geometrical model is first constructed in Bioimpedance Simulator. Model elements are specified, including current sources, an extracellular medium (plasm), an

Table 9.3 Circuit Elements in Bioimpedance Simulator

Symbol	Circuit Element
R_C	Cytoplasm resistivity/(2 × pixel thickness)
R_P	Plasm resistivity/(2 × pixel thickness)
R_{GAP}	Gap junction resistance/(pixel size × pixel thickness)
R_{PT}	Plasm resistivity × pixel size/(2 × pixel thickness × junction width)
R_{PL}	Plasm resistivity × junction width/(2 × pixel thickness × pixel size)
R_M	Membrane resistance/(pixel size × pixel thickness)
C_M	Membrane capacitance × pixel size × pixel thickness

(a)

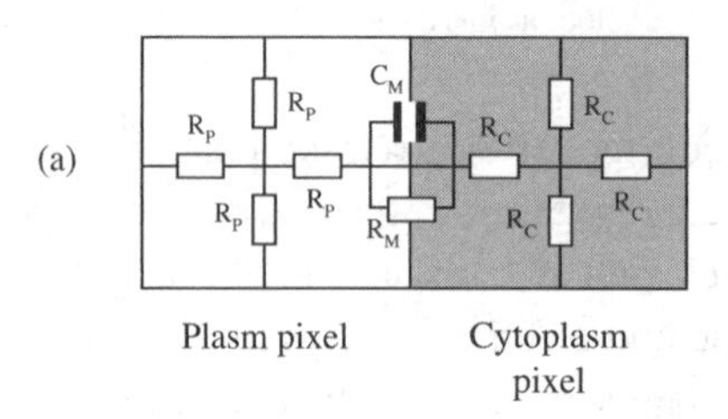

(b)

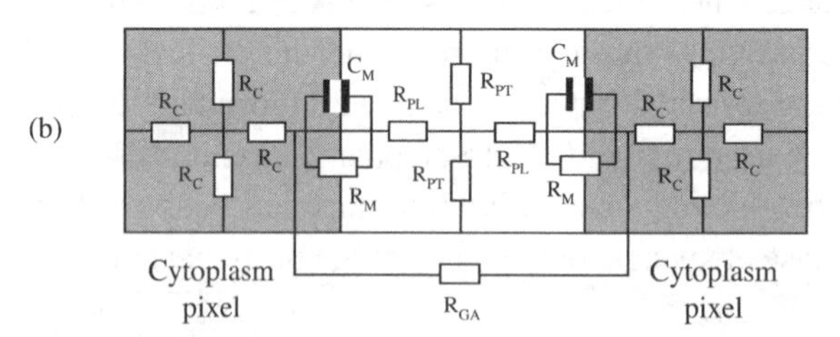

FIGURE 9.17 Equivalent circuit models of (a) plasm and cytoplasm pixels separated by membrane resistance and capacitance elements and (b) cytoplasm pixels separated by a gap junction pixel and circuit elements in the Bioimpedance Simulator.

intracellular medium (cytoplasm), and gap junctions. The geometries of the intra- and extracellular regions and gap junctions are pixilated with user-specified pixel width. A key feature of the simulator is that membranes are automatically specified at the interfaces between extracellular and intracellular regions and are not separately pixilated. Table 9.3 lists various circuit elements used to construct the model. Once the model is constructed, the frequency range of analysis is then specified, and the simulation is executed. SPICE input files called netlists are generated. Various impedance quantities can be plotted versus frequency or on Cole–Cole plots.

Figure 9.17 shows equivalent circuit models of (a) plasm and cytoplasm pixels separated by membrane capacitance and resistance and (b) the circuit element corresponding to a gap junction separating cytoplasm regions.

Impedance and Cole–Cole plots are shown in Figures 9.18 and 9.19 of a whole cell modeled with and without organelles, respectively.

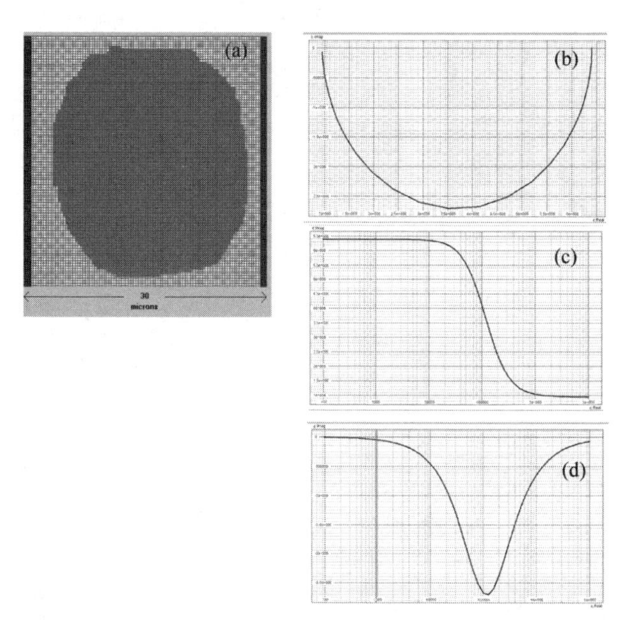

FIGURE 9.18 Simulation results obtained with the Bioimpedance Simulator of a whole cell without organelles. (a) Problem geometry. (b) Cole–Cole plot of imaginary vs. real impedance components. (c) Real part of epsilon vs. frequency. (d) Imaginary part of epsilon vs. frequency.

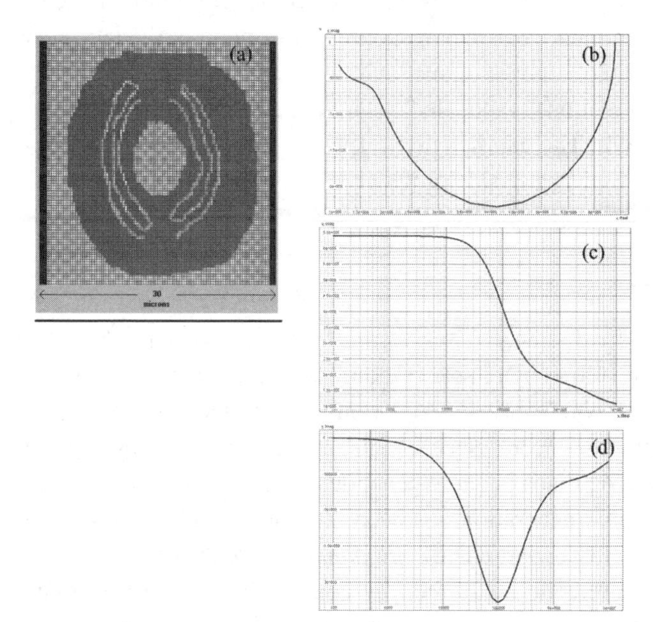

FIGURE 9.19 Simulation results obtained with the Bioimpedance Simulator of a whole cell with organelles. (a) Problem geometry. (b) Cole–Cole plot of imaginary vs. real impedance components. (c) Real part of epsilon vs. frequency. (d) Imaginary part of epsilon vs. frequency.

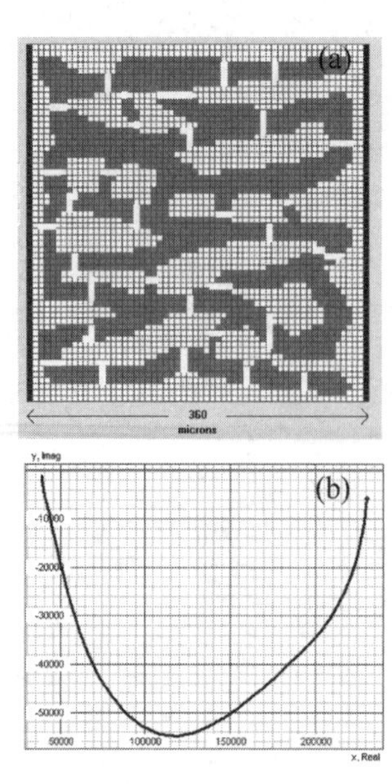

FIGURE 9.20 Simulation results obtained with the Bioimpedance Simulator of a tissue region including gap junctions. (a) Problem geometry. (b) Cole–Cole plot of imaginary vs. real impedance components.

The plots in Figures 9.18 and 9.19 show impedance variations at high frequencies due to organelles. Note that the high-frequency responses correspond to the left-hand sides of the Cole–Cole plots.

A tissue region with gap junctions is modeled in Figure 9.20, showing a distorted semicircle in the Cole–Cole plot. The gap junction response is prominent at low frequencies corresponding to the right-hand side of the Cole–Cole plot.

9.6 Nonlinear Effects

At sufficiently low excitation field strengths $\approx$1 V/cm, the current increases linearly with voltage in a cell sample resulting in a linear response with some harmonic distortion due to the electrode–electrolyte interface. Excitations greater than a few volts per centimeter can generate harmonics from electric field interactions with membrane proteins below $\approx$100 Hz. Electric fields begin to penetrate the plasma membrane near kHz frequencies where harmonic generation has been attributed to field interaction with the photosynthetic electron transport chain in *Rhodobacter capsulatus*.

Nonlinear response measurement are limited by confounding electrode polarization effects at electrode/electrolyte interfaces that give rise to sparious harmonic generation. Efforts to eliminate sparious harmonics include the use of dual cell configurations and remote field sensors.

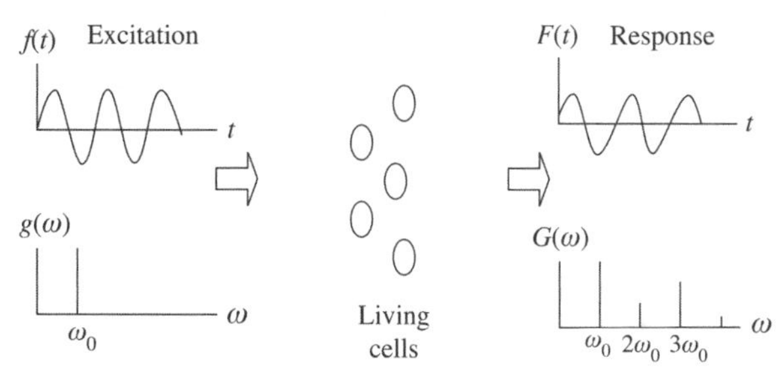

FIGURE 9.21 Schematic illustrating the technique of nonlinear harmonic analysis. (Adapted from Claycomb, 2009. Harmonic analysis of neuronal membranes and tissue using SQUIDs. *IEEE Transactions on Applied Superconductivity.* 17: 812–815.)

Figure 9.21 illustrates the technique of nonlinear harmonic analysis. A sinusoidal electric field $f(t)$ with excitation frequency ω_0 is applied to a suspension of living cells. The Fourier transform of the excitation is $g(\omega) \sim \delta(\omega - \omega_0)$. The response $F(t)$ is a periodic function with Fourier transform $G(\omega) \sim \delta(\omega - n\omega_0)$ consisting of integer multiples of ω_0.

EXERCISES

Exercise 9.1 The equation developed by Maxwell and Fricke describing the low frequency conductivity σ of a cell suspension with volume fraction p is

$$\frac{\sigma - \sigma_a}{\sigma + \gamma\sigma_a} = p\frac{\sigma_i - \sigma_a}{\sigma_i + \gamma\sigma_a}$$

where σ_a is the conductivity of the extracellular space and σ_i is the cytoplasm conductivity. Make a plot of σ versus with $\gamma = 2$ for spheres.

Exercise 9.2 For water $\alpha \approx 0$, $\varepsilon_s \approx 80$, $\varepsilon_\infty \approx 6$, and $\tau \approx 10$ ps. Create a Cole–Cole plot of the dielectric response of water potting ε' versus ε'' where $\varepsilon^* = \varepsilon' - i\varepsilon''$.

Exercise 9.3 The impedance of a test cell with cross-sectional area $A = 10^{-4}\,\text{m}^2$ is measured to be $Z = 2 + 3i\ \Omega$ at $f = 50$ Hz ($\omega = 2\pi f$). What are the corresponding admittance, susceptance, conductance, conductivity, and permittivity values of the sample?

Exercise 9.4 Plot the $\text{Re}\Delta U_e$ versus $\text{Im}\Delta U_e$ in Equation 9.23 as a function of frequency for different values of τ.

Exercise 9.5 QuickField Electrostatics Simulation. Create a model consisting of two regions with charge density ρ_1 and ρ_2 separated by a plasma membrane with thickness $\delta = 7$ nm using x–y symmetry. Create a vector plot of the electric field and plot the potential along the dotted contour shown in Figure 9.22.

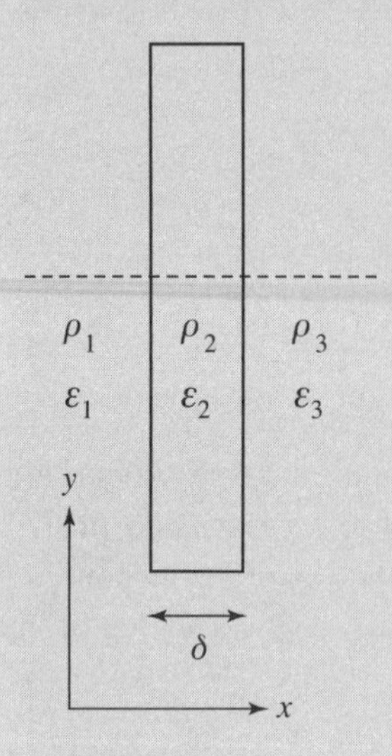

FIGURE 9.22

Exercise 9.6 QuickField Electrostatics Simulation. Create a model consisting of two regions with charge density ρ_1 and ρ_2 separated by a cylindrical axonal membrane with thickness $\delta = 7$ using r-z symmetry. Create a vector plot of the electric field and plot the potential along the center of the axon shown in Figure 9.23. Perform your simulation for both myelinated and unmyelinated axons, taking the thickness of the myelin sheath ~0.5 μm.

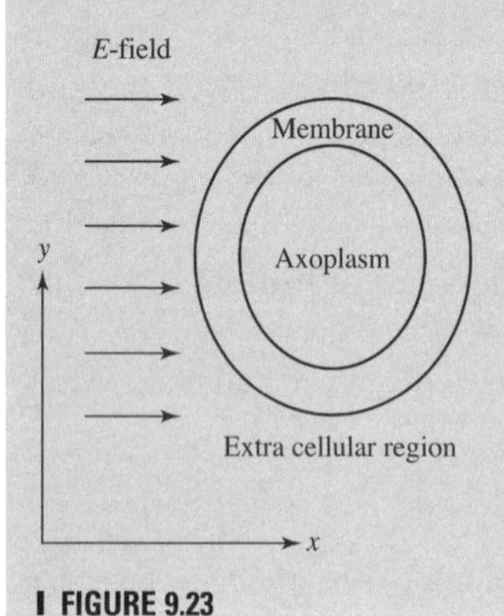

FIGURE 9.23

Exercise 9.7 QuickField AC Conduction Simulation. Create a model of a spherical liposome with $(\varepsilon_1, \sigma_1)$ inside the liposome, $(\varepsilon_2, \sigma_2)$ inside the membrane, and $(\varepsilon_3, \sigma_3)$ outside the liposome with thickness $\delta = 7$ nm. Use the QuickField LabelMover feature to plot the electric field at the center of the liposome as a function of excitation frequency. Create a vector plot of the electric field and plot the potential along the z axis shown in Figure 9.24. Exploit symmetry about the plane bisecting the liposome to create a finer mesh.

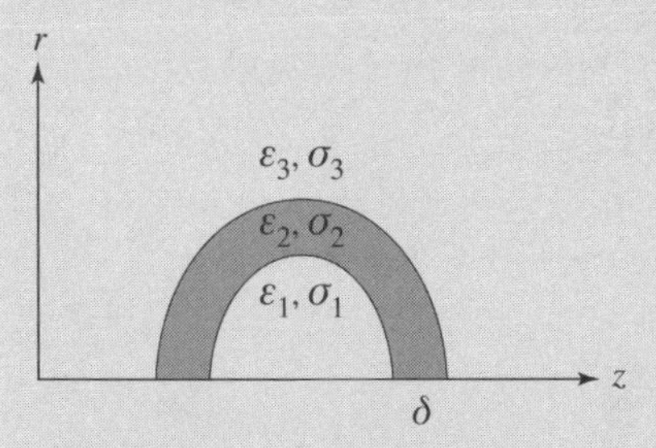

FIGURE 9.24

Exercise 9.8 Create a double-shell version of Exercise 9.8 by placing a second membrane a radial distance of 7 nm outside the inner membrane. Calculate the membrane potential across the inner membrane as a function of excitation frequency.

Exercise 9.9 Calculate the potential and electric field in an electrolyte between two coaxial long cylinders of radii R_a and R_b held at opposite potentials $U_e(R_a) = V$ and $U_e(R_a) = -V$, respectively. Use the Helmholtz form of the Poisson–Boltzmann equation in the low field limit. Assume that the potential can vary only in the radial direction.

Exercise 9.10 Repeat Exercise 9.9, calculating the potential and electric field in an electrolyte outside a single cylinder with $U_e(R_a) = V$.

Exercise 9.11 Calculate the potential and electric field in an electrolyte between two concentric spheres of radii R_a and R_b withheld at opposite potentials $U_e(R_a) = V$ and $U_e(R_a) = -V$, respectively. Use the Helmholtz form of the Poisson–Boltzmann equation in the low field limit. Assume that the potential can vary only in the radial direction.

Exercise 9.12 Repeat Exercise 9.11, calculating the potential and electric field in an electrolyte outside a single sphere with $U_e(R_a) = V$.

Exercise 9.13 Calculate the potential and electric field in an electrolyte outside of a thin needle with potential V. Use the Helmholtz form of the Poisson–Boltzmann equation in the low field limit. Perform your calculation using prolate spheroidal coordinates.

Exercise 9.14 Calculate the potential and electric field in an electrolyte outside of a thin disk with potential V. Use the Helmholtz form of the Poisson–Boltzmann equation in the low field limit. Perform your calculation using oblate spheroidal coordinates.

10 Nerve Conduction

10.1 Nerve Impulses

The human nervous system is made up of ~10^{11} neurons, roughly the same as the number of stars in the Milky Way galaxy. The central nervous system consists of the brain and the spinal cord. The peripheral nervous system relays messages between the central nervous system and the rest of the body.

The three types of neurons include sensory, association, and motor neurons. Sensory neurons receive input stimuli from the environment and subsequently activate association neurons (or interneurons) that transmit signals between nerve cells. Association neurons are the most common type of neuron in complex animals. Integration centers that perform cognitive functions in higher life forms are composed of these neurons. Motor neurons receive input from association neurons to initiate muscle contraction and activate glands.

Nerve cells are a few microns in diameter consisting of a long axon that terminates on the cell body surrounding the nucleus as shown in Figure 10.1. Axons can be up to a meter in length. Dendrites are shorter, tentacle-like branches a few millimeters in length surrounding the nerve cell body and are essential in signaling between neurons.

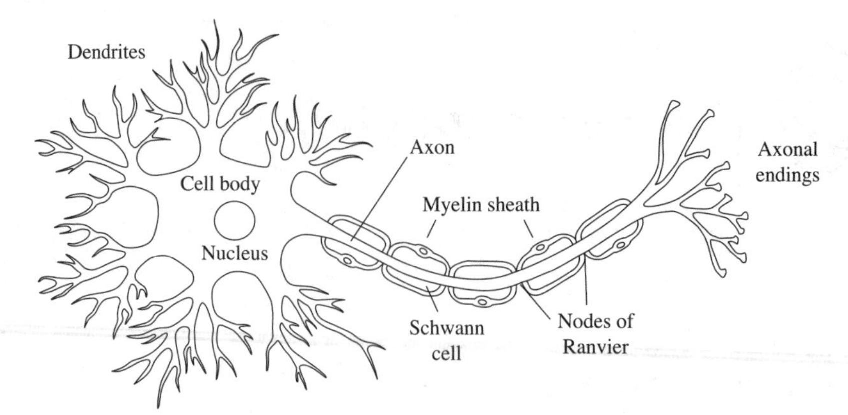

FIGURE 10.1 Schematic of a nerve cell showing the cell body, dendrites, nucleus, axon, Schwann cells, myelin sheath with nodes of Ranvier surrounding the axon, and axonal endings. (Drawing created using Motifolio®.)

Active signal transport occurs along the axon by propagating action potentials. Some axons are surrounded by Schwann cells that supply nutrients to the nerve cell and assist in signal propagation. The plasma membrane of the Schwann cell is known as the myelin sheath. Signal propagation is between 0.5 and 2.5 m/s in unmyelinated axons compared with up to 130 m/s in myelinated axons because of the increased capacitance provided by the myelin sheath.

The increased signal propagation speed is accompanied by a decrease in signal amplitude along the ~1-mm-long myelinated segments. Ion channels located at gaps between the Schwann cells called nodes of Ranvier serve to boost the amplitude of the nerve signals. The velocity of action potential is reduced at these nodes. However, since the nodes are only ~2 μm long, the average signal velocity is hardly affected. This relay type of signal along the myelinated axon is known as saltatory conduction.

10.2 Neurotransmitters and Synapses

Action potentials can pass between neurons across electrical or chemical synapses. Electrical synapses pass signals directly by the exchange of ions through neuronal gap junctions. Chemical synapses operate by the exchange of neurotransmitter molecules. Neurotransmitter diffusion between axonal endings and dendrites is slower than direct ion exchange through gap junctions, but it can be chemically regulated.

More than 10 different types of neurotransmitters have been discovered, including acetylcholine, dopamine, and serotonin. Figure 10.2 illustrates the transport of vesicles containing neurotransmitter molecules to the presynaptic membrane. Vesicle transport to the presynaptic membrane can also be facilitated by microtubules. There are no microtubules to facilitate the diffusion of neurotransmitters in the synapse, however.

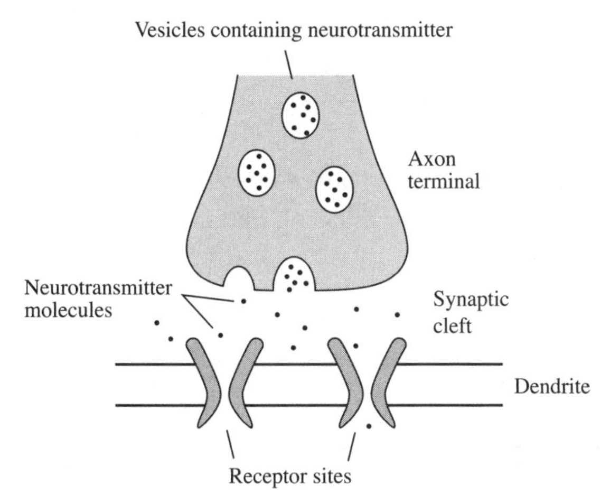

FIGURE 10.2 Diffusion of neurotransmitter molecules across a synapse.

The time it takes neurotransmitter molecules to diffuse across the synaptic cleft of width $\delta \approx 20$ nm can be estimated from the one-dimensional diffusion equation

$$R_{\text{rms}} = \sqrt{2Dt} \tag{10.1}$$

with $\delta = R_{\text{rms}}$ giving the diffusion time $t = \delta^2/2D = 2 \cdot 10^{-7}$ s.

10.3 Passive Transport in Dendrites

Both active and passive transport is required for the propagation of nerve impulses. Dendrites, which receive signals from other nerve cells, do not have ion channels and therefore respond passively to signal voltages. These passive dendritic signals travel roughly 50 times slower than action potentials. However, the dendrite signals need to travel just a few millimeters to the cell nucleus compared to action potentials that travel up to a meter in length.

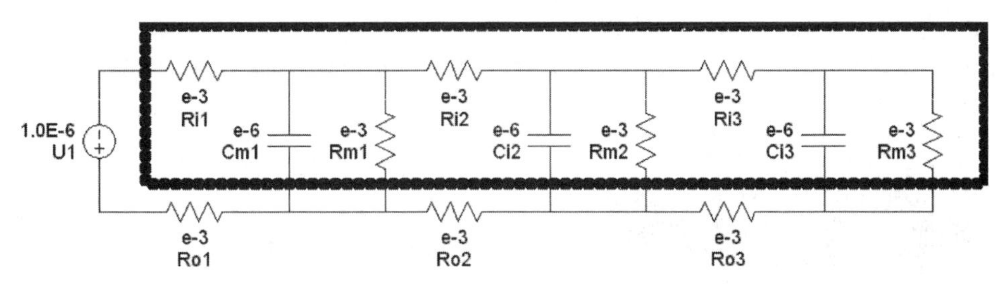

FIGURE 10.3 Passive circuit model of a dendrite (contained in the boxed region) where R_i is the resistance per unit length, R_m and C_m are the membrane resistance and capacitance, respectively, per unit length. R_0 is the resistance outside the dendrite.

Resistor–capacitor networks can be used to model the passive response of nerve cells to electrical stimuli. Figure 10.3 shows an equivalent circuit of a dendrite where R_i is the resistance per unit length, R_m and C_m are the membrane resistance and capacitance, respectively, per unit length. R_0 is the resistance outside the dendrite. Circuit elements corresponding to the dendrite are contained in the boxed region in this figure.

Only three segments are included in this simplified model. For an infinite chain, the voltage $v(x, t)$ is described by the cable equation

$$\frac{R_m}{R_i}\frac{\partial^2 v(x,t)}{\partial x^2} - R_m C_m \frac{\partial v(x,t)}{\partial t} = v(x,t). \tag{10.2}$$

This equation can be written simply as

$$\lambda^2 \frac{\partial^2 v(x,t)}{\partial x^2} - \tau \frac{\partial v(x,t)}{\partial t} = v(x,t). \tag{10.3}$$

The parameter $\lambda = (R_m/R_i)^{1/2} = (r\rho_m/2\rho_i)^{1/2}$ where ρ_i and ρ_m have units of resistivity and resistivity times length, respectively. The time constant is given by $\tau = R_m C_m$. The solution of Equation 10.3 can be written as a product of exponential and complementary error functions

$$v(x,t) = \frac{1}{2}\begin{pmatrix} V_0 \\ \lambda R_i I_0 \end{pmatrix}\left\{ e^{-x/\lambda}\mathrm{erfc}\left(\frac{x/\lambda}{2\sqrt{t/\tau}} - \sqrt{t/\tau}\right) \pm e^{x/\lambda}\mathrm{erfc}\left(\frac{x/\lambda}{2\sqrt{t/\tau}} + \sqrt{t/\tau}\right)\right\} \tag{10.4}$$

where the upper and lower prefactors and signs correspond to a unit step voltage V_0 or step current I_0 applied at $(x, t) = (0, 0)$, respectively.

10.4 Active Transport and the Hodgkin–Huxley Equations

Sodium and potassium voltage-gated ion channels in the axonal membrane, depicted in Figure 10.4, facilitate active transport in neurons. A single neuron receives inputs from ~10^4 synapses on average. An action potential is initiated that propagates down

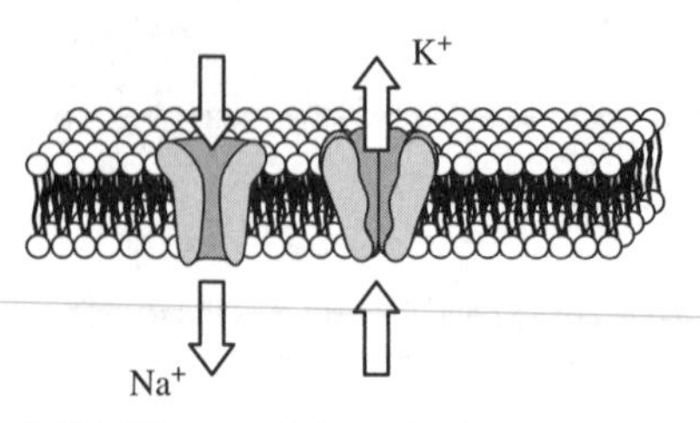

FIGURE 10.4 Schematic of sodium and potassium ion channels in the axonal membrane.

the length of the axon if the total integrated stimulus voltage exceeds a threshold of about 65 mV. This all-or-nothing behavior results in an action potential of constant magnitude for triggering voltages greater than the threshold.

In 1952 Hodgkin and Huxley (H–H) developed an equivalent circuit model shown in Figure 10.5 describing time-dependent action potentials mediated by voltage-gated ion channels.

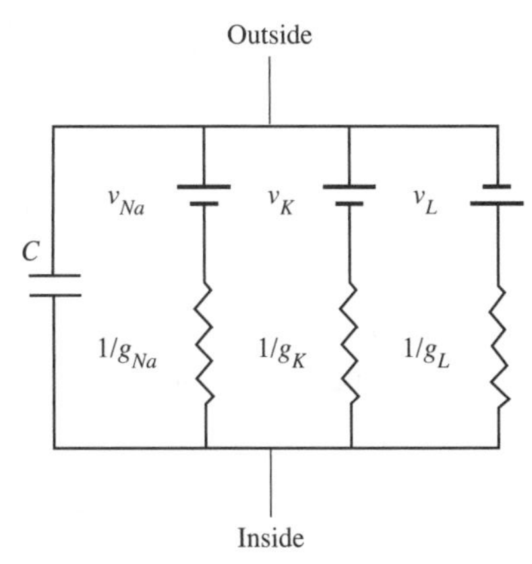

FIGURE 10.5 Hodgkin–Huxley equivalent circuit model.

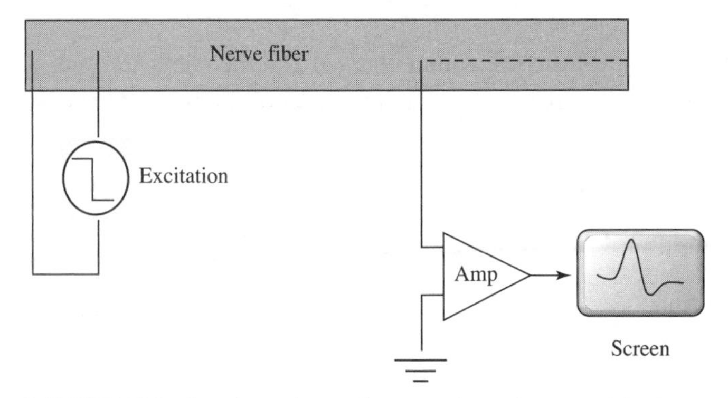

FIGURE 10.6 Experimental setup for measuring action potentials along a nerve fiber.

Figure 10.6 illustrates measurement of the action potential in the giant squid axon. The H–H model, originally developed for the giant squid axon, remains the standard for computational neuroscience today. The model includes potential sources for each of the ion channels along with resistivity and capacitance of the axon.

Channel conductances dependent on membrane potential regulate the flow of sodium and potassium ions through channels in the neuronal membrane. The ionic current and axon membrane potential v obeys the differential equation

$$C\frac{dv}{dt} + I_{\text{ion}} = 0 \tag{10.5}$$

where C is the membrane capacitance. The total ionic current consists of contributions from individual species:

$$I_{\text{ion}} = g_K(v - v_K) + g_{Na}(v - v_{Na}) + g_L(v - v_L) \tag{10.6}$$

where (g_K, g_{Na}, g_L) and (v_K, v_{Na}, v_L) are the potassium, sodium, and leakage conductances, and equilibrium potentials, respectively. The equilibrium potentials are relative to the resting potential of the axon. The equilibrium voltage is obtained with $dv/dt = 0$, giving

$$v_{\text{eq}} = \frac{g_K v_K + g_{Na} v_{Na} + g_L v_L}{g_K + g_{Na} + g_L}. \tag{10.7}$$

The leakage potential v_L is calculated from the membrane equilibrium condition. The sodium and potassium conductances depend on the activities of the sodium and potassium channels:

$$g_{Na} = G_{Na} m^3 h \tag{10.8}$$

$$g_K = G_K n^4 \tag{10.9}$$

while g_L is constant. The activity n of the potassium channel and the respective activity m and inactivity h of the sodium channels obey the first-order differential equations:

$$\frac{dn}{dt} = \alpha_n(1 - n) - \beta_n n \tag{10.10}$$

$$\frac{dm}{dt} = \alpha_m(1 - m) - \beta_m m \tag{10.11}$$

$$\frac{dh}{dt} = \alpha_h(1 - h) - \beta_h h \tag{10.12}$$

where m, h, and n range between zero and one. These derivatives are zero in the resting state so that

$$n_\infty = \frac{\alpha_n}{\alpha_n + \beta_n} \quad m_\infty = \frac{\alpha_m}{\alpha_m + \beta_m} \quad h_\infty = \frac{\alpha_h}{\alpha_h + \beta_h}. \tag{10.13}$$

Equations 10.13 provide equilibrium conditions for numerical simulations. The rate constants α_m, β_m, α_n, β_n, α_h, and β_h are functions of the membrane potential experimentally determined by fitting to the measured giant squid axon potential. These six rate constants are

$$\alpha_m(v) = \frac{0.1(25 - v)}{\exp(2.5 - 0.1v) - 1} \tag{10.14}$$

$$\beta_m(v) = 4\exp\left(\frac{-v}{18}\right) \tag{10.15}$$

$$\alpha_n(v) = \frac{0.01(10 - v)}{\exp(1 - 0.1v) - 1} \tag{10.16}$$

$$\beta_n(v) = 0.125\exp\left(\frac{-v}{80}\right) \tag{10.17}$$

$$\alpha_h(v) = 0.7\exp\left(\frac{-v}{20}\right) \tag{10.18}$$

$$\beta_h(v) = \frac{1}{1 + \exp(3 - 0.1v)}. \tag{10.19}$$

In these equations, the respective units of potential, current, and time are mV, mA, and ms. The sign convention of the membrane potential is used here with respect to the outside of the axon or $v = v_{in} - v_{out}$. Table 10.1 shows parameter values for the giant squid axon.

Table 10.1 Hodgkin–Huxley Parameters for the Giant Squid Axon

Parameter in Hodgkin–Huxley Equation	Value for Giant Squid Axon
Membrane Capacitance	$C = 1\ \mu F/cm^2$
Potassium Channel Conductance	$G_K = 36$ mS/cm^2
Sodium Channel Conductance	$G_{Na} = 120$ mS/cm^2
Leakage Channel Conductance	$G_L = 0.3$ mS/cm^2
Potassium Equilibrium Potential	$v_K = -12$ mV
Sodium Equilibrium Potential	$v_{Na} = 115$ mV
Leakage Equilibrium Potential	$v_L = 10.5988$ mV

10.5 Simulation of Action Potential

10.5.1 Excitation Threshold

The excitation threshold exhibited by action potentials can be demonstrated in an H–H equation by applying stimuli I_{app} near the threshold and solving

$$C\frac{dv}{dt} + I_{ion} = I_{app}. \tag{10.20}$$

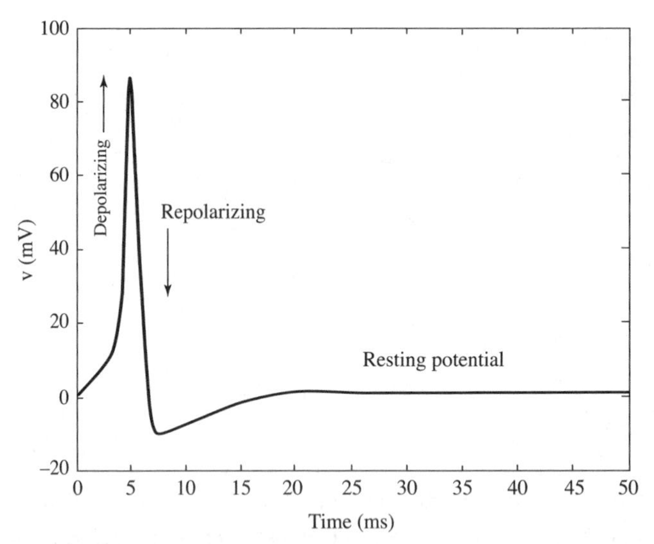

FIGURE 10.7 Time-dependent action potential at one position simulated with the H–H equations for I_{app} = 1.0 mA. The depolarizing rising part of the action potential occurs with the motion of sodium ions into the cell. The action potential drops after about 5 ms in the repolarizing phase with the motion of potassium ions outside of the cell.

Figure 10.7 shows the action potential traces with an excitation of I_{app} = 1.0 mA. The electric field and equipotentials near the leading edge of the action potential are plotted in Figure 10.8.

10.5.2 Neuronal Refractoriness

Once the action potential has been initiated, there is finite time before a second potential can fire called the refractory period. The absolute refractory period is the shortest time interval between two action potentials.

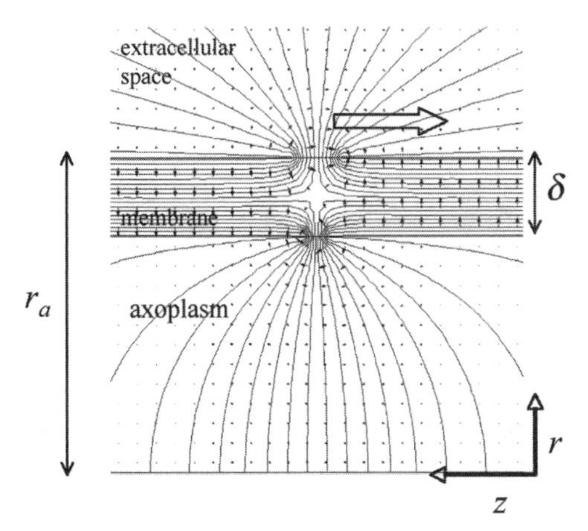

FIGURE 10.8 Finite element calculation showing the equipotentials and electric vectors near the leading edge of an action potential propagating along a nerve cell. The axon radius r_a is 48 μm with a 6-μm membrane thickness δ. The model has rotational symmetry about the z-axis horizontal at the bottom of the figure.

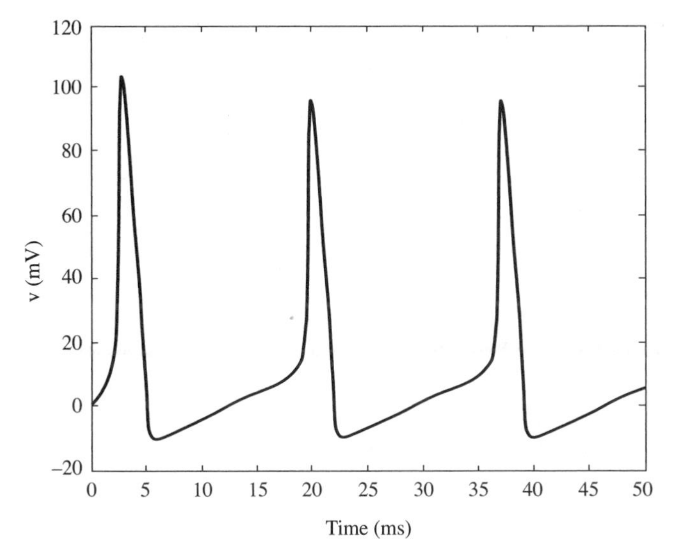

FIGURE 10.9 Time-dependent action potential simulated with the H–H equations for $I_{app} = 7.0$ mA showing repetitive spiking.

10.5.3 Repetitive Spiking

Regular spiking events are observed in neurons for excitations greater than $I_{app} \approx 6$ mA/cm^2. This behavior is simulated in Figure 10.9 with excitation amplitude of $I_{app} = 7.0$ mA.

To solve the H–H equations using the fourth-order Runge–Kutta method, the following MATLAB code is saved as an M-file named HH_RK.m:

```
function yp=HH_RK(t,y)

% Resting potentials
VNa=115; VK=-12.0; VL=10.6;

% Conductances and capacitance
gNa=120; gK=36; gL=0.3; Cm=1.0;

% Applied current
Iapp=7;

v=y(1);
m=y(2);
n=y(3);
h=y(4);

am=(2.5-0.1*v)/(exp(2.5-0.1*v)-1);
bm=4*exp(-v/18);
ah=0.07*exp(-v/20);
bh=1./(exp(3-0.1*v)+1);
an=(.1-0.01*v)/(exp(1-0.1*v)-1);
bn=0.125*exp(-v/80);

INa =(m^3)*h*gNa*(v-VNa);
IK=(n^4)*gK*(v-VK);
IL=gL*(v-VL);

% Derivative vector

yp=[ (Iapp-INa-IK-IL)/Cm;
       (am*(1-m)-bm*m);
       (an*(1-n)-bn*n);
       (ah*(1-h)-bh*h) ];
```

The system is solved by entering at the command prompt

```
>> [t,y]=ode45(@HH_RK,[0,30],[.1, .1, .1, .1]);
```

with trial initial conditions `[.1,.1,.1,.1]` and with `Iapp = 0.0` in the M-file above.

The above simulation will converge to a steady state where the last point is taken as the new initial condition

```
>> x0=y(end,:)
```

The simulation is now executed with the initial conditions x0 and with a nonzero Iapp

```
>> [t,y]=ode45(@HH_RK,[0,30],x0);
```

The action potential is now plotted

```
>> plot(t,y(:,1))
```

Iapp may also be defined as a global variable so that it can be changed from the command prompt.

10.6 FitzHugh–Nagumo Model

FitzHugh developed a model with dynamics similar to the H–H system but with only two variables instead of four. FitzHugh noticed that the time-dependent curves n and $-h$ in the H–H model had a similar shape where $n + h \approx 0.85$. It turns out that FitzHugh's equations are not derivable from the H–H system but are a modification of van der Pol's model of a relaxation oscillator of the form

$$\frac{dv}{dt} = \frac{1}{c}\left(v - \frac{v^3}{3} - w + I_{\text{ext}}\right) = F_v(v, w) \tag{10.21}$$

$$\frac{dw}{dt} = c(v - aw + b) = F_w(v, w) \tag{10.22}$$

where v is attributed to membrane potential and the excitability of the sodium channel and w is a recovery variable. Typical parameter values are $c = 0.2$, $a = 0.8$, and $b = 0.7$. Because the model has only two variables, its solutions can be viewed in the (v, w) plane unlike the four-dimensional H–H equations. FitzHugh originally solved these equations using an analog computer. Later, Nagumo developed the circuit model in Figure 10.10 that employed a tunneling diode to reproduce the cubic behavior in the first equation.

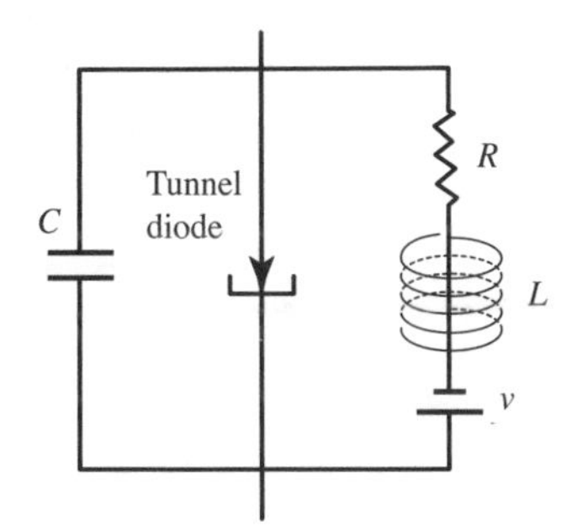

FIGURE 10.10 FitzHugh–Nagumo (F–N) equivalent circuit model of nerve impulses. The tunnel diode has a cubic current-voltage characteristic that appears in the F–N equations.

The (v, w) plane can be partitioned into four regions by two intersecting curves called nullclines where $dv/dt = 0$ (v-nullcline) and $dw/dt = 0$ (w-nullcline). F_v is positive below and negative above the v-nullcline. F_w is positive below and negative above the w-nullcline. The nullclines corresponding to $I_{\text{ext}} = 0$ are given by

$$v - \frac{v^3}{3} - w = 0 \quad \text{and} \quad aw - v - b = 0. \tag{10.23}$$

These curves are plotted in Figure 10.11 along with the vector field.

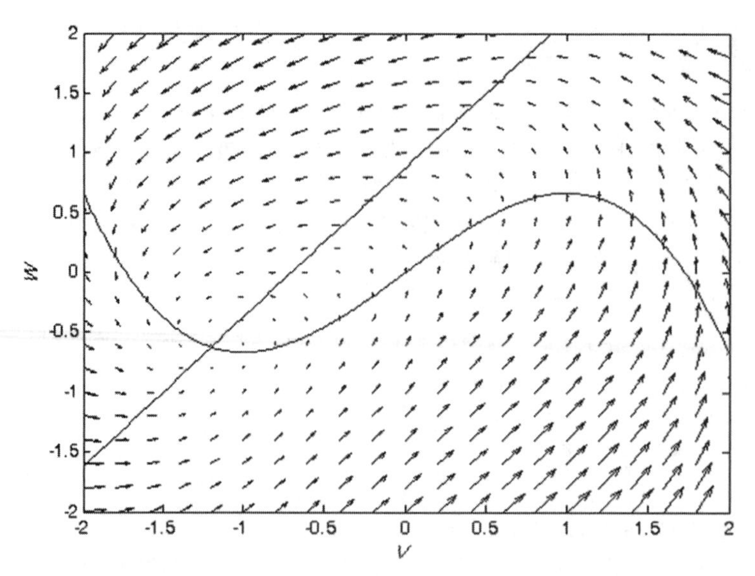

FIGURE 10.11 Vector field plot of the FitzHugh–Nagumo model superimposed on the nullclines with zero applied current.

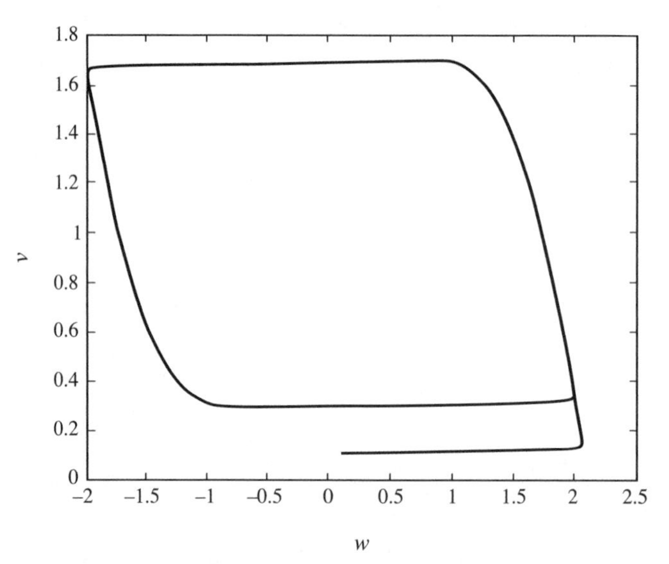

FIGURE 10.12 Phase plot of the v–w variables calculated from FitzHugh–Nagumo equations with $I_{app} = 1.0$ mA. The orbit is a stable limit cycle corresponding to repeated spiking of the action potential.

The w nullcline is a straight line whereas the v nullcline is an N-shaped curve. The nullclines intersect at a singular point that represents the resting state. The v nullcline will shift with increasing values of I_{ext}. The vector field is mapped out by plotting vectors with components (F_v, F_w) over a grid covering the (v, w) plane. Phase space trajectories will be tangent to these vectors with orbits depending on the initial conditions. For values of I_{ext} below ≈ 1, trajectories will terminate on the fixed point. The all-or-none dynamic of the H–H model is contained the F–N model where weak stimuli fail to generate action potentials. The phase space orbits will transverse a stable limit cycle for $1 < I_{ext} < 2$ as shown in Figure 10.12.

The limit cycle trajectory corresponds to current over threshold in the H–H model. Figure 10.13(a) shows a time series for excitation current under the spiking threshold with spiking behavior shown in Figure 10.13(b) for excitation above threshold.

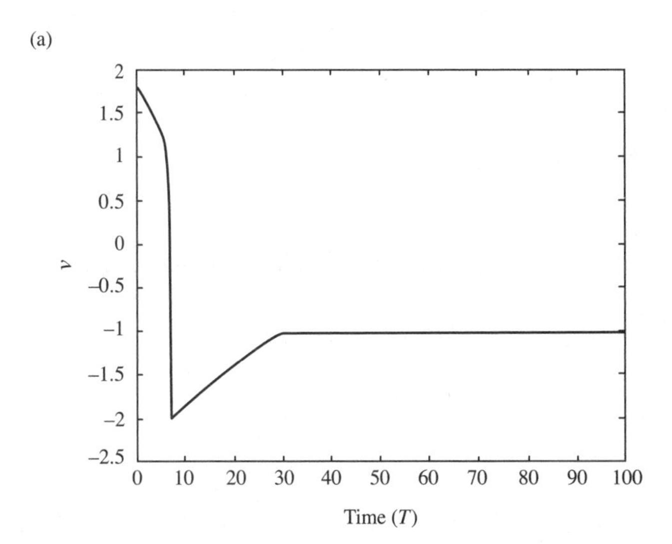

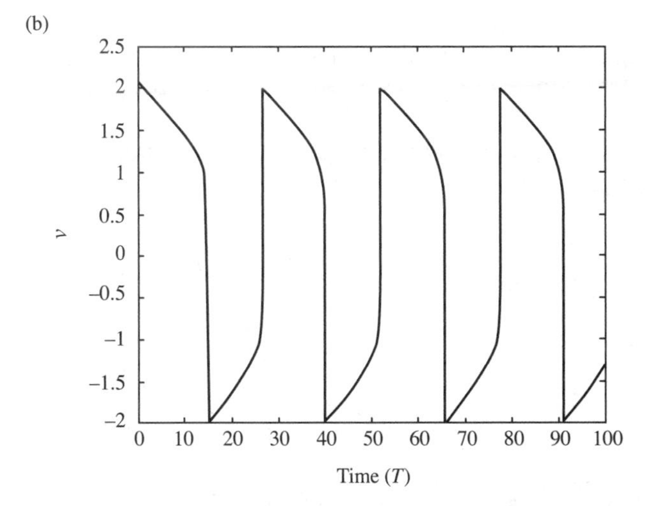

FIGURE 10.13 v–t time series calculated from the FitzHugh–Nagumo equations with (a) $I_{app} = 0.25$ mA and with (b) $I_{app} = 1.0$ mA.

The following MATLAB code plots the vector field and nullclines of the FitzHugh–Nagumo model as shown in Figure 10.11:

```
v=-2:.2:2;
w=-2:.2:2;

 [V,W]=meshgrid(v,w);

x= V-V.^3/3-W;
y= V-0.8*W+0.7;

quiver(V,W,x,y)

hold on

plot(v,-v.^3/3+v,v,(0.7+v)/0.8)

axis([-2, 2, -2, 2])

title('FitzHugh-Nagumo Nullclines & Vector Field')
xlabel('V')
ylabel('W')
```

The following M-file used to integrate the F–N differential equations is saved as nagumo.m:

```
function yp=nagumo(t,y)

c=0.08;
a=0.8;
b=0.7;
Iext=1;

yp= [(1/c)*(y(1)-(y(1)^3)/3-y(2)+Iext);
                c*(y(1)-a*y(2)+b)];
```

The system is solved by entering the command prompt

```
>> [t,y]=ode45(@nagumo,[0,30],[.5,.5]);
```

where different initial conditions will correspond to different phase space trajectories. Time series are plotted with commands:

```
>> plot(t,y(:,1))
```

or

```
>> plot(t,y(:,2))
```

Phase plots are generated accordingly

```
>> plot(y(:,1),y(:,2))
```

10.7 Action Potentials in the Earthworm Nerve Fiber

Measurements of action potential in the giant squid axon require careful setup and dissection procedures. Simple measurement of the action potential can be carried out with various species of earthworm such as *Lumbricus terrestris* by dissecting the worm and making direct contact with the nerve fiber. Action potentials can also be measured by surface-contacting electrodes without dissection. This provides an excellent neurophysiology laboratory exercise with minimal setup. Earthworm samples may be obtained at a local bait store.

Earthworm preparation is a straightforward procedure that can be conducted by physics students with little dissection experience. The worm is first rinsed and anesthetized in a dilute ethanol solution. The worm is then pinned out on a dissection dish and a 3–4 cm incision is made along the dorsal side of the animal exposing the nerve cord. Recording electrodes are placed under the nerve cord and connected to an oscilloscope. Electrical spiking of the nerve cord can be observed even in the absence of external stimulation. Two pin electrodes connected to a signal generator are placed through the head end of the worm a few millimeters apart. As the square pulse stimulus voltage is increased, the medial giant action potential is first observed on the oscilloscope with a peak voltage of roughly 100 mV. Further increase in stimulus voltage activates the lateral giant fibers, and a second peak appears on the oscilloscope trace.

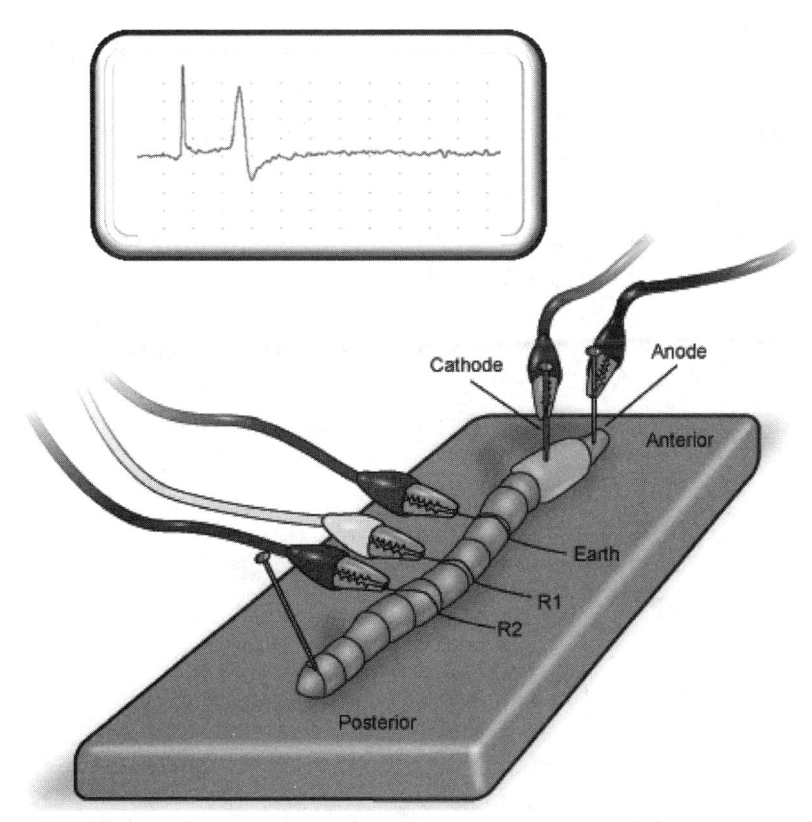

FIGURE 10.14 Experimental setup for measuring the action potential of an earthworm using external electrodes. (Courtesy of ADInstruments Inc.)

The refractory period between action potentials can be observed by decreasing the time interval between stimulus pulses.

Earthworm action potentials are detected at the surface of the worm using the four-electrode configuration shown in Figure 10.14. During the spiking of the action potential, sodium ions enter the nerve fiber resulting in a transient potential drop in the tissue surrounding the nerve fiber. Potential differences at the surface of the earthworm ~0.1 mV are roughly a thousand times smaller than near the nerve fiber requiring signal amplification. The ADInstruments LabTutor® experiments use this setup and include the all-or-nothing response, refractory period, and action potential conduction velocity.

EXERCISES

Exercise 10.1 The relative permeability of the lipid bilayer comprising the axon membrane is $\varepsilon_r = 10$ with a thickness of $d = 6$ nm. Because of the thickness of the membrane, we can approximate its capacitance with that of a parallel plate

$$C = \frac{\varepsilon_0 \varepsilon_r A}{d}.$$

Myelinated axons are nearly 1 micron thick: $d = 1\ \mu$m because of the surrounding myelin sheath. Compare the capacitance per unit area for both myelinated and unmyelinated axons.

Exercise 10.2 Write a program to numerically solve the passive transport using Equation 10.3 with unit step voltage $v(0, t) = V_0$. Make a surface plot of $v(x, t)$.

Exercise 10.3 Write a program to numerically solve the passive transport using Equation 10.3 for a dendrite with a square wave excitation. Plot the potential $v(x, t)$ at several positions x as a function of time.

Exercise 10.4 Plot the spiking rate as a function of I_{app} in the Hodgkin–Huxley model.

Exercise 10.5 Plot the spiking rate as a function of I_{app} in the FitzHugh–Nagumo model.

Exercise 10.6 The ratio of the number of ion-gating molecules in open O and closed C configurations is given by the Boltzmann factor

$$\frac{O}{C} = \exp\left(\frac{zev_m - w}{k_B T}\right)$$

in the presence of a membrane potential v_m with ionic charge $q = ze$. The difference in energy levels between the open and closed configurations is $w = E_O - E_C$. Calculate the probability that the ion channel is open, $P(O)$, using the fact that $P(O) + P(C) = 1$.

11 Mechanical Properties of Biomaterials

11.1 Elastic Moduli

11.1.1 Young's Modulus

Thomas Young developed a theory for characterizing material bodies deformed by external forces. In one dimension, the stress σ is the force applied to a body divided by its area

$$\sigma = \frac{F}{A}. \tag{11.1}$$

The dimensionless strain ε is the change in length ΔL divided by the original length L_0, or

$$\varepsilon = \frac{\Delta L}{L_0}. \tag{11.2}$$

Young's modulus E, often referred to as the elastic modulus, is defined as the ratio of the stress divided by the strain

$$E = \frac{\sigma}{\varepsilon} \tag{11.3}$$

for a body in tension or compression. Young's modulus is an intrinsic property of a given material that is independent of its shape. The units of E are force divided by area

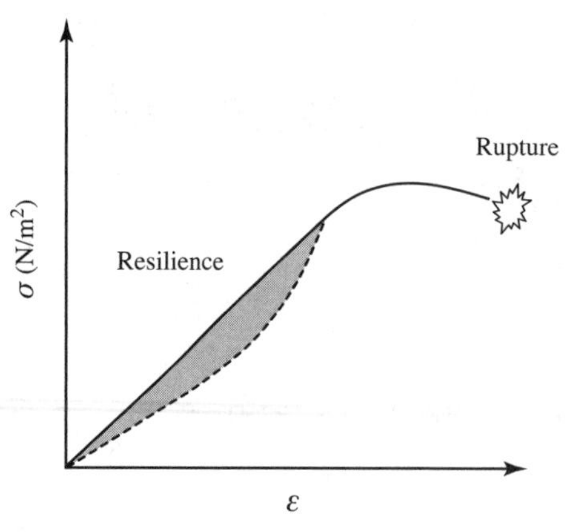

FIGURE 11.1 Schematic of a biomaterial stress–strain curve. The gray area is the resilience, or the maximum recoverable energy per unit volume that can be elastically stored in the biomaterial. (Adapted from Glaser, R. 2001. *Biophysics* Heidelberg, Germany.)

N/m^2 or Pascals (Pa). The relation between Young's modulus and Hooke's law $F = k\Delta L$ relating the linear stretch of a body with force constant k under a given load for sufficiently small ΔL may be obtained considering the ratio

$$E = \frac{F/A}{\Delta L/L_0}. \tag{11.4}$$

Solving for F gives

$$F = \left(\frac{EA}{L_0}\right)\Delta L, \tag{11.5}$$

so that the familiar spring constant is Young's modulus times the cross-sectional area divided by the initial length. Figure 11.1 illustrates a characteristic stress–strain curve for a biomaterial that obeys Hooke's law over some range of elongation.

The Young's moduli of biological materials are usually anisotropic depending on the direction loading is applied. Cell membranes have moduli on the order of magnitude $E \sim 10^7 - 10^8$ Pa with a larger resistance to stretching compared to bending. The viscoelastic properties of cell membranes are very different than rubber membranes that can be stretched in plane without tearing. The plasma membrane will rupture if its area is expanded by more than 1–2% because the phospholipid molecules have little ability to expand in the radial direction.

11.1.2 Shear Modulus

The shear modulus S is defined by the shear stress $\sigma_\perp = F_\perp/A$ divided by the shear strain $\varepsilon_\perp = \gamma$ in radians:

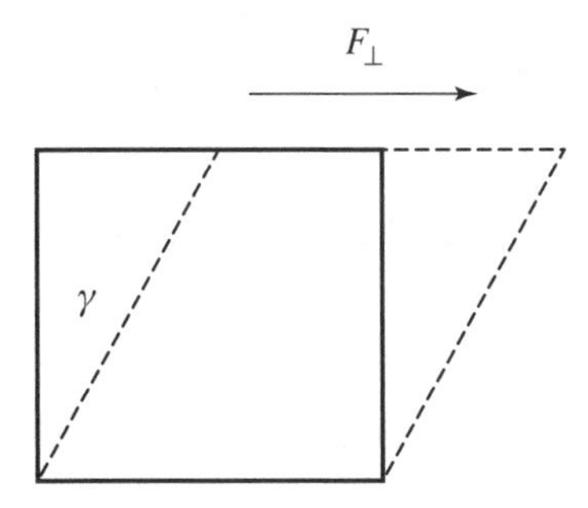

FIGURE 11.2 Illustration of shear stress. The dotted line is the deformed boundary under a shear force.

$$S = \frac{\sigma_\perp}{\varepsilon_\perp} = \frac{F_\perp/A}{\gamma}. \tag{11.6}$$

The shear strain is illustrated in Figure 11.2. The shear modulus is roughly two to three times smaller than Young's modulus in most materials. The shear modulus of human cortical bone is 3.5 GPa compared to its Young's modulus of 9.6 GPa (longitudinal) and 17.4 GPa (transverse). Young's moduli of several biomaterials are tabulated in Appendix 9.

11.1.3 Poisson's Ratio

Material bodies pulled in one direction generally tend to contract in the orthogonal direction as shown in Figure 11.3. The negative transverse contraction strain perpendicular to the load divided by the axial, or extension strain in the direction of the load gives Poisson's ratio:

$$\nu = -\frac{\varepsilon_{\text{trans}}}{\varepsilon_{\text{axial}}}. \tag{11.7}$$

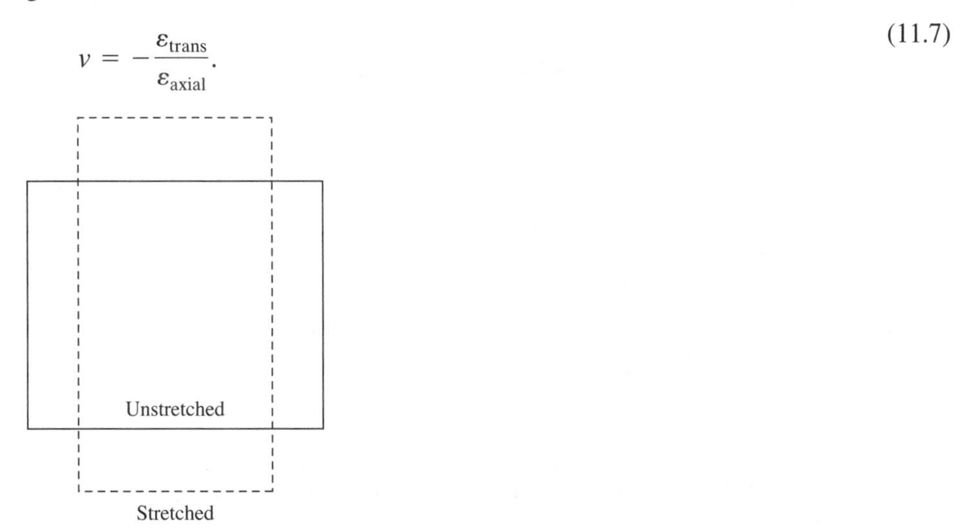

FIGURE 11.3 Transverse and axial strains of a material stretched in the vertical direction illustrating Poisson's ratio.

Most materials have v values between 0 and 0.5, with many metals $v = 0.27–0.35$. The Poisson ratio of bone is $v \approx 0.36$.

11.2 Electric Stresses in Biological Membranes

Electric forces may be included in the stress analysis problems calculated in QuickField. The von Mises stress distribution and the deformed boundary of a liposome in a 20-V/m static electric field is shown in Figure 11.4. The liposome is extended along the direction of the electric field. This simulation is performed in axial symmetry with the z-axis horizontal to the page. The coupled stress and electrostatic problems share the same model file where a link is created in the stress analysis problem properties to the electrostatics problem file.

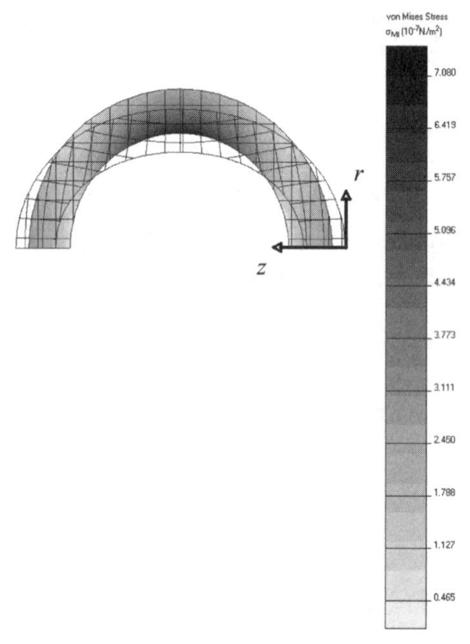

FIGURE 11.4 Finite element calculation of the von Mises stress distribution and the deformed boundary ($\times 10^{15}$) of a liposome in an electric field with magnitude 20 V/m. The simulation is performed using coupled stress and electric field analyses in QuickField.

11.3 Mechanical Effects of Microgravity during Spaceflight

Several studies have been performed to ascertain the effects of microgravity on the human body during space flight. Mechanical reconditioning experienced in

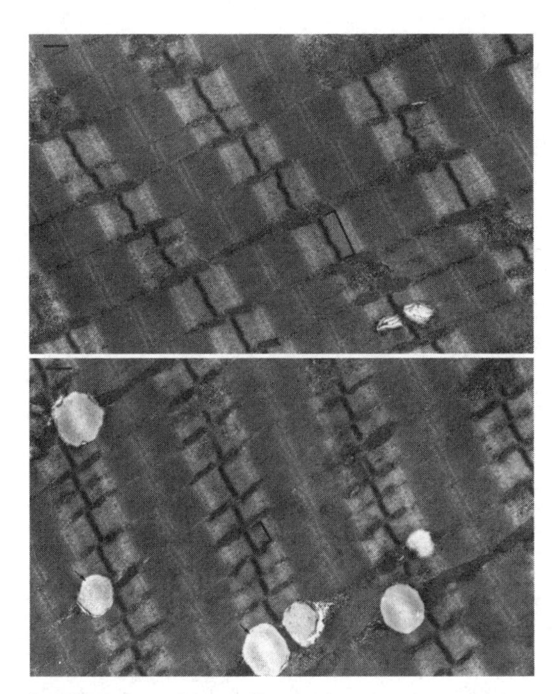

FIGURE 11.5 Muscle fiber section taken from an astronaut before spaceflight (top) and a muscle section taken from the same astronaut after spaceflight (bottom). Lipid droplets are found in the post flight sample. (Courtesy of NASA Marshall Space Flight Center Collection/Courtesy of nasaimages.org.)

low-gravity environments includes bone loss, muscle atrophy, and reduced cardiovascular function. Adipose pockets evidently form in the *bulk* of the muscle during spaceflight as shown in Figure 11.5. Bone loss in space evidently results from the adaptation of the body to the low-gravity environment with a corresponding reduced production of bone. The rate of bone loss can vary between 1% and 2% per month and is most pronounced in lower-body skeletal elements. This presents a challenge for astronauts embarked on long-duration space missions. The reduced bone mass increases blood calcium levels, increasing the risk of developing kidney stones. Solving the problem of bone loss in space may someday advance treatments for osteoporosis on Earth. Future progress may include increasing the resistance of on-flight exercises using Bowflex®–type machines to reduce bone loss and muscle atrophy, and to improve cardiovascular functionality.

EXERCISES

Exercise 11.1 QuickField Coupled Electric and Stress Analyses: Calculate the stress distribution in a cell membrane punctured by a charged microelectrode (r–z symmetry).

Exercise 11.2 QuickField Stress Analysis: Model the stress distribution in a bone segment containing a fracture under transverse loading shown in Figure 11.6. Take Young's modulus of bone ~17 GPa with zero Young's modulus in the fracture. Repeat the calculation with an anisotropic modulus of 9.6 GPa (y-direction) and 17.4 GPa (x-direction). Perform your simulations using x–y symmetry.

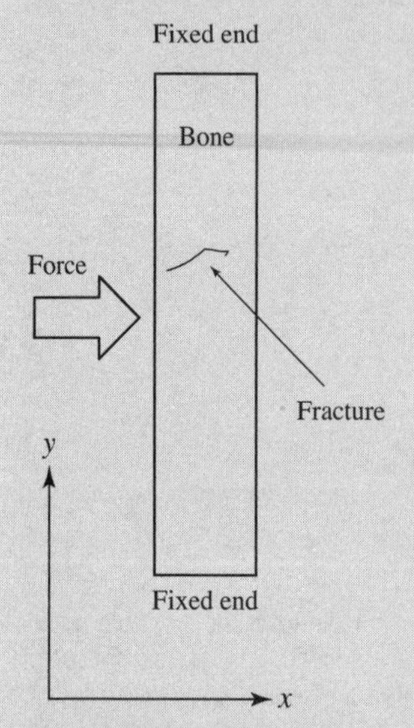

FIGURE 11.6 Bone segment under transverse loading.

Exercise 11.3 QuickField Stress Analysis: Model the stress distribution in a cylindrical bone segment containing a longitudinal fracture with zero Young's modulus under transverse loading shown in Figure 11.7. Take Young's modulus of bone ~17 GPa.

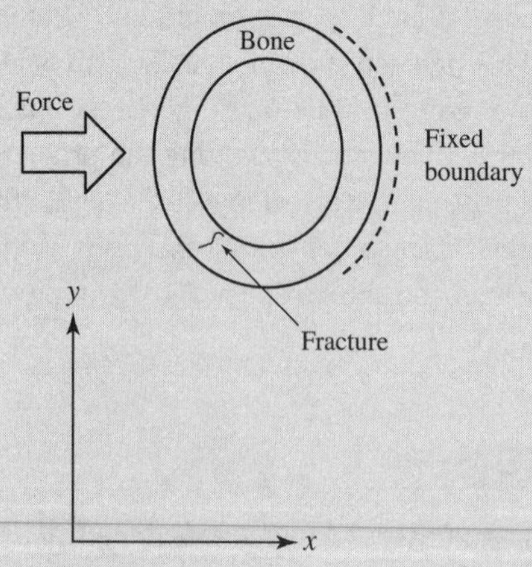

FIGURE 11.7 Bone segment under transverse loading.

Exercise 11.4 QuickField Stress Analysis: Model the stress distribution in a cylindrical bone segment containing a radial fracture with zero Young's modulus under tension and compression as shown in Figure 11.8. Take Young's modulus of bone ~17 GPa. Repeat the calculation with an anisotropic modulus of 9.6 GPa (z-direction) and 17.4 GPa (r-direction). Create the model using r–z symmetry.

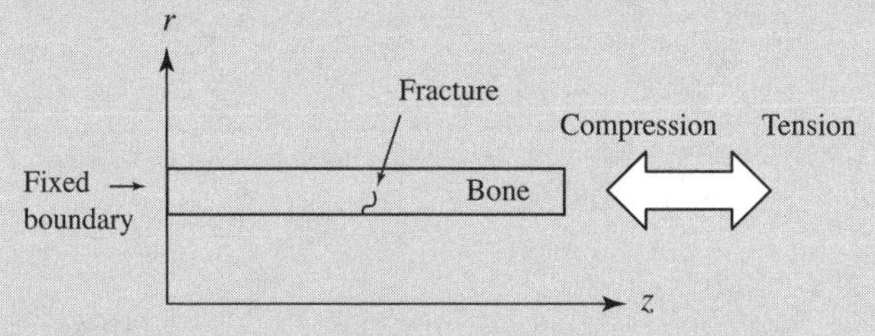

FIGURE 11.8 Bone segment under longitudinal loading.

Exercise 11.5 QuickField Coupled AC Current Flow and Stress Analyses: Calculate the stress distribution in a liposome as a function of frequency. Create the model using r–z symmetry.

Exercise 11.6 QuickField Coupled Electrostatics and Stress Analyses: Model the stress distribution in a plasma membrane surrounding a membrane protein with dipole moment p perpendicular to the plane membrane. How does the stress distribution change if an electric field is applied parallel to the membrane and perpendicular to the dipole moment? Create the model using x–y symmetry.

12 Biomagnetism

12.1 Biomagnetic Field Sources

Bioelectric currents generated by propagating action potentials give rise to electrical potentials that can be detected using contacting electrodes. Electrocardiography (ECG) is a standard clinical technique employing surface electrodes to measure heart signals. Electroencephalography (EEG) can be used to detect potentials due to the propagation of cortical action potentials in the brain.

The same bioelectric currents that produce measurable electrical potentials also generate magnetic fields that can be detected using superconducting quantum interference device (SQUID) magnetometers. The magnetic analogs of ECG and EEG are MCG (magnetocardiography) and MEG (magnetoencephalography), respectively. Relations between bioelectric and biomagnetic field sources and their measurements are illustrated in Figure 12.1.

Because the relative permeability of living tissue is nearly that of free space $\mu_r \approx 1$, the body does not significantly affect magnetic signals generated by bioelectric

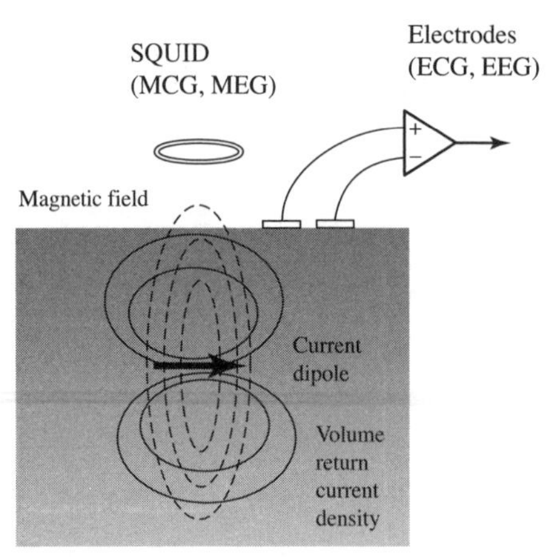

FIGURE 12.1 Relation between bioelectric and biomagnetic field sources and measurement techniques. Bioelectric current sources generate volume return currents (solid lines) that induce surface potentials that can be measured using contacting electrodes (ECG and EEG). Biomagnetic fields (dashed lines) are generated both by source and return currents and are detected outside the body without contacting electrodes using SQUID magnetometers (MCG and MEG).

current sources. Electrical signals, on the other hand, are attenuated by the inhomogeneous conductivity distribution of the body. Insulating components such as bone and air spaces can reduce potentials at the surface of the skin. Another example discussed below is the insulating vernix caseosa surrounding the fetus beginning in about the third trimester of pregnancy that precludes fetal electrocardiograms.

Magnetic vector fields enable greater source localization compared to scalar electrical potentials. Biomagnetic measurements of action potentials, bioelectric currents, and intrinsic magnetization are noninvasive, requiring no electrical contact with the subject, and can provide both temporal (real-time) and spatial information. Magnetic measurements can be used to detect magnetization from hepatic iron deposits in the liver or intestinal ischemia, where lack of oxygen due to poor blood supply to an organ can result in ischemic tissue damage. These biomagnetic sources could not be detected using electrical techniques.

Although biomagnetic signals are unattenuated by the body and offer useful physiological information, they are extraordinarily weak. A given biomagnetic source can be orders of magnitude lower than typical background fields encountered in the laboratory or clinical environment. Noise factors and difficulties associated with handling liquid cryogens required to cool SQUIDs to their operating temperatures have hampered the use of SQUIDs in medicine. Figure 12.2 plots several biomagnetic signals and environmental noise sources with their corresponding frequency ranges.

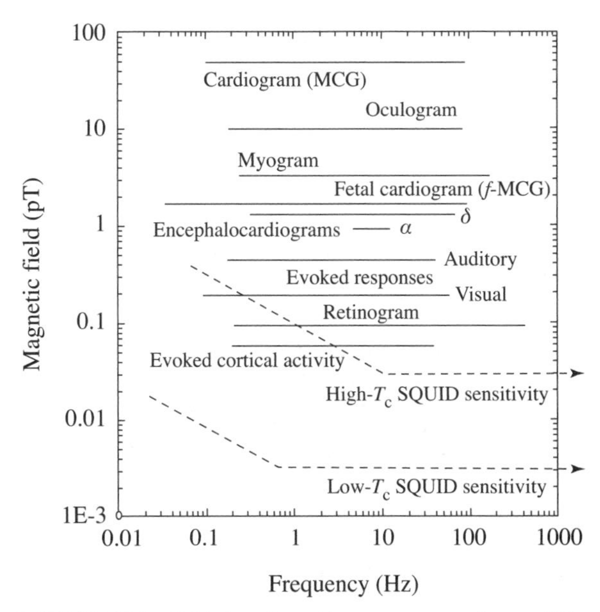

FIGURE 12.2 Field strength and frequency range of several biomagnetic sources ranging between roughly 0.1 pT–100 pT and 0.1 Hz–1 kHz. The highest sensitivities of high- and low-T_c SQUIDs are shown by comparison. (Adapted from Wikswo, J. P. 2000. Applications of SQUID magnetometers to biomagnetism and nondestructive evaluation. In *Applications of superconductivity*, ed. H. Weinstock, 139–228.)

12.1.1 Current Dipole Model

Most biomagnetic sources result from charge transport during physiological processes. The Biot–Savart law gives the magnetic field at the location **r** due to bioelectric current density **J** at **r**′:

$$\mathbf{B}(\mathbf{r}) = \frac{\mu_0}{4\pi} \iiint\limits_{\text{vol}} \frac{\mathbf{J}(\mathbf{r}') \times (\mathbf{r} - \mathbf{r}')}{|\mathbf{r} - \mathbf{r}'|^3} \, dv' \tag{12.1}$$

where the integral is evaluated over the volume containing the source current. Current dipoles are often used to model biomagnetic field distributions. A current dipole can be thought of as a current source and sink separated by a displacement **d** pointing from source to sink. For current sources localized in small regions, the volume integral can be approximated as

$$\mathbf{B}(\mathbf{r}) = \frac{\mu_0}{4\pi} \mathbf{Q} \times \frac{(\mathbf{r} - \mathbf{r}')}{|\mathbf{r} - \mathbf{r}'|^3} \tag{12.2}$$

where $\mathbf{Q} = I\mathbf{d}$ is the current dipole moment with the current I propagating in the direction of **d**. The magnetic field of a current dipole is qualitatively similar to the field of a

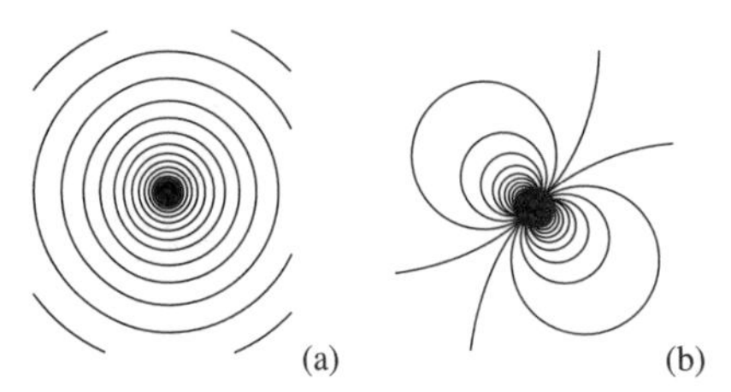

FIGURE 12.3 Magnetic flux lines due to (a) current dipole source modeled as a line current normal to the page and (b) magnetic dipole source with a 45° orientation with respect to the horizontal.

line current plotted in Figure 12.3(a). As a comparison, a magnetic dipole field is plotted in Figure 12.3(b). Magnetic dipole fields are generated by intrinsic and induced magnetization distributions and current loops. In general, biomagnetic sources will have both current and magnetic dipole moments, as well as higher multiple moments.

12.1.2 Vector Potential Formulation

The magnetic field is derivable from the curl of the vector potential

$$\mathbf{B} = \nabla \times \mathrm{A}. \tag{12.3}$$

Since $\nabla \times \mathbf{B} = \mu_0 \mathbf{J}$, the vector potential satisfies Poisson's equation

$$\nabla^2 \mathbf{A} = -\mu_0 \mathbf{J}. \tag{12.4}$$

The vector potential points in the same direction as the current density and is given by the volume integral of the source current

$$\mathbf{A}(\mathbf{r}) = \frac{\mu_0}{4\pi} \iiint_{\text{vol}} \frac{\mathbf{J}(\mathbf{r}')}{|\mathbf{r} - \mathbf{r}'|} dv' \tag{12.5}$$

where the source current is specified at $\mathbf{r}'$. The vector potential due to a current dipole source located at $\mathbf{r}'$ is approximated as

$$\mathbf{A}(\mathbf{r}) = \frac{\mu_0}{4\pi} \frac{\mathbf{Q}}{|\mathbf{r} - \mathbf{r}'|}. \tag{12.6}$$

12.2 Nerve Impulses

The first measurement of the magnetic field produced by a single axon was made by Wikswo, Barach, and Freeman (1980). A peak field of 100 pT was recorded at a distance of 1.3 mm from an isolated frog sciatic nerve using a toroidal pickup coil inductively coupled to a SQUID magnetometer. Figure 12.4 depicts the field distribution resulting from propagating action potentials along the axon. The magnetic field has both

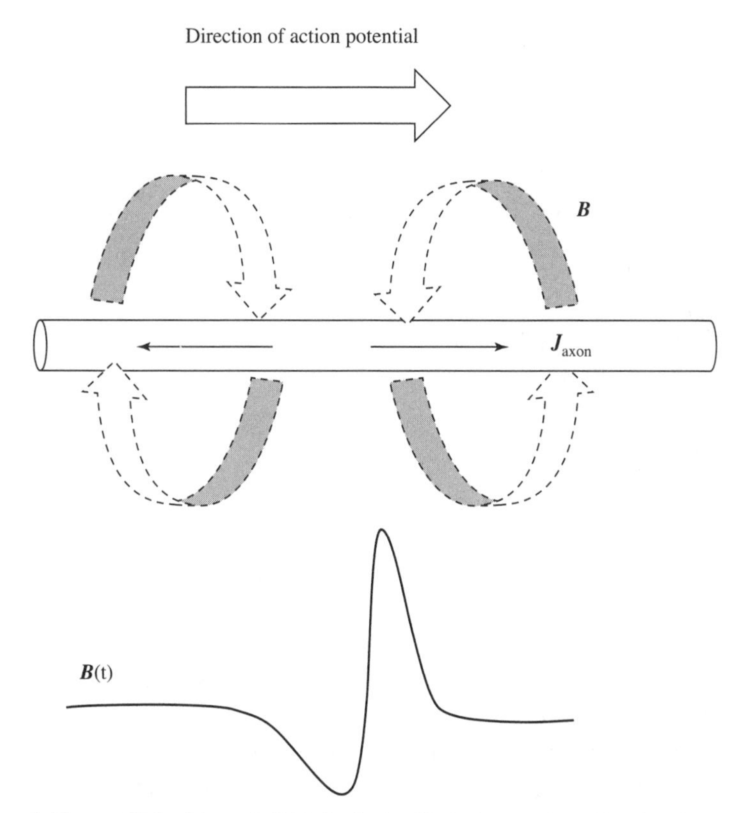

Figure 12.4 Magnetic field distribution (dashed arrows) resulting from intercellular axon currents (solid arrows). The time-dependent magnetic field at a given location is sketched in the bottom trace.

counterclockwise and clockwise orientations around the axon because of the changing polarity of the action potential charge density. The magnetic field measured at a single location thus will change signs as the action potential passes by the magnetometer.

12.3 Magnetotactic Bacteria

Magnetotactic bacteria form chains of magnetic crystals, enabling them to align themselves with the Earth's magnetic field. Two types of crystals Fe_3O_4 (magnetite) and Fe_3S_4 (greigite) are formed by these bacteria. Originally discovered in bog sediments, the ability to magnetically align with the Earth's field is believed to assist these organisms in seeking out regions of lower oxygen concentration.

Magnetite crystalline remains of these bacteria preserved as magnetofossils have been identified dating back to the Cretaceous Era. Similar magnetic crystals have been found in the Martian meteorite ALH84001, although it has not been conclusively proven that these crystals are biological in origin.

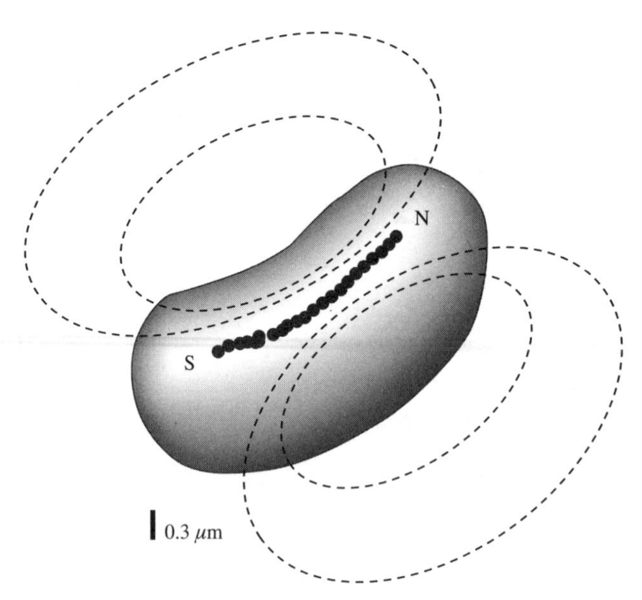

FIGURE 12.5 Magnetotactic bacteria with a chain of magnetite crystals. (Drawing based on structural data.)

Figure 12.5 illustrates the alignment of magnetite crystals in a magnetotactic bacterium and the resulting field distribution.

The total dipole moment $\boldsymbol{\mu}$ is the sum of dipole moments of each crystal. The bacterium will experience a net torque $\boldsymbol{\tau}$ tending to align its magnetic moment with an external magnetic field **B** according to the cross product

$$\boldsymbol{\tau} = \boldsymbol{\mu} \times \mathbf{B}. \tag{12.7}$$

The magnetic interaction energy W_B is obtained from the dot product

$$W_B = -\boldsymbol{\mu} \cdot \mathbf{B} = -\mu B \cos\theta \tag{12.8}$$

so that W_B is minimal (maximal) when $\boldsymbol{\mu}$ is parallel (antiparallel) to **B**. The field distribution of a linear string of magnetite crystals is calculated in Figure 12.6. SQUID measurements of magnetotactic bacteria samples give an estimated magnetic moment $\mu = 1.8 \pm 0.4 \cdot 10^{-12}$ emu with an average magnetic-to-thermal energy ratio $\mu B_{\oplus} / k_B T \approx 9$, corresponding to a local value of Earth's magnetic field $B_{\oplus} \approx 0.2$ G.

12.4 SQUID Magnetometry

Superconductivity is characterized by a resistanceless flow of electrical current below a critical transition temperature T_c that depends on the superconducting material. The

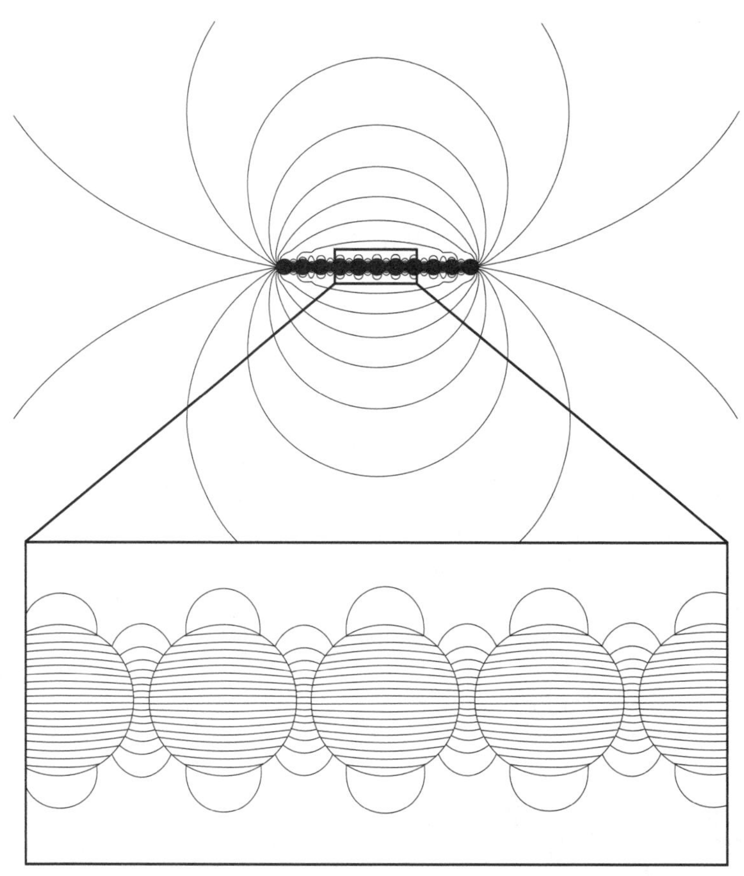

FIGURE 12.6 Finite element calculation of the field distribution of a linear array of 11 magnetite crystals. Each crystal has a diameter of 0.1 μm with a separation of 0.01 μm between crystals. (Calculated using QuickField Magnetostatic Analysis.)

absence of electrical resistivity offers many technological advantages. One of the most promising applications is the development of magnetic sensors. SQUIDs are currently the world's most sensitive magnetic flux to voltage transducers.

SQUID magnetometers are based on three principle phenomena: the physics of the superconducting state, quantization of magnetic flux in a superconducting loop, and quantum tunneling of superconducting Cooper pair electrons. The quantum mechanics of superconductivity is remarkable in that a single wave function ψ describes the superconducting electrons in a bulk or thin-film superconductor. Superconductivity is a *macroscopic* quantum phenomenon. The superconducting wave function, also known as the order parameter ψ, can be written as a product of a spatially dependent magnitude and phase ϕ at a location $\mathbf{r}$ inside the superconductor:

$$\psi(\mathbf{r}) = |\psi(\mathbf{r})| \exp[i\phi(\mathbf{r})]. \quad (12.9)$$

The current density $\mathbf{J}_s$ inside a superconductor is proportional to the square of the wave function magnitude times the vector potential:

$$\mathbf{J}_s(\mathbf{r}) = -\frac{e^2}{m}|\psi(\mathbf{r})|^2\mathbf{A}(\mathbf{r}). \tag{12.10}$$

A SQUID is constructed of a superconducting loop interrupted by one or two insulating regions through which this supercurrent can flow by quantum tunneling. DC SQUIDs have two junctions, whereas RF SQUIDs have only one tunnel junction. Superconducting tunnel junctions were theoretically predicted by Brian Josephson soon before their development in the mid-1960s.

12.4.1 Josephson Effect

Cooper pair electrons pass without electrical resistance through the bulk of the superconductor and are quantum mechanically coupled on either side of a Josephson tunnel junction shown in Figure 12.7(a). This Josephson junction (JJ), or weak link, can be an insulating oxide layer forming a normal gap. Using photolithography techniques, a superconducting thin film can be deposited on a bicrystal substrate consisting of two

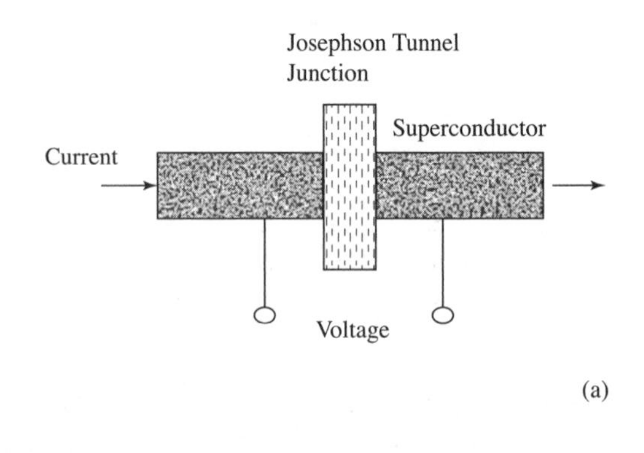

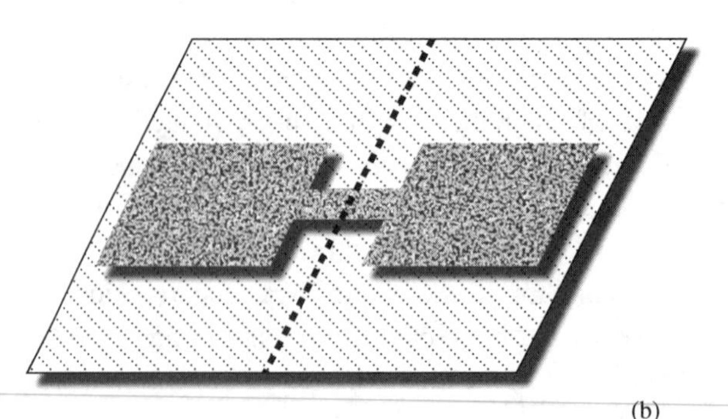

FIGURE 12.7 (a) Schematic of a Josephson junction. (b) Josephson junction formed by depositing a high-T_c superconducting thin film onto a bicrystal substrate.

pieces of $SrTiO_3$ fused with a misorientation in crystal structure at the grain boundary as shown in Figure 12.7(b). A superconducting $YBa_2Cu_3O_7$ thin film deposited on the bicrystal mimics the crystal structure of the underlying substrate, thereby forming a weak link at the grain boundary.

The supercurrent through the barrier is related to the phase difference between the superconducting order parameters on the left and right sides of the junction, with

$$\psi_L = \sqrt{\rho_L}\exp(i\phi_L) \quad \text{and} \quad \psi_R = \sqrt{\rho_R}\exp(i\phi_R) \tag{12.11}$$

where $\rho = |\psi|^2$. The tunneling current is given by

$$I = I_c \sin\delta \tag{12.12}$$

where I_c is the critical current of the junction and $\delta = \phi_L - \phi_R$ is the phase difference in the order parameter across the junction. If a finite voltage V exists across the junction, the phase difference evolves according to the AC Josephson relation

$$\frac{d}{dt}\delta = \frac{2e}{\hbar}V. \tag{12.13}$$

For two JJ's interrupting a closed loop, the resulting field-modulated critical current is represented by

$$I_c \approx 2I_0\left|\cos\frac{\pi\Phi_L}{\Phi_0}\right| \tag{12.14}$$

where $\boldsymbol{\phi}_L = BA$ is the magnetic flux passing through the loop of area A. This equation can be viewed as the superconducting magnetic analog of a double-slit diffraction pattern. The DC SQUID is thus a two-junction interferometer. Single-junction RF SQUIDs are coupled to a tank circuit while DC SQUIDs, which consist of two Josephson junctions in parallel, are supplied with a DC bias current as shown in Figure 12.8. SQUIDs are usually operated in a null feedback mode where a voltage is measured that is proportional to the externally applied flux.

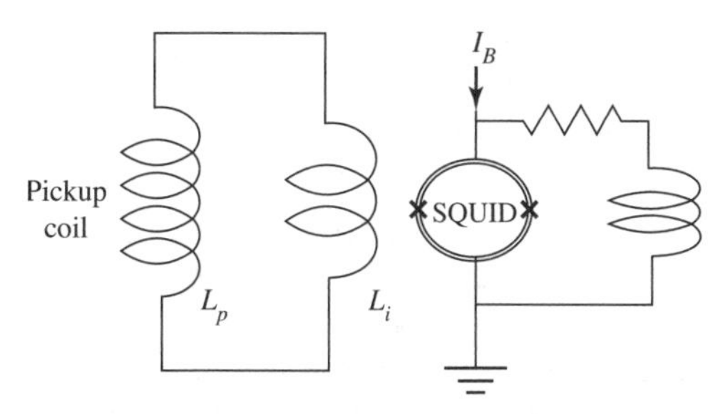

FIGURE 12.8 Schematic of a DC SQUID with pickup and input coil.

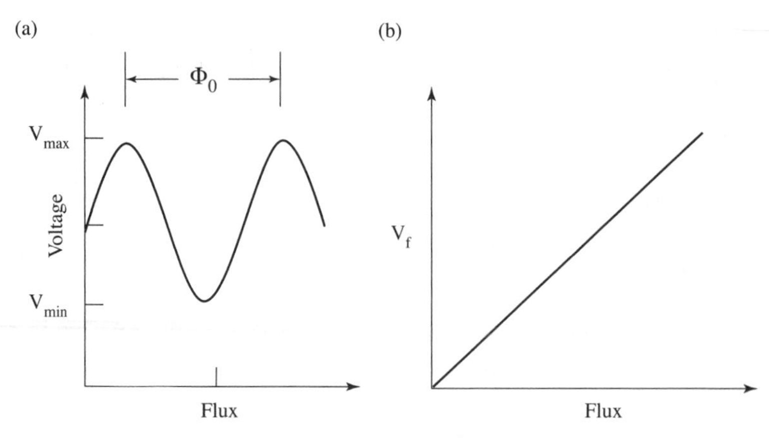

FIGURE 12.9 (a) V-Φ curve of a DC SQUID for constant bias current I_b. (b) Linear output voltage of the flux-locked loop vs. applied magnetic flux.

In general, the SQUID's electrical properties depend on the externally applied magnetic flux. These electrical properties are exploited in constructing highly sensitive magnetometers. The V–Φ curves of the DC SQUID are plotted in Figure 12.9(a), illustrating that a bare SQUID behaves as a nonlinear flux to voltage transducer.

12.4.2 Flux-Locked Loop

To linearize the SQUID's output to the externally applied flux, the SQUID is operated in a flux-locked loop (FLL) as shown in Figure 12.10.

Here, feedback electronics keep flux in the SQUID at a maximum or a minimum on the V–Φ curve. First, an AC-modulation flux is applied to the SQUID through a modulation coil. The SQUID output is then fed to a lock-in amplifier referenced to the modulation frequency. In the absence of an external applied field, the modulation flux traces over an extrema of the V–Φ curve. This results in a rectified voltage output at twice the modulation frequency that is not detected by the lock-in amplifier.

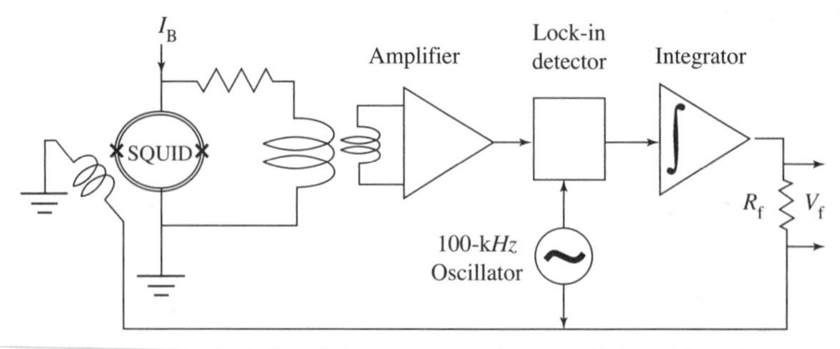

FIGURE 12.10 DC SQUID operated in a flux-locked loop. (Adapted from Jenks W. G. et al. 1997. SQUIDs for nondestructive evaluation. *Journal of Physics D: Applied Physics* 30: 293–323.)

An external magnetic field skews the modulation flux resulting in a voltage output at the modulation frequency that is detected by the lock-in amplifier. The nonzero output voltage from the lock-in amplifier is then fed through a feedback resistance that supplies a current to the modulation coil that cancels the external applied flux. A voltage is then measured across the feedback resistor R_f that is a linear function of applied magnetic field as shown in Figure 12.9(b). Hence the SQUID is operated as a linear flux to voltage transducer.

$YBa_2Cu_3O_7$ (YBCO) and niobium are the most common materials with which SQUIDs are fabricated. Usually, a superconducting flux transformer is used to couple flux into the SQUID washer. This flux transformer consists of a pickup coil that senses the external magnetic field and an input coil that couples flux into the SQUID. Low-T_c niobium SQUIDs often utilize wire-wound superconducting input and pickup coils or axial gradiometers, although some now employ integrated, thin-film, planar gradiometers. High-T_c YBCO SQUIDs utilize a thin-film pickup coil on the same chip as, and directly coupled to the SQUID or, alternatively, on a separate chip in a "flip-chip" configuration.

12.4.3 Intrinsic Noise Factors

Intrinsic noise factors in the SQUID arise from fluctuations in the JJ's critical current density as well as the thermally activated hopping of flux vortices in the SQUID washer. Fluctuations in critical current are due to trapping and release of tunneling electrons at the location of defects or impurities in the junction. This process results in $1/f$ noise at low frequencies. Noise produced by the critical current fluctuations is substantially reduced with a bias-reversal scheme whereby the bias current across the SQUID is periodically reversed. Another source of $1/f$ noise in the power spectrum arises from the motion of flux vortices in the superconducting transformer circuit. Cooling the SQUIDs in a magnetically shielded enclosure is necessary to reduce trapped flux in the SQUID.

12.4.4 Extrinsic Noise Factors

Extrinsic noise factors encompass all forms of environmental electromagnetic interference (EMI). Traditionally, external EMI is reduced by enclosing the entire instrument in an expensive magnetically shielded room. Further noise reduction is obtained by using low-T_c wire-wound first- or second-order gradiometers that couple flux into a SQUID surrounded by a superconducting shield. Gradiometers measure only magnetic field gradients such as those produced by local dipole sources while rejecting uniform noise fields. Low-T_c gradiometers are configured to measure field gradients such as dB_z/dz as shown in Figure 12.11(a). Other gradiometer configurations measure d^2B_z/dz^2, $dB_\rho/d\rho$, and so on. Currently, no high-T_c wire-wound gradiometers are

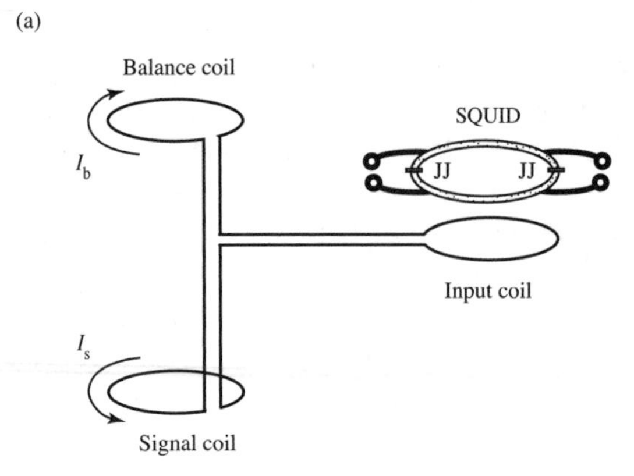

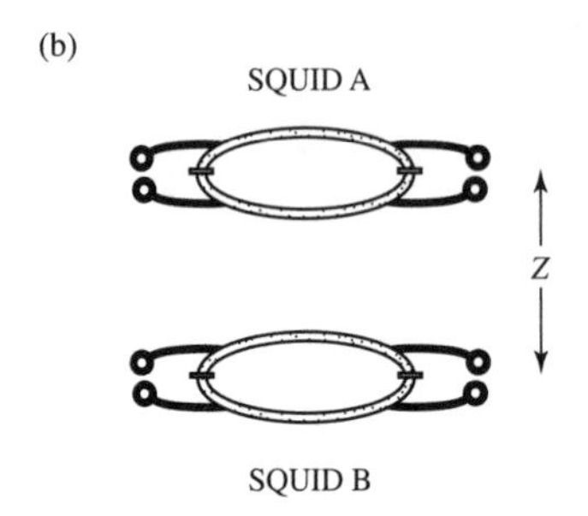

FIGURE 12.11 (a) First-order wire wound and (b) electronic gradiometers that measure dB_z/dz.

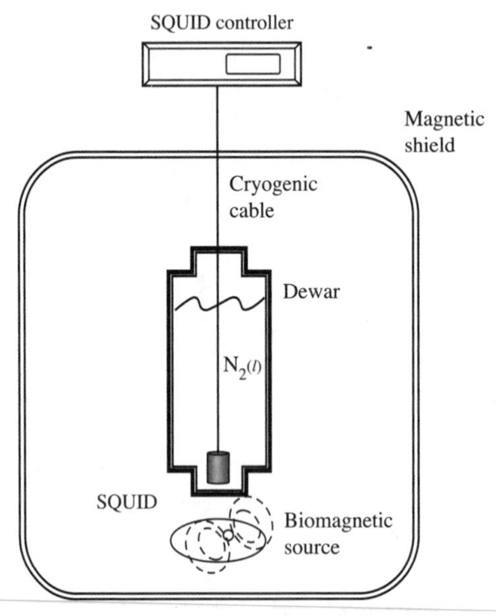

FIGURE 12.12 High-T_c SQUID measurement system consisting of a SQUID in a liquid nitrogen Dewar suspended over a biomagnetic source. A magnetic shield surrounds the sensor and the source while the system electronics are located outside of the magnetic shield.

manufactured, although recent developments in thin-film planar gradiometers appear promising. In Figure 12.11(b), a first-order electronic gradiometer is made by subtracting the output of two SQUIDs, thereby measuring dB_z/d_z. This procedure requires careful mechanical balancing of the SQUIDs that are still exposed to background noise fields.

Phase-sensitive detection serves to eliminate EMI from AC SQUID measurements. This is a useful technique for AC magnetic susceptibility measurements and impedance magnetocardiography where a lock-in amplifier is used to measure conductivity and permeability variations at a given frequency and phase angle. Other noise-reduction schemes include signal processing and localized magnetic shielding simulated in Figure 12.12 that can be highly permeable or superconducting.

Highly permeable shielding channels noise fields through the bulk of the shield thereby creating a magnetically silent region in Figure 12.13(a), whereas a superconductor shields by expelling flux in Figure 12.13(b).

12.4.5 Digital Filters

Digital signal processing can be used to remove unwanted electromagnetic noise sources such as 60-cycle power line noise (50 Hz in Europe), fluctuations in the Earth's magnetic field, and electromagnetic interference generated by electronic equipment. One technique employs a fast Fourier transform (FFT) algorithm where specific unwanted frequency components are deleted in the frequency domain. The filtered signal is then obtained by performing an inverse Fourier transform of the modified spectrum.

12.5 Magnetocardiography

Human MCG signals were first recorded by Baule and McFee in 1963. The first magnetocardiogram was recorded by Zimmerman in 1970. The peak magnetic field at the surface of the human torso is approximately 50 pT. By comparison, the Earth's magnetic field is a million times greater $B_{\oplus} \approx 50\ \mu$T, or 0.5G. It is therefore necessary to perform MCG measurements in magnetically shielded enclosures, or to use gradiometry and signal-processing techniques.

Spatial maps of various magnetic field components B_x, B_y, and B_z are obtained from multichannel SQUID systems. These maps can be used to reconstruct the underlying current distribution by solving the magnetic inverse problem. This is an advantageous technique for locating sources such as those resulting from Wolff–Parkinson–White (WPW) syndrome. Cardiomagnetic field maps can also be measured in posterior regions where ECG signals are attenuated because of the high electrical resistivity of the lungs. Because MCG is noncontacting, the technique holds

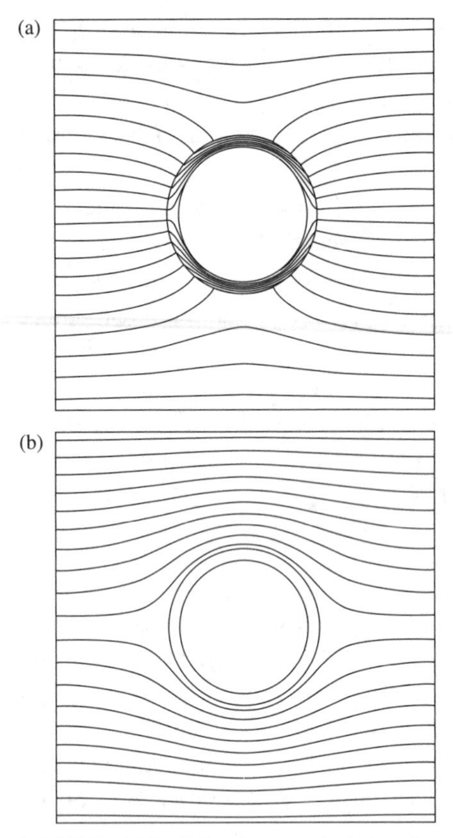

FIGURE 12.13 Finite element calculation of magnetic shielding by (a) highly permeable magnetic shield and (b) superconducting shield. Biomagnetic measurements are conducted in the shielded regions in (a) and (b) for helmet-type MEG systems and for shielding SQUIDs coupled to gradiometers outside the magnetic shield. (Calculated using QuickField Magnetostatic Analysis.)

the promise of measuring heart signals outside of a space suit where it is not possible to make direct electrical contact with the body.

12.6 Fetal Magnetocardiography

Fetal magnetocardiography (FMCG) measures the magnetic fields produced by the fetal heartbeat. The peak magnetic field resulting from fetal heart activity is about one-tenth that of the human heart ~5 pT at the surface of the maternal abdomen. Electrical signals from the fetal heart can be recorded providing useful physiological information such as the presence of arrhythmias and other pathological heart rhythms. However, electrical signals are attenuated by the maternal abdomen, amniotic fluid, and electrically insulating waxy substance (vernix caseosa) that coats on the fetal skin

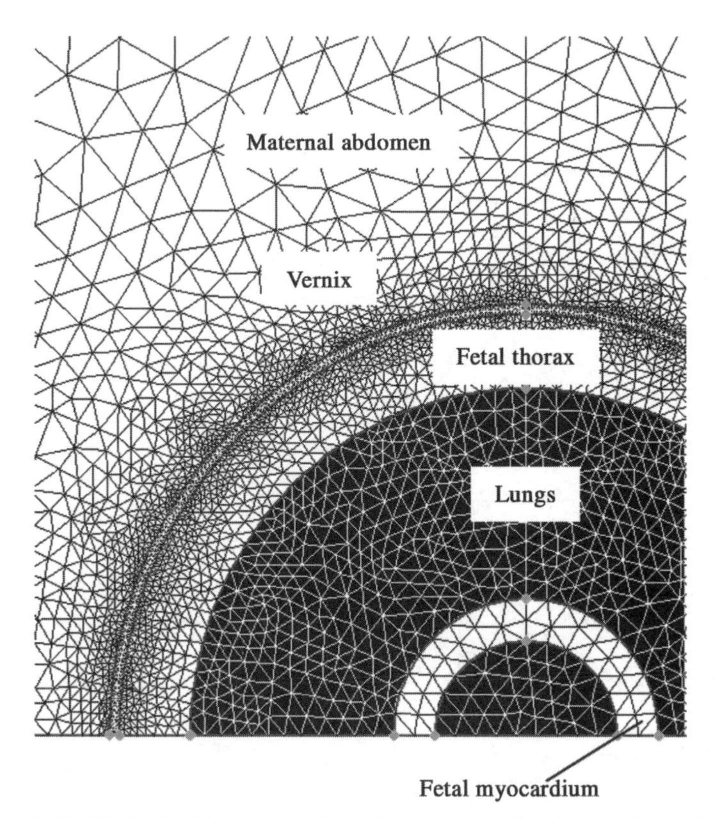

FIGURE 12.14 Nested sphere finite element model of fetal conducting regions including the myocardium, lungs, thorax, vernix caseosa, and maternal abdomen. Model created in QuickField.

during the third trimester. The insulating vernix severely attenuates the fetal-ECG amplitude, making temporal and morphological changes difficult to identify, whereas magnetic fields are virtually unattenuated by this layer. A finite element nested sphere model is shown in Figure 12.14, illustrating various conducting regions in and surrounding the fetus, including the vernix.

A FMCG is recorded in Figure 12.15, showing (a) the raw SQUID data taken over 14 cycles and (b) the averaged PQRST wave at 33 weeks gestation. Peak magnetic fields are between 3 pT and 5 pT in this figure. Note that it would not be possible to record the fetal ECG at this stage of pregnancy because of the vernix.

12.7 Magnetoencephalography

The first MEG was recorded by David Cohen in 1968 using conventional pickup coils in a magnetically shielded room. Magnetic fields produced by bioelectric currents in the brain are considerably weaker than cardiomagnetic fields. Typical MEG field

(a)

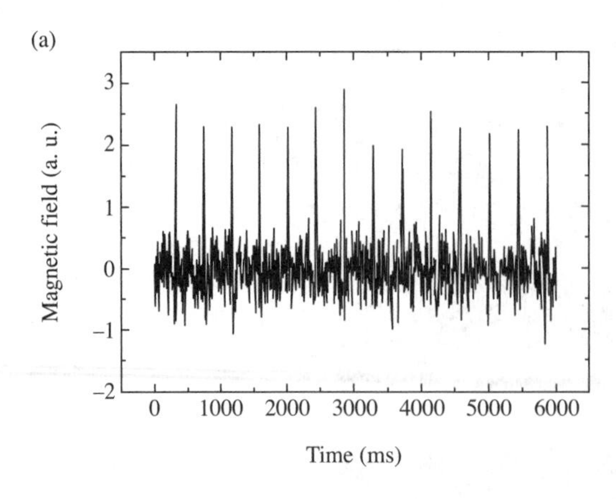

(b)

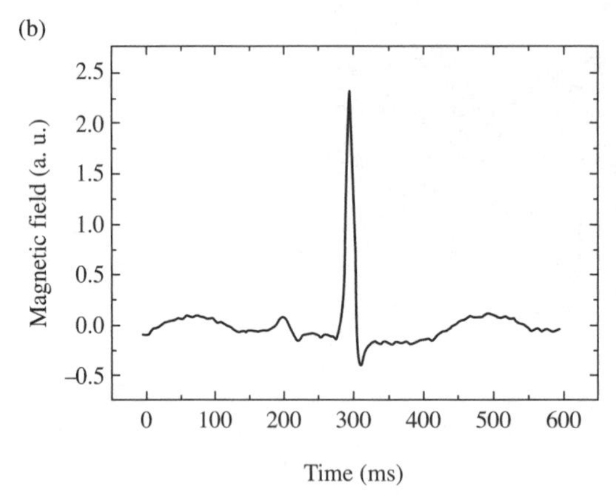

FIGURE 12.15 Fetal magnetocardiogram showing (a) the raw SQUID data taken over 14 cycles and (b) the averaged PQRST wave. (Courtesy of Audrius Brazdeikis at the University of Houston, Texas Center for Superconductivity.)

strengths are less than 10 fT, corresponding to the collective activation of ~10^4 neurons. For source localization, multiple sensors are necessary to construct a field map. This can involve SQUID arrays with as many as 300 sensors in a helmet-type configuration. Once a field map is obtained, various techniques can be implemented in solving the magnetic inverse problem to determine the underlying current distribution. The current dipole model is useful for modeling biomagnetic sources such as epileptic foci in the cortex. Volume return currents illustrated in Figure 12.16 forming closed loops will accompany dipole sources in a conducting tissue.

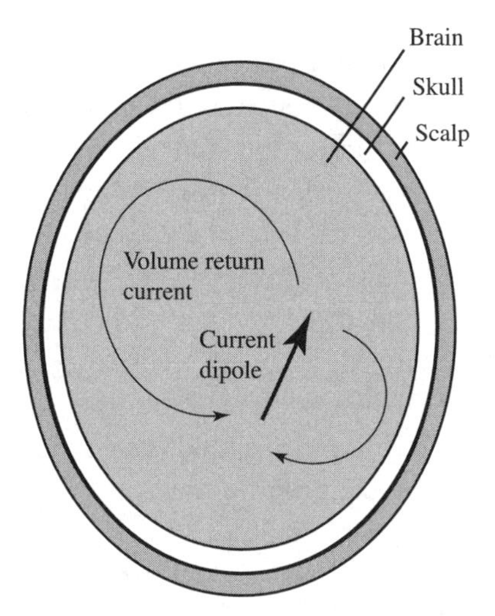

FIGURE 12.16 Schematic of a current dipole (solid arrow) inside of a spherical tissue region with volume return current (dashed arrows).

EXERCISES

Exercise 12.1 A population of magnetotactic bacteria, each with dipole moment μ, is located in a region of space with uniform magnetic induction in the z-direction. Calculate the average value of the cosine of the angle that the bacterial dipole moments make with respect to the incident field $\langle \cos\theta \rangle$.

Exercise 12.2 QuickField AC Magnetics Simulation: Construct a nested sphere model of the human torso including blood, myocardium, lung, and thorax regions as illustrated in Figure 12.17. Use resistivity values in Appendix 9 for each tissue region. Place a circular drive coil outside the torso to induce AC currents in the thorax. Calculate the change in magnetic field on the opposite side of the torso for both end-systole and end-diastole phases. Construct the model using axial symmetry with sphere diameters of 4.3 cm, 7.3 cm, 22 cm, and 28 cm, from innermost to outermost. Use the QuickField LabelMover feature to sweep the excitation frequency of the drive coil.

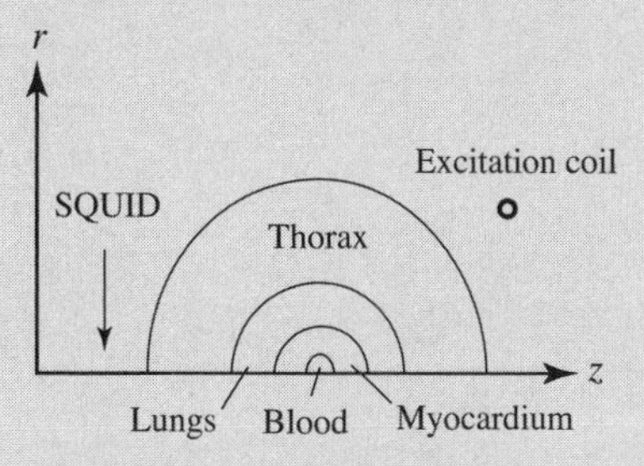

FIGURE 12.17 Nested sphere model of the thorax under AC excitation.

Exercise 12.3 QuickField DC Current Flow Simulation: Construct a nested sphere model of the human head including brain, skull, and scalp regions similar to Figure 12.16. Place a dipole source in the brain and calculate the volume return currents. Model the dipole as two charges separated by a small distance. Use conductivity values in Appendix 9 for the brain and tissue regions. Make contour plots of the electric potential on the outer surface of the scalp and on the outer surface of the brain. Construct the model using axial symmetry.

Exercise 12.4 QuickField AC Magnetics Simulation: Model a permeable conducting cylindrical magnetic shield in a z-directed AC magnetic field as illustrated in Figure 12.18. Use the QuickField LabelMover feature to plot the field at a point inside the shield as a function of frequency. Repeat the calculation with an air gap near one end of the model corresponding to a lid of the magnetic shield. Construct the model using axial symmetry. Take the shield conductivity $\sigma = 3 \cdot 10^7$ *S*/m and relative permeability $\mu_r = 10^4$, similar to that of commercially available Mumetal.

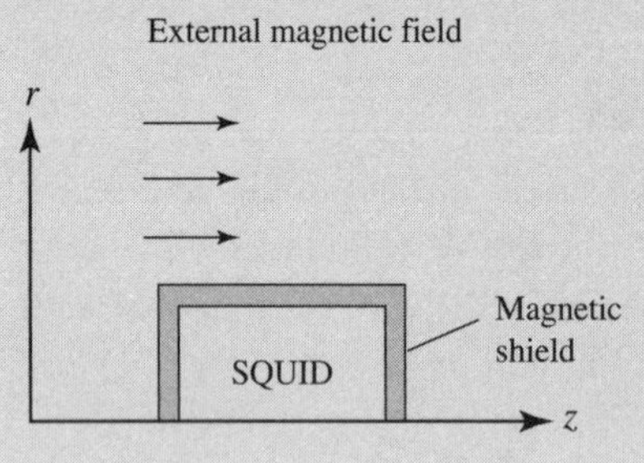

FIGURE 12.18 Cylindrical magnetic shield in a z-directed external noise field.

13 Nonlinearity and Chaos in Biological Systems

13.1 Chaotic Dynamics

Many simple physical systems with few interacting degrees of freedom exhibit erratic and unpredictable behavior. Chaotic behavior is observed in a diverse array of systems including orbital motion involving three or more bodies, certain chemical reactions such as the Belousov–Zhabotinsky (B–Z) reaction, intensity fluctuations in laser cavities, and fluid dynamics. The simplist example of an undriven chaotic system is the

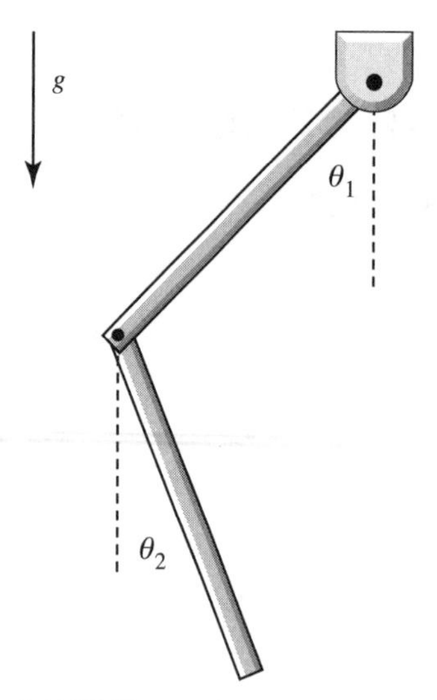

FIGURE 13.1 Simple double pendulum consisting of two connected rods that are free to pivot at their upper ends. This system will exhibit chaotic oscillations when set in motion with sufficiently large initial values of θ_1 and θ_2.

double pendulum illustrated in Figure 13.1. Chaotic systems are often described by systems of nonlinear differential equations that do not have analytical solutions. Physical processes that are purely random, such as radioactive decay, are not chaotic. In biology, complex oscillations can occur in cellular metabolism, population dynamics, heart rhythms, and driven nerve impulses. This chapter highlights several biological systems that exhibit nonlinear or chaotic behavior.

13.1.1 Characteristics of Chaotic Dynamics

Chaos occurs in systems whose time evolution is described by nonlinear differential equations. Nonlinearity does not necessarily imply chaotic behavior, however. Chaos never occurs in linear systems or systems with analytical solution. One prerequisite for chaos is at least three degrees of freedom such as spatial coordinates, momenta, reactant concentrations, and explicit time dependence. For example, the nonlinear, coupled system proposed by Sel'kov describing glycolysis,

$$\begin{aligned} \frac{\partial x}{\partial t} &= -x + ay + x^2 y \\ \frac{\partial y}{\partial t} &= b - ay - x^2 y \end{aligned} \tag{13.1}$$

with kinetic parameter a and b, is not chaotic because it has only two degrees of freedom: x and y. This system contains the nonlinear term x^2y where x = [ADP] is the adenosine diphosphate concentration and y = [F6P] is the concentration of fructose-6-phosphate. The time dependence is implicit in these autonomous differential equations so that time is not included in the number of degrees of freedom. If we were to specify explicit time dependence in this system, say by modulation of the kinetic parameter $b = b_0\sin(\omega t)$, then there would be three degrees of freedom including time, and chaos would be possible.

13.1.2 Sensitive Dependence on Initial Conditions

It is not possible to forecast the state of chaotic systems for long durations because of their extreme sensitive dependence on initial conditions. Small uncertainties in initial conditions result in complete uncertainty in future states of the system. Also, slight changes in initial conditions can result in dramatic changes later on. This phenomenon is known as the butterfly effect.

13.1.3 Phase Space

A dynamical system follows a trajectory in a multidimensional phase space where each degree of freedom corresponds to an axis in the phase space. The trajectory of a particle moving in one dimension may be plotted in a two-dimensional phase plane with position and momentum coordinates x and p_x, respectively. For example, the orbit of an object exhibiting simple harmonic motion such as a block on a spring would be a closed loop in x and p_x space.

A famous model exhibiting chaotic behavior was first developed in 1963 by Edward Lorenz to describe complex atmospheric convection. The Lorenz model consists of the set of first-order, autonomous, nonlinear differential equations:

$$\begin{aligned}\frac{dx}{dt} &= -\sigma(x+y)\\ \frac{dy}{dt} &= -xz + rx - y\\ \frac{dz}{dt} &= xy - bz\end{aligned} \tag{13.2}$$

where σ is proportional to the Prandtl number given by the fluid velocity divided by the thermal conductivity. The parameter r is the Rayleigh number, and b is a geometry-dependent factor. The solution of this system generates a three-dimensional phase-space orbit shown in Figure 13.2 for $\sigma = 10$, $r = 28$, and $b = 8/3$. The Lorenz equations represent an extreme truncation of the Navier–Stokes equations describing fluid flow between the two boundaries held at different temperatures representing the Earth's surface and the upper atmosphere.

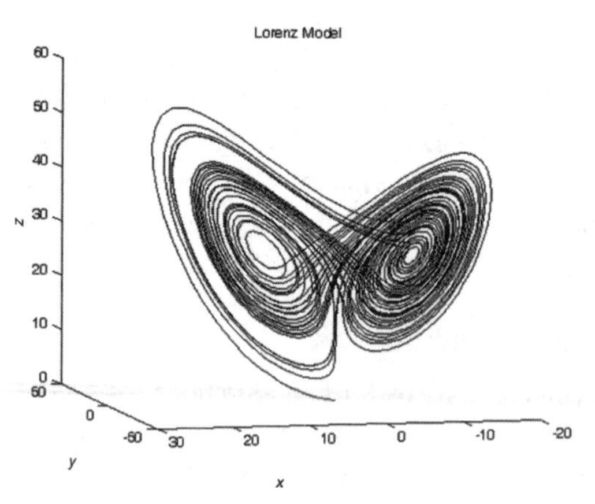

FIGURE 13.2 3-D phase plot of the Lorenz attractor.

13.1.4 Phase Space Reconstruction

In practice, it may not be possible to obtain separate recordings of all of the dynamical variables associated with a dynamical system. Often only a single-channel measurement such as an electrochemical potential or reactant concentration is available. It is possible to qualitatively reconstruct the phase space from a single variable $x(t)$ by using time delay coordinates, however. The reconstruction is performed in two dimensions by plotting $x(t)$ vs. $x(t - \tau)$ where τ is a suitably chosen time delay. A reconstructed orbit may be plotted in three dimensions with coordinates $x(t)$, $x(t - \tau)$, and $x(t - 2\tau)$. Figure 13.3 illustrates a 3-D reconstruction of the Lorenz attractor in Figure 13.2 from a single time series using two time delays. The plot is qualitatively similar to the actual attractor where details of the phase space are revealed that would not be evident by inspection of the time series.

The following MATLAB code creates the 3-D reconstruction of the Lorenz model in Figure 13.3.

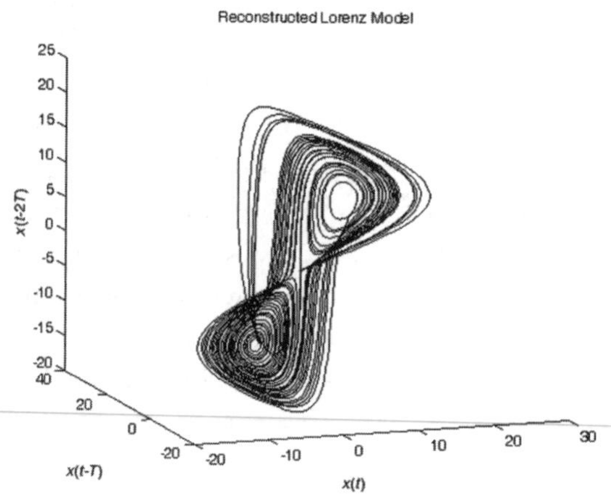

FIGURE 13.3 Lorenz attractor reconstructed from the x-time series using time-delay coordinates.

```
rho=28;
sigma=10;
beta=8/3;
dt=.01;

% create function handles

f=@(x,y,z)sigma*(y-x);
g=@(x,y,z)rho*x-y-x*z;
h=@(x,y,z)x*y-beta*z;

nmax=2^12;

x=zeros(nmax,1); x1=zeros(nmax,1);
y=zeros(nmax,1); x2=zeros(nmax,1);
z=zeros(nmax,1); x3=zeros(nmax,1);

x(1)=.1; y(1)=.1; z(1)=.1;

% integrate the differential equations using
% a first order Euler scheme

for n=1:nmax

x(n+1)=x(n)+f(x(n),y(n),z(n))*dt;
y(n+1)=y(n)+g(x(n),y(n),z(n))*dt;
z(n+1)=z(n)+h(x(n),y(n),z(n))*dt;

end

% perform the reconstruction from the x-time series
% using two time delays

for n= 15:nmax
      x1(n)=x(n);
      x2(n)=x(n-7);
      x3(n)=x(n-14);
end

plot3(x1,x2,x3,'k')

title('Reconstructed Lorenz Model')
xlabel('x(t)')
ylabel('x(t-T)')
zlabel('x(t-2T)')
```

13.1.5 Poincaré Sections

Poincaré sections, also know as strobe plots, are cross sections of the phase space where trajectories intersect a plane in phase space where one of the variables is constant. Plotting a point each time the trajectory crosses the plane forms the section. The resulting plot may consist of a single point, a curve, or a fractal set corresponding to periodic, quasiperiodic, or chaotic systems, respectively. The Poincaré section is a useful tool for investigating chaotic dynamics. One can think of the Poincaré section as analogous to an MRI scan revealing structure within the phase space.

13.1.6 Lyapunov Exponents

The Lyapunov exponent λ gives an indication of the sensitive dependence on initial conditions of a dynamical system as the divergence of two nearby trajectories. Given two initial conditions in a coordinate x separated by Δx_0, the separation in trajectories will evolve according to

$$\Delta x(t) = \Delta x_0 \exp(\lambda t). \tag{13.3}$$

In general, a dynamical system will have as many Lyapunov exponents as degrees of freedom. An initial volume or block of coordinates in phase space $V_0 = \Delta x_0 \Delta y_0 \Delta z_0$ has time dependence

$$V(t) = V_0 \exp\left(\sum_{i=1}^{3} \lambda_i t\right). \tag{13.4}$$

If the sum of Lyapunov exponents is negative, the system is dissipative, and trajectories will remain confined to some finite volume of phase space. If the sum of Lyapunov exponents is positive, then the system is divergent, and V expands with time. Conservative systems preserve phase-space volume elements although V may be stretched and folded many times as the system evolves. Chaotic systems can be conservative or dissipative but must have at least one positive Lyapunov exponent.

13.1.7 Power Spectra

Fourier's theorem states that a periodic waveform can be represented as a superposition of sine and cosine functions. A function with period T

$$F(t) = F(t + nT) \qquad n = 0, 1, 2, 3 \ldots \tag{13.5}$$

where n is an integer is thus represented as

$$F(t) = \frac{A_0}{2} + \sum_{n=1}^{\infty}\left(A_n \cos\left(n\frac{2\pi}{T}t\right) + B_n \sin\left(n\frac{2\pi}{T}t\right)\right) \tag{13.6}$$

where A_n and B_n are the Fourier coefficients. Equation 13.6 can be written in a more compact form in terms of complex exponentials

$$F(t) = \sum_{n=-\infty}^{\infty} a_n \exp(in\omega_0 t) \tag{13.7}$$

where $\omega_0 = 2\pi/T$ and $f_0 = 1/T$ is the fundamental frequency. A function with arbitrary time dependence can be represented by the Fourier integral

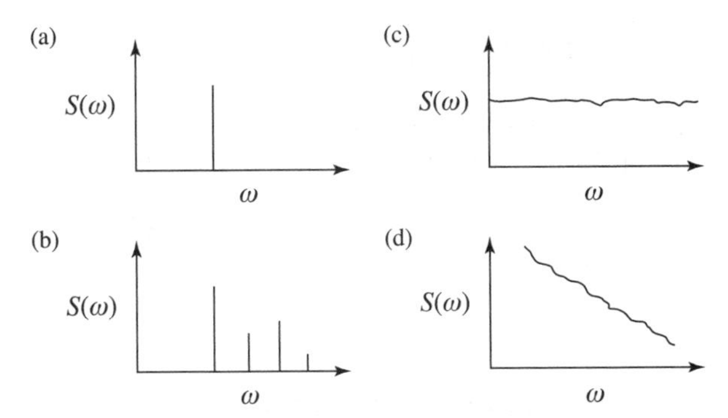

FIGURE 13.4 Power spectra of (a) a pure sine wave, (b) a periodic signal, (c) white noise, and (d) pink noise.

$$F(t) = \frac{1}{\sqrt{2\pi}} \int_{-\infty}^{\infty} a(\omega) \exp(i\omega t)\, d\omega. \tag{13.8}$$

The inverse Fourier transform solves for the complex coefficients

$$a(\omega) = \frac{1}{\sqrt{2\pi}} \int_{-\infty}^{\infty} F(t) \exp(-i\omega t)\, dt. \tag{13.9}$$

The power spectrum is obtained from the squared magnitude of the $a(\omega)$

$$S(\omega) = |a(\omega)|^2. \tag{13.10}$$

Figure 13.4 shows a sketch of different types of power spectra of both periodic and noisy signals. The power spectrum of a pure sine wave in (a) is a single peak, while the spectrum of an arbitrary periodic signal in (b) consists of multiple peaks, or harmonics of the fundamental frequency. White noise (c) is characterized by a flat power spectrum with roughly equal power over a broad frequency range. Pink noise (d) is often referred to as $1/f$ noise and has stronger lower-frequency components.

The Fast Fourier Transform

The fast Fourier transform (FFT) is an algorithm that can be applied to a discrete time series sampled from an arbitrary waveform. Commercial spectrum analyzers employ this algorithm to display and record power spectra in real time. In order to perform the FFT, the time series must have 2^n elements. A time series of arbitrary length will have to be truncated or padded with zeros.

13.2 Population Growth in a Limited Environment

Species interaction in a limited environment can result in nonlinear oscillations in the populations of each species. We first consider the population growth of a single species with the number of individuals changing in proportion to the population size N:

$$\frac{dN}{dt} = rN \tag{13.11}$$

where the proportionality constant $r = b - d$ is equal to the birthrate b minus the mortality rate d. The solution to this equation is

$$N(t) = N_0\exp(rt). \tag{13.12}$$

If the mortality rate exceeds the birthrate, then r is negative, and the population declines into extinction. If r is positive, overcrowding will eventually ensue in a finite environment. To include environmental limitations, we introduce a carrying capacity K equal to the maximum number of organisms that can be sustained by the environment. The right-hand side of Equation 13.11 is then multiplied by $1 - N/K$, giving the logistic equation

$$\frac{dN}{dt} = rN\left(1 - \frac{N}{K}\right) \tag{13.13}$$

so that $dN/dt = 0$ when $N = K$. The solution to this equation is

$$N(t) = \frac{KN_0e^{rt}}{K + N_0(e^{rt} - 1)} \tag{13.14}$$

where $N(0) = N_0$ and $N(t) \to K$ as $t \to \infty$ for $r > 0$. We can write the logistic equation in a more succinct form dividing by K:

$$\frac{d}{dt}\frac{N}{K} = r\frac{N}{K}\left(1 - \frac{N}{K}\right) \tag{13.15}$$

and then letting $x = N/K$, giving the reduced form

$$\frac{dx}{dt} = rx(1 - x) \qquad x \in [0, 1]. \tag{13.16}$$

We investigate a discrete form of this equation in Section 13.4 that gives rise to complex behavior and chaos.

13.3 Predator–Prey Models of Population Dynamics

The simplest predator–prey system models the interaction of two species such as cheetahs (predators) and baboons (prey). The predators multiply as they consume the prey, resulting in a reduction of the prey population. The predators begin to starve off as their food supply is diminished, causing a rebound of the prey population. Oscillations in the population of each species are then established that can be modeled by the system

$$\begin{aligned} \frac{dN_1}{dt} &= pN_1 - qN_1N_2 \\ \frac{dN_2}{dt} &= -rN_2 + sN_1N_2 \end{aligned} \tag{13.17}$$

where N_1 is the prey population and N_2 is the predator population. The prey mortality rate qN_1N_2 is proportional to the number of predators N_2. The growth rate of the predator population is given by sN_1N_2. The parameters q and s govern interaction between the species. Without species interaction, the prey population grows exponentially $N_1 \sim \exp(pt)$ assuming limitless food supply, while the predators starve $N_2 \sim \exp(-rt)$. This system of first-order equations has periodic solutions without analytical form and therefore requires numerical solution. The following MATLAB program solves these equations using a fourth-order integration scheme and plots the results.

The following code block is first saved in an M-file named Pred_Prey.m

```
function yp = Pred_Prey(t,x)
S=.01; R=2; P=1; Q=.05;
yp = [ P*x(1)-Q*x(1)*x(2);
S*x(1)*x(2)-R*x(2)];
```

The following instructions are then entered at the command prompt to solve the differential equations and plot the population time series and phase plot

```
>> [t,x]=ode45(@Pred_Prey,[0,10],[100,20]);
>> plot(t,x)
>> plot(x(:,1),x(:,2))
```

Here the initial number of prey is 100 with 20 predators. The first plot command graphs the time evolution of both N_1 and N_2 in Figure 13.5.

In this figure, there is a time lag between peaks in the predator and prey populations. This time lag is referred to as a phase difference. Because of this phase difference, phase-space orbits formed by graphing N_1 vs. N_2 are closed loops in Figure 13.6 (second plot command). The phase-space orbits would be open curves if the predator and prey populations were exactly in phase (both peaking at the same time) or 180° out of phase (with predator population maxima corresponding to prey population minima,

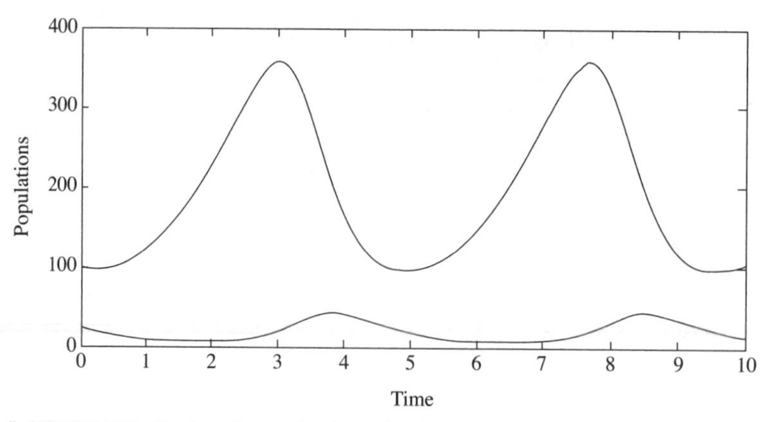

FIGURE 13.5 Predator (bottom) and prey (top) population time series.

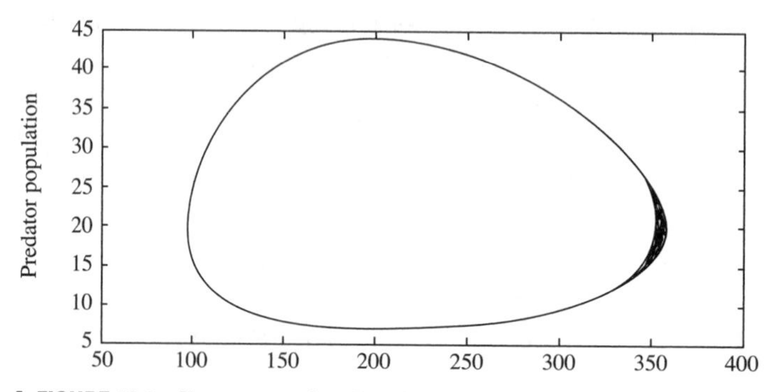

FIGURE 13.6 Phase space plot of predator vs. prey populations.

and vice versa). This predator–prey model (shown above) does not include environmental influences, interaction between with other species, or the evolutionary adaption of species to predatation.

Nearly out-of-phase oscillations were experimentally observed in microorganism populations of rotifer *Brachionus calyciflorus* (predator) and unicellular green algae (prey). These oscillations were attributed to the rapid evolution of the prey. Evidently algae clones that are more resistant to depletion by rotifers represent a larger fraction of the rebounding prey population. The less-resistant algae clones rebound later in the cycle when the rotifer population is close to its minimum.

13.4 Discrete Logistic Equation

In the simple case of a linear population growth, we have that the number of organisms alive in the next generation $n + 1$ is proportional to the number of organisms in the current generation n:

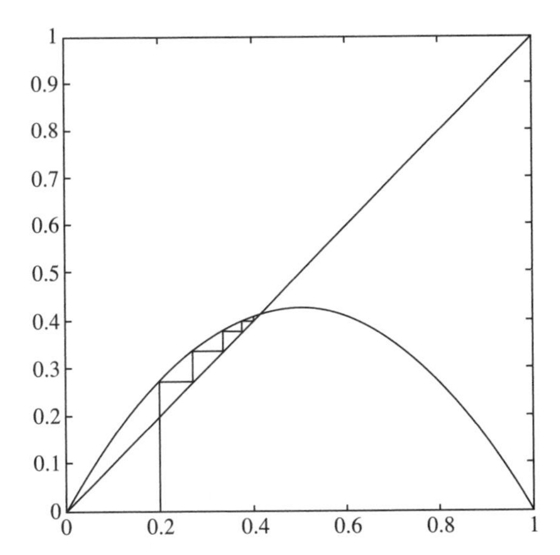

FIGURE 13.7 Iterations of the logistic map with the parameter $\mu = 1.7$. The initial condition is 0.2, with the map rapidly converging to a fixed point.

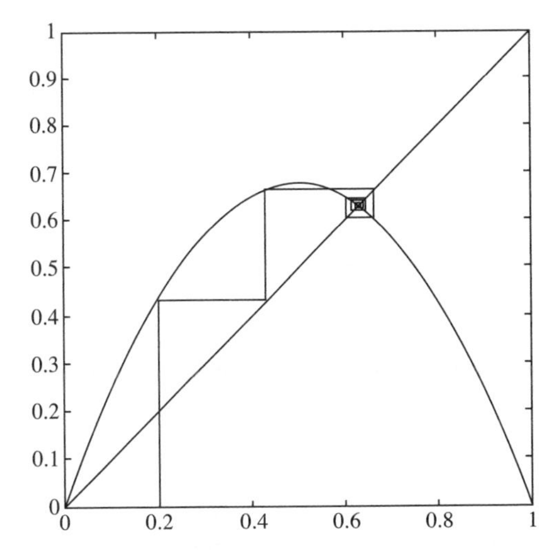

FIGURE 13.8 Iterations of the logistic map with the parameter $\mu = 2.7$. The initial condition is 0.2, with the map rapidly converging to a fixed point.

$$x_{n+1} = \mu x_n \qquad \mu > 0 \tag{13.18}$$

where μ represents a growth rate. The population remains constant if $\mu = 1$ and declines if $0 < \mu < 1$. In a limited environment, the discrete time–logistic equation is given by the difference equation

$$x_{n+1} = \mu x_n(1 - x_n) \qquad x_n \in [0, 1] \tag{13.19}$$

where the first term μx_n corresponds to the unimpeded growth of the population. The second term $-\mu x_n^2$ represents a decrease in growth resulting from overcrowding or disease. Periodic and chaotic oscillations in x_n can result, depending on the value of μ.

13.4.1 Period-Doubling Route to Chaos

For small values of μ, the system reaches a "fixed point" after several steps where the number of organisms is constant. Figures 13.7 and 13.8 show iterates of the logistic map for μ values equal to 1.7 and 2.7, respectively.

In both of these figures, the trajectory approaches a fixed point on the map. As μ increases the system goes into a two-cycle alternating between two populations as shown in Figure 13.9, with μ equal to 3.2.

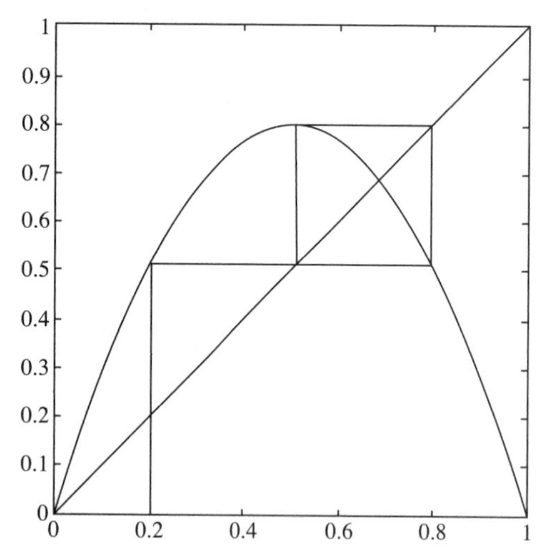

FIGURE 13.9 Iterations of the logistic map with the parameter $\mu = 3.2$. The initial condition is 0.2, with the map entering into a limit cycle.

With further increase in μ, a period-doubling cascade ensues where the system progresses through 4-, 8-, 16-, and 2^n-cycles leading to chaotic population fluctuations. This type of cascade resulting from the adjustment of a system parameter is characteristic of the period-doubling route to chaos. Figures 13.10 and 13.11 show iterates of the logistic map for μ values equal to 3.83 and 3.9, respectively, in turn corresponding to periodic and chaotic oscillations.

Figure 13.12 shows the Fourier transform of x_n sequences with (a) $\mu = 2.7$ (b) $\mu = 3.83$, and (c) $\mu = 3.9$.

The logistic map also exhibits extreme sensitivity with respect to small changes in initial conditions in the chaotic regime. Two time series of the logistic map are plotted in Figure 13.13 with slightly different initial conditions, illustrating the butterfly

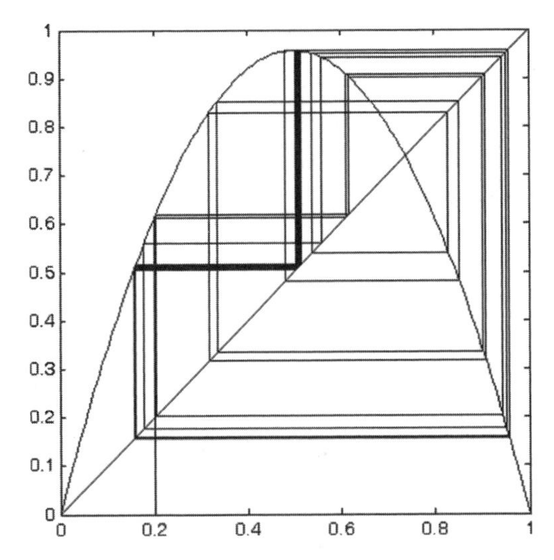

FIGURE 13.10 Iterations of the logistic map with the parameter $\mu = 3.83$.

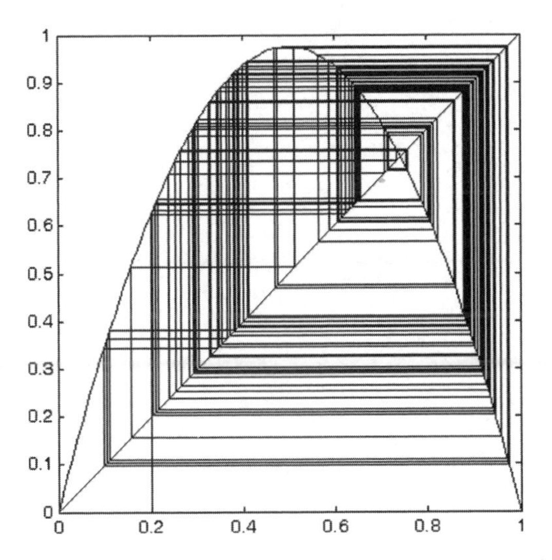

FIGURE 13.11 Iterations of the logistic map with the parameter $\mu = 3.9$. The map is chaotic.

effect for $\mu = 3.9$. The x_n sequences initially coincide and then become completely uncorrelated.

The following MATLAB code generates iterates of the logistic map:

```
mu=3.2;                 % parameter
x0=0.2;                 % initial condition
Nmax=100;               % number of iterations
x(1) = x0;
```

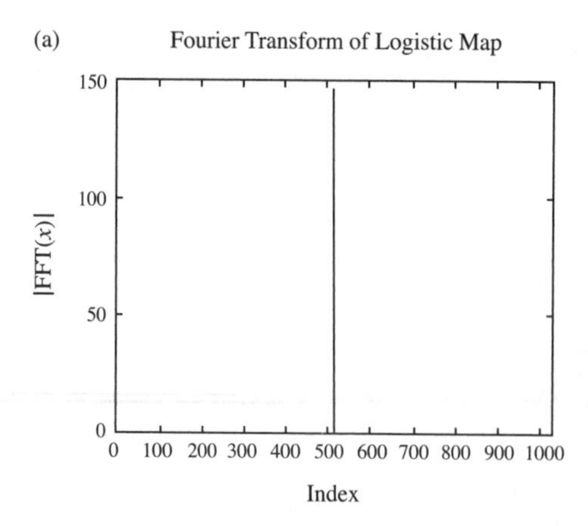

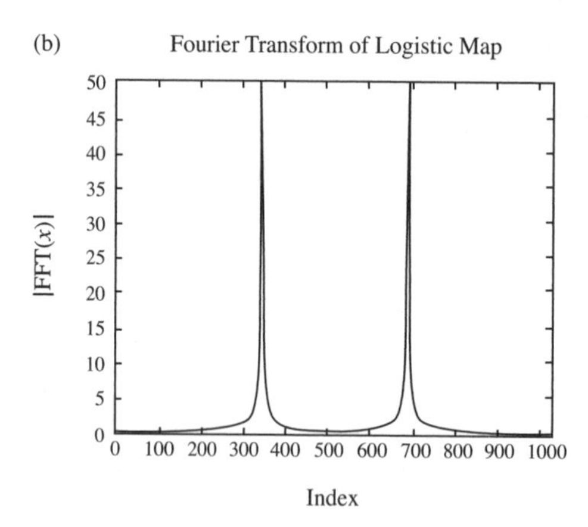

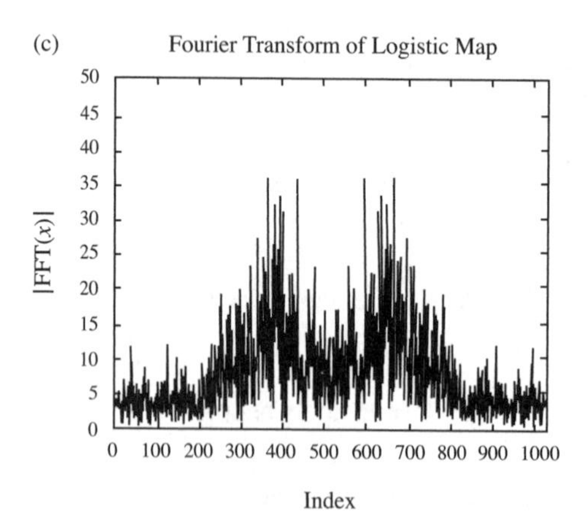

FIGURE 13.12 Fast Fourier transform (FFT) of logistic map time series with (a) $\mu = 2.7$, (b) $\mu = 3.83$, and (c) $\mu = 3.9$.

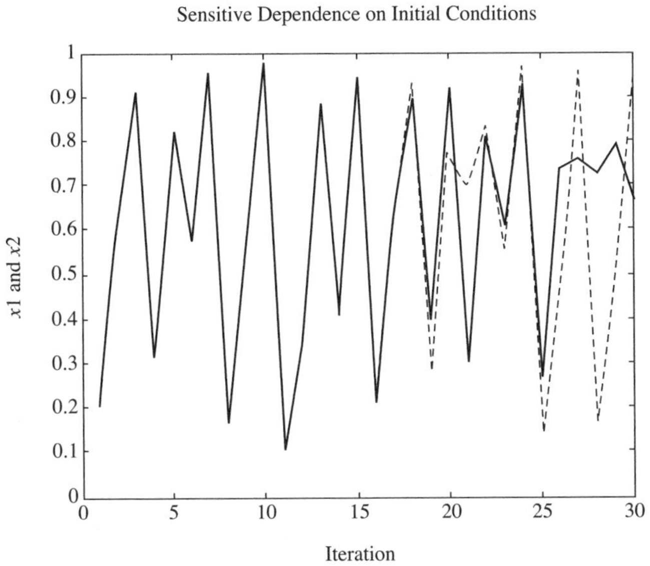

Sensitive Dependence on Initial Conditions

FIGURE 13.13 Two time series of the logistic map with initial conditions differing by 10^{-4} with $\mu = 3.9$. The plots initially track for roughly 20 iterations and then become completely uncorrelated.

```
% iterate the map

for i=1:Nmax
  x(i+1) = mu*x(i)*(1-x(i));
end

% plot the function and the line y=x

hold on

t = 0:0.01:1;
plot(t,mu*(t.*(1-t)),'r');
plot(t,t,'k');

line([0 0],[1 1])

% draw the map

line([x(1) x(1)], [0 x(2)])

 for i=1:Nmax-1
    line([x(i) x(i+1)],[x(i+1) x(i+1)])
    line([x(i+1) x(i+1)],[x(i+1) x(i+2)])
  end

hold off
axis equal
```

The following MATLAB code calculates the FFT of the logistic map:

```
Nmax = 2^10;      % Length of signal

x=zeros(Nmax:1);
```

```
R=3.83;
y=0.2;

% initial transients

for i=1:100

    y=R*y*(1-y);
end

% begin recording after 100 initial steps

x(1)=y;

for n=1:Nmax-1
    x(n+1)=R*x(n)*(1-x(n));
end

Y=fft(x);

plot(abs(Y),'k')
axis ([0 Nmax 0 50])
title('Fourier Transform of Logistic Map')
xlabel('index')
ylabel('|FFT(x)|')
```

The following MATLAB code demonstrates sensitive dependence on initial conditions in the logistic map:

```
Nmax = 30;
x=zeros(Nmax:1);
y=zeros(Nmax:1);

t=linspace(1,Nmax,Nmax);

R=3.9;
y(1)=0.2;

for n=1:Nmax-1
    y(n+1)=R*y(n)*(1-y(n));
end

% introduce small change in initial condition

x(1)=0.20001;

for n=1:Nmax-1
    x(n+1)=R*x(n)*(1-x(n));
end

plot(t,x,':k',t,y,'k','LineWidth',2)

title('Sensitive Dependence on Initial Conditions')
xlabel('iteration')
ylabel('x1 and x2')
```

13.4.2 Bifurcation Diagrams

Period-doubling bifurcations occur at specific values of μ_m as can be seen in the bifurcation diagram in Figure 13.14, plotting x_n vs. μ varied between 1 and 4.

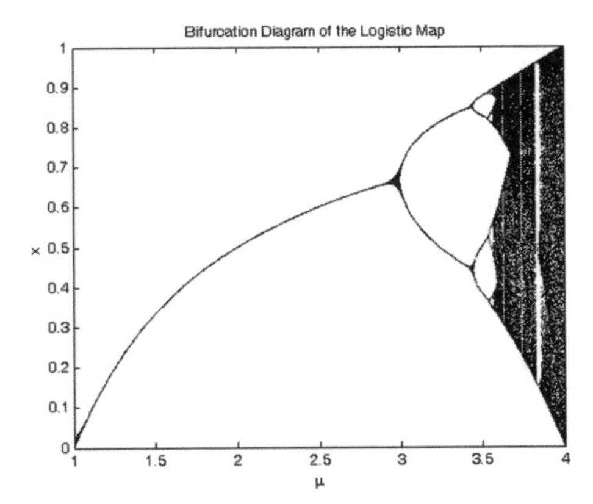

FIGURE 13.14 Bifurcation diagram of the logistic map with the parameter μ varied between 1 and 4.

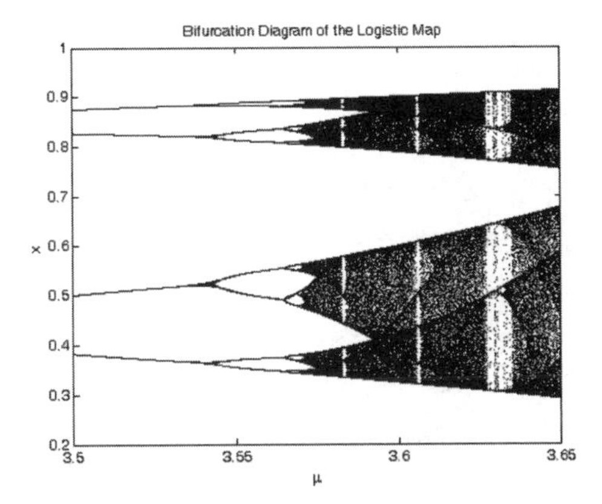

FIGURE 13.15 Bifurcation diagram of the logistic map with the parameter μ varied between 3.5 and 3.65.

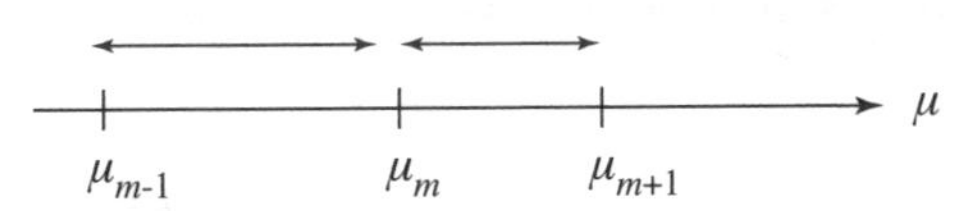

FIGURE 13.16 Intervals in μ between period-doubling bifurcations of the logistic map.

Figure 13.15 shows a close up of the bifurcation diagrams with μ between 3.5 and 3.65. Evidently the ratio R of subsequent intervals in μ between bifurcations,

$$R = \frac{\mu_m - \mu_{m-1}}{\mu_{m+1} - \mu_m}, \tag{13.20}$$

approaches the Feigenbaum number $R = 4.6692\ldots$ in the limit $m \to \infty$.

Two intervals between μ_m are shown schematically in Figure 13.16. This number was first discovered by Michael Feigenbaum then at Los Alamos National Laboratory, New Mexico. The Feigenbaum number is said to be universal since it has the same value for all chaotic maps with quadratic maxima such as the logistic map. Equation 13.20

can also be used to predict a sequence of μ values where bifurcations occur given the first three values of μ_m.

The following MATLAB code plots the bifurcation diagram of the logistic map:

```
N = 900;
R_init = 3.5;
R_fin = 4;
R = linspace(R_init, R_fin, N);

trans = 40;
total = 300;
results=zeros(length(R),total);

for j = 1:length(R)
x = 0.4;

for i = 1:trans
        x = R(j) * x * (1-x);
end

for i = 1: total
x = R(j) * x * (1-x);
results(j,i) = x;
end

end

plot(R, results,'.k','MarkerSize',1);
xlabel('\mu');
ylabel('x');
title('Bifurcation Diagram of the Logistic Map')
```

13.4.3 Lyapunov Exponent of the Logistic Equation

The Lyapunov exponent for the logistic map may be calculated from the formula

$$\lambda = \lim_{N\to\infty} \frac{1}{N} \sum_{i=0}^{N-1} \ln|f'(x_i)| \tag{13.21}$$

where $f'(x_i) = \mu(1 - 2x_i)$. The following MATLAB code calculates λ for the logistic map and plots the results shown in Figure 13.17 for μ between 2.8 and 4.0:

```
Jmax=4000;
Jtrans=100;
y=zeros(Jmax);


r_init=2.8;
r_fin=4.0;
pts=1000;
delta_r=(r_fin-r_init)/pts;


Lyap=zeros(pts);
```

```
R=zeros(pts);

n=0;
r=r_init;

while r<r_fin

r= r+ delta_r;
n=n+1;

y(1)=.5;

for j=2:Jmax;

y(j)=r*y(j-1)*(1-y(j-1));

end

sum = 0;

for j=Jtrans:Jmax;

sum = sum + log(abs(r*(1.0-2.0*y(j))));

end

Lyap(n)= sum/(Jmax-Jtrans);

R(n)=r;

end

plot(R,Lyap,'k');
xlim([r_init,r_fin]);
hold on
plot([r_init,r_fin],[0.0,0.0],'k','EraseMode','none');
hold off
xlabel('\mu');
ylabel('\lambda');
title('Lyapunov Exponent of the Logistic Map')
```

13.4.4 Shannon Entropy

The Shannon entropy may be calculated by binning the unit interval into cells and calculating the probability $p_i = n_i/N$ that each cell is occupied. The entropy is then

$$S = -\sum_{i=1}^{N} p_i \ln p_i. \tag{13.22}$$

The following MATLAB code calculates the Shannon entropy of the logistic map and plots the results shown in Figure 13.18 with μ varied between 2.6 and 4.0.

```
numR = 400;
startR = 2.6;
endR = 4;
R = linspace(startR, endR, numR);
```

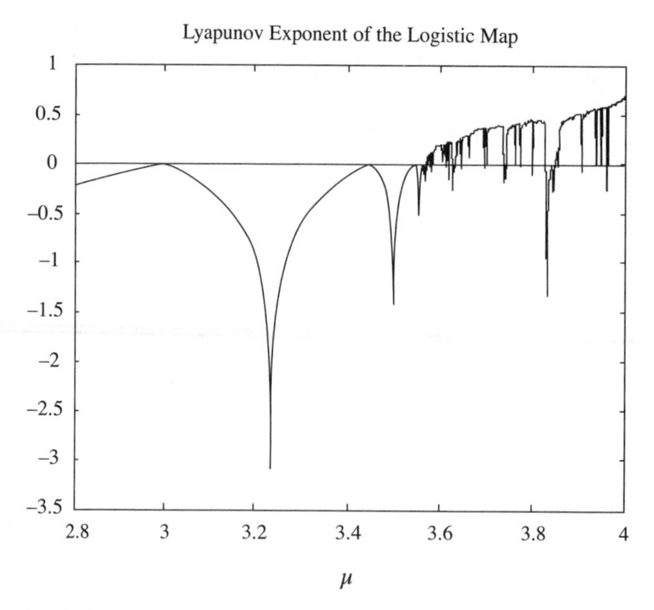

FIGURE 13.17 Variation of the Lyapunov exponent of the logistic map with the parameter μ varied between 2.8 and 4.0. The map is chaotic for positive values of λ above the horizontal line in the figure. The Lyapunov exponent drops precipitously in the periodic windows of the map.

```
skipnum = 40;
num = 2000;
bin=40;
dx=1/bin;

results=zeros(num,length(R));
P=zeros(bin, length(R));
S=zeros(length(R),1);

for j = 1:length(R)
x = 0.4;
for i = 1:skipnum
x = R(j) * x * (1-x);
end;

for i = 1:num
x = R(j) * x * (1-x);
results(j,i) = x;
end;
end;

for j=1:length(R)
    for i=1:num
        for n=1:bin
            if (n-1)*dx<=results(j,i) && results(j,i)< n*dx
                P(n,j)=P(n,j)+1;
```

```
            end
        end
    end
end

P=P/num;

for j=1:length(R)

        for n=1:bin
            if P(n,j)>0
                S(j)=S(j)-P(n,j)*log(P(n,j));
            end
        end
end

plot(R,S,'k');

xlabel('\mu')
ylabel('Entropy')
title('Entropy of the Logistic Map')
```

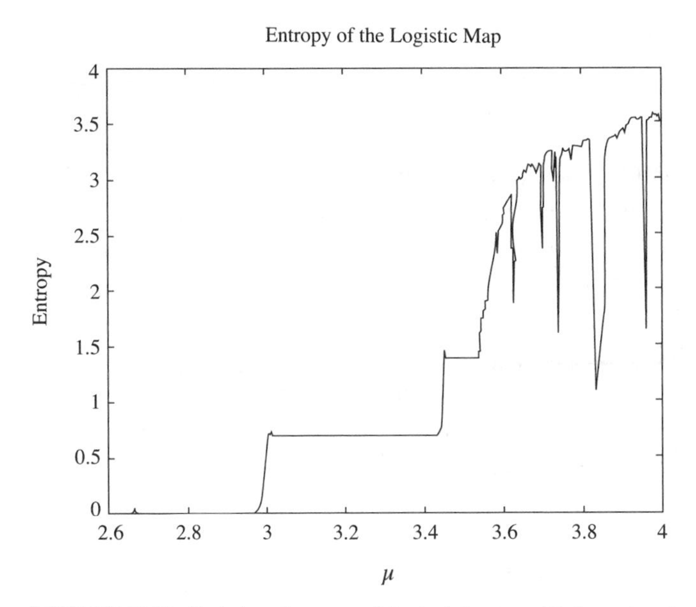

FIGURE 13.18 Variation of entropy of the logistic map with the parameter μ varied between 2.6 and 4.0. The entropy drops precipitously in the periodic windows of the map.

13.5 Chaos in the Heart

The tools of time series analysis including phase-space reconstruction and power spectra have been applied to the study of recorded ECG signals of the human heart. This analysis provides useful physiological information that can indicate the presence of pathology and heart disease. Healthy hearts beat more chaotically with $1/f$-like power spectra compared with pathological heart rhythms that are more periodic, exhibiting a strong frequency response near 50 Hz.

13.6 Reaction Diffusion Equations

The temporal and spatial variation of a chemical species v due to both reaction and diffusion is described by the single-component equation

$$\frac{\partial v}{\partial t} = \underbrace{f(v)}_{\text{reaction}} + \underbrace{D\frac{\partial^2 v}{\partial^2 x}}_{\text{diffusion}}. \tag{13.23}$$

Systems undergoing both reaction and diffusion may form complex patterns and traveling waves. The two-component reaction diffusion equations

$$\begin{aligned} \frac{\partial v}{\partial t} &= f(v, w) + D_v \frac{\partial^2 v}{\partial^2 x} \\ \frac{\partial w}{\partial t} &= g(v, w) + D_w \frac{\partial^2 w}{\partial^2 x} \end{aligned} \tag{13.24}$$

describe processes such as inhibitor and promoter reactions in one spatial and one time dimension. The behavior of this system will depend on the functions f and g, including system parameters and rate constants. If the diffusion constant $D_w \gg D_v$, then the concentration of w can be treated as uniform.

13.6.1 Action Potentials

A model describing the two-dimensional action potential in excitable heart tissue is obtained by adding the spatially varying term $\nabla^2 v$ to the first FitzHugh–Nagumo equation, resulting in the reaction diffusion system

$$\begin{aligned} \frac{\partial v}{\partial t} &= \nabla^2 v + \frac{1}{c}\left(v - \frac{v^3}{3} - w + I_{\text{ext}}\right) \\ \frac{\partial w}{\partial t} &= c(v - aw + b). \end{aligned} \tag{13.25}$$

This system may be used to simulate spiral waves such as those associated with arrhythmias in the heart.

13.6.2 Fertilization Calcium Waves

The propagation of fertilization calcium waves in eggs is described by the reaction diffusion equations

$$\frac{\partial[\mathrm{Ca}^{2+}]}{\partial t} = D_C\nabla^2[\mathrm{Ca}^{2+}] - k_+[B][\mathrm{Ca}^{2+}] + k_-[\mathrm{Ca}\cdot B] + J_{\mathrm{Ca}}$$
$$\frac{\partial[\mathrm{B}_{\mathrm{total}}]}{\partial t} = D_B\nabla^2[\mathrm{B}_{\mathrm{total}}] \tag{13.26}$$

in the presence of a buffer B such as a Ca^{2+} indicator dye where $[B_{total}] = [B] + [Ca \cdot B]$ is the free-plus-bound buffer concentration. D_C and D_B are free calcium and buffer diffusion constants, respectively. J_{Ca} is the Ca^{2+} flux into and out of the endoplasmic reticulum.

13.6.3 Pattern Formation

Reaction diffusion equations describe pattern formation in certain fish, seashells, and other animals. A two-dimensional example is given by the equation

$$\frac{\partial u}{\partial t} = D\nabla^2 u + \gamma u(1 - u) \tag{13.27}$$

with $u = u(x, y)$. The following MATLAB code generates the pattern formed in Figure 13.19.

```
Nmax =200;    % time steps
n =100;       % grid size
D=0.2;        % diffusion constant
gamma = 0.5;

reaction ='u.*(1-u)';

% initialize u and grad arrays
u=rand(n); grad= zeros(n);

i= 2:n-1; j=2:n-1;

for k = 1:Nmax

       grad(i,j)=u(i,j-1)+u(i,j+1)+u(i-1,j)+u(i+1,j);
       u=(1-4*D)*u+D*grad+gamma*eval(reaction);

end

pcolor(u);
shading interp;
```

13.7 Dynamics of the Driven Hodgkin–Huxley System

The Hodgkin–Huxley equation describing the time development of action potential is

$$C\frac{dv}{dt} + I_{\mathrm{ion}} = I_{\mathrm{ext}}(t) \tag{13.28}$$

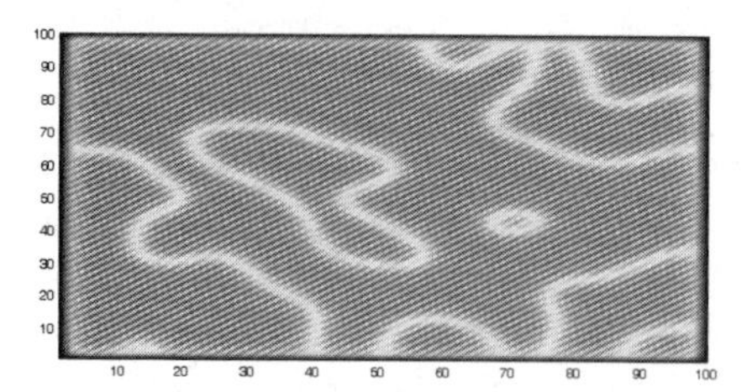

FIGURE 13.19 Pattern formation simulated by reaction diffusion Equation 13.27. (Adapted from Yang X-S. 2006. *An introduction to computational engineering with MATLAB.* Cambridge International Science Publishing, Cambridge, U.K. Chapter 7: Pattern Formation.)

with I_{ion} given by Equation 10.6. Types of behavior observed in the harmonically driven H–H system with $I_{\text{ext}}(t) = A\sin(\omega t)$ include phase locking, quasiperiodic, chaotic, and transient chaotic behavior. Phase locking, also referred to as entrainment, occurs when the action potential synchronizes with the external driving. The natural frequency of sustained-action potential spiking can be a fractional multiple of the driving frequency. This corresponds to the ratio n/m of synchronized oscillations where the action potential is activated n times in m periods of the external driving. For example, a 4/5 oscillation would correspond to four spiking events in five periods of the excitation. This type of synchronization can give rise to fractional harmonics and higher subharmonics in the power spectrum observed in both the H–H equations and experimentally in the harmonically driven giant squid axon.

The parameters A and ω can be continuously adjusted from some initial values so that the *route* to *chaos* can be from synchronized oscillations, successive period doubling, or intermittency. These routes to chaos can be mapped along trajectories in $A - \omega$ parameter space.

13.8 Models of DNA Motility

The structure of DNA consists of two helical strands. DNA bases adenine (A), guanine (G), thymine (T), and cytosine (C) run parallel with bases connected by hydrogen bonds with two types of base pairs A–T and G–C. Angular fluctuation of DNA bases around the sugar–phosphate chains can lead to large angle excursions of the base pairs rupturing weakly interacting hydrogen bonds. Resulting open states can form as shown in Figure 13.20.

Englander et al. (1980) proposed a model of DNA rung fluctuations analogous to the motion of coupled pendulums in a gravitational potential. The dynamics are described by the equation

$$\underbrace{I\frac{\partial^2\phi_n}{\partial t^2}}_{\text{intertial term}} = \underbrace{K(\phi_{n+1} - 2\phi_n + \phi_{n-1})}_{\text{coupling between bases}} - \underbrace{mg\ell\sin\phi_n}_{\text{hydrogen bond attraction}} \tag{13.29}$$

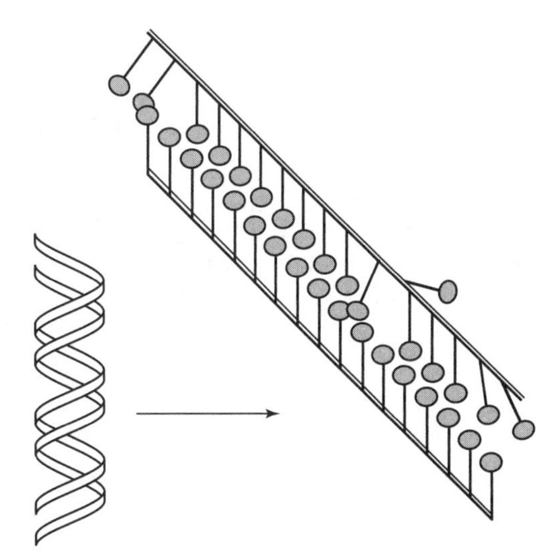

FIGURE 13.20 DNA ladder chain modeled as an array of coupled pendulums. Each pendulum in the upper chain makes an angle θ_n with respect to the normal of the sugar–phosphate backbone.

where ϕ_n is the angular deviation of the nth pendulum from its equilibrium position. The net rotational torque acting on a DNA base is provided by coupling between neighboring bases and attractive hydrogen bonding to the lower ladder. In the continuum approximation, we replace the difference terms in the base coupling by the second derivative:

$$\frac{(\phi_{n+1} - 2\phi_n + \phi_{n-1})}{a^2} \rightarrow \frac{\partial^2 \phi}{\partial x^2} \tag{13.30}$$

where a is the distance between pairs on the chain. We then have the sine–Gordon equation:

$$\frac{I}{mg\ell}\frac{\partial^2 \phi}{\partial t^2} = \frac{Ka^2}{mg\ell}\frac{\partial^2 \phi}{\partial x^2} - mg\ell \sin\phi \tag{13.31}$$

whose solutions are well known. This equation can be written in a compact form by rescaling the time and position coordinates

$$t\sqrt{\frac{mg\ell}{I}} \rightarrow t' \text{ and } x\frac{mg\ell}{Ka^2} \rightarrow x' \tag{13.32}$$

so that

$$\frac{\partial^2 \phi}{\partial t'^2} = \frac{\partial^2 \phi}{\partial x'^2} - \sin\phi. \tag{13.33}$$

The effects of damping with an externally applied force F can be included:

$$\frac{\partial^2\phi}{\partial t^2} = \frac{\partial^2\phi}{\partial x^2} - \sin\phi - \alpha\frac{\partial\phi}{\partial t} - F \tag{13.34}$$

where α represents the strength of the damping. We discard the primes, understanding that we are working with dimensionless coordinates. The solution to this equation is that of a traveling wave of the form $\phi(x, t) = \phi(x - vt)$:

$$\phi(x, t) = 4\tan^{-1}\left[\exp\left(\pm\frac{x - x_0 - vt}{\sqrt{1 - v^2}}\right)\right] \tag{13.35}$$

where the plus and minus signs correspond to kink and antikink solutions, respectively, with velocity

$$v = \left[1 + \left(\frac{4\alpha}{\pi F}\right)^2\right]^{-1/2}. \tag{13.36}$$

These solutions describe corkscrew-type waves moving to the left or right. Other solutions to the sine–Gordon equation include so-called breather solutions that are smaller amplitude ripples propagating down the chain.

The above model can be extended by allowing base pairs in the second chain to rotate about the lower sugar–phosphate backbone. We now have two systems of coupled, second-order, differential equations:

$$\begin{aligned} I\frac{\partial^2\phi_n^1}{\partial t^2} &= K\left(\phi_{n+1}^1 - 2\phi_n^1 + \phi_{n-1}^1\right) - mg\ell\left[2\sin\phi_n^1 - \sin\left(\phi_n^1 + \phi_n^2\right)\right] \\ I\frac{\partial^2\phi_n^2}{\partial t^2} &= K\left(\phi_{n+1}^2 - 2\phi_n^2 + \phi_{n-1}^2\right) - mg\ell\left[2\sin\phi_n^2 - \sin\left(\phi_n^1 + \phi_n^2\right)\right] \end{aligned} \tag{13.37}$$

where ϕ_n^1 and ϕ_n^2 are the angular deviations of the upper and lower chain. The $\sin\left(\phi_n^1 + \phi_n^2\right)$ terms represent coupling between the upper and lower chains. The effects of impurities and inhomogeneities can be modeled in the discrete equations, with coefficients of the sine terms dependent on n. In the continuum limit, making the substitution (Equation 13.32) we have

$$\begin{aligned} \frac{\partial^2\phi_1}{\partial t^2} &= \frac{\partial^2\phi_1}{\partial x^2} - 2\sin\phi_1 + \sin(\phi_1 + \phi_2) \\ \frac{\partial^2\phi_2}{\partial t^2} &= \frac{\partial^2\phi_2}{\partial x^2} - 2\sin\phi_2 + \sin(\phi_1 + \phi_2), \end{aligned} \tag{13.38}$$

where damping and driving can be included as before.

EXERCISES

Exercise 13.1 Plot the nullclines corresponding to

$$\frac{\partial x}{\partial t} = 0 \quad \text{and} \quad \frac{\partial y}{\partial t} = 0$$

in the Sel'kov glycolysis model given by Equation (13.1) with kinetic parameters $(a, b) > 0$. Also plot the vector field and determine the fixed point where the two nullclines cross.

Exercise 13.2 Consider the discrete model of flour beetle (*Tribolium castaneum*) population dynamics with three distinct age classes including larvae, pupae, and adults. Adults and pupae can cannibalize larvae, resulting in complex population dynamics. The model consists of three finite difference equations where the populations of each age class at a time $n + 1$ are calculated from the previous populations at a time n:

$$L_{n+1} = bA_n \exp(-c_{ea}A_n - c_{el}L_n)$$

$$P_{n+1} = L_n(1 - \mu_l)$$

$$A_{n+1} = P_n \exp(-c_{pa}A_n) + A_n(1 - \mu_a).$$

Table 13.1

Adult Population	A_n ($A_1 = 100$)
Adult Mortality Rate (other than by cannibalism)	$\mu_A \approx 0.04 - 0.96$ (variable)
Larvae Population	L_n ($L_1 = 250$)
Larvae Mortality Rate (other than by cannibalism)	$\mu_l = 0.513$
Pupae Population	P_n ($P_1 = 5$)
Number of Larval Recruits	$b = 11.68$
Probability Pupae Are Not Eaten by Adults	$\exp(-c_{ea}A_n)$ ($c_{ea} = 0.011$)
Probability Pupae Are Not Eaten by Larvae	$\exp(-c_{el}L_n)$ ($c_{el} = 0.013$)
Survival Probability of Pupae	$\exp(-c_{pa}A_n)$ ($c_{pa} = 0.017$)

Refer to Table 13.1 to answer the following:

(a) Compare discrete time series of each population L, P, and A.

(b) Create power spectra for different values of μ_a.

(c) Construct a bifurcation diagram plotting the number of larvae versus the adult mortality rate μ_a.

Exercise 13.3 Investigate the sensitive dependence with respect to small changes in initial conditions in the flour beetle model in the chaotic regime by plotting time series of the same variable together for slightly different initial conditions; for example, $A_1 = 100$ and $A_1 = 101$. The time series should coincide initially and then diverge.

Exercise 13.4 Calculate the entropy of the larvae, pupae, and adult populations as a function of μ_a in the flour beetle model.

Exercise 13.5 Complex periodic patterns including stripe formations such as those found on the angelfish (*Pomacanthus*) can be modeled by the coupled reaction diffusion equations

$$\frac{\partial A}{\partial t} = c_1 A + c_2 I + c_3 - D_A \frac{\partial^2 A}{\partial x^2} - g_A A$$

$$\frac{\partial I}{\partial t} = c_4 A + c_5 - D_I \frac{\partial^2 I}{\partial x^2} - g_I I.$$

Investigate the spatial–temporal evolution and pattern formation by numerically solving these equations. The variables and parameter values as given by Kondo and Asai are listed in Table 13.2.

Table 13.2

Diffusion Constant for Activator Molecule	$D_A = 0.007$
Activator Decay Constant	$g_A = 0.03$
Concentration of Activator Molecules	A
Diffusion Constant for Activator Molecule	$D_I = 0.1$
Inhibitor Decay Constant	$g_I = 0.06$
Concentration of Inhibitor Molecules	I
Constraint on Synthesis Rate of Activator	$0 < c_1 A + c_2 I + c_3 < 0.18$
Constraint on Synthesis Rate of Inhibitor	$0 < c_4 A + c_5 < 0.5$

Exercise 13.6 The driven Hodgkin–Huxley equation obeys the differential equation

$$C\frac{dv}{dt} + I_{\text{ion}} = I_{\text{app}}$$

where I_{ion} is given by Equations 10.6–10.19. Plot the $v(t)$ time series with AC excitation $I_{\text{app}} = A\sin(\omega t)$ for different driving amplitudes A and angular frequencies $\omega = 2\pi f$.

Exercise 13.7 Plot the $v(t)$ time series in the pulse-driven H–H system with $I_{\text{app}} = A\,\delta(t - nT)$ where T is the pulsing period, A is the pulse amplitude, and n is an integer.

Exercise 13.8 Create a time-delay plot of $v(t)$ vs. $v(t - \tau)$ in either the pulse-driven or the ac-driven H–H system. Experiment with different values of τ. Plots with values of τ that are too small will be a straight line. Values of τ that are too large will produce plots that resemble "pick-up sticks." Such phase-space reconstructions are useful when analyzing experimental data where only one time series is recorded.

Exercise 13.9 Create a Poincaré section, or strobe plot, of the Lorenz model, plotting points in the x–y plane when z passes through zero. Strobe plots can be made in any of the coordinate planes by plotting a point where the third variable is zero (or by plotting a point periodically in driven systems). It is important to plot points without enabling line connection.

Exercise 13.10 Create a return map of the Lorenz model by plotting subsequent maxima of one of the variables, for example, z_M vs. z_{M+1}. It is important to plot subsequent maxima without enabling line connection.

Exercise 13.11 Create a plot of R vs. m in the logistic map where R is the ratio defined by Equation 13.20 and m is the number of bifurcations that have occurred increasing the parameter μ.

Exercise 13.12 Investigate pattern formations in the following system of reaction diffusion equations:

$$\frac{\partial A}{\partial t} = D_A\left(\frac{\partial^2 A}{\partial x^2} + \frac{\partial^2 A}{\partial y^2}\right) + \gamma f(A, B)$$
$$\frac{\partial B}{\partial t} = D_B\left(\frac{\partial^2 B}{\partial x^2} + \frac{\partial^2 B}{\partial y^2}\right) + \beta g(A, B)$$

where

$$D_A = 0.2, D_B = 0.1, \gamma = 0.5, \beta = 0.2$$

and

$$f(A, B) = A(1 - A)$$

$$g(A, B) = A(1 - AB).$$

Apply periodic boundary conditions to the sides of an N×N grid where, on the top, for example, with $\Delta x = \Delta y = \Delta$

$$\nabla^2 A(N, j) = \frac{1}{\Delta^2}\left(A(N-1, j) + A(1, j) + A(N, j+1) + A(N, j-1) - 4A(N, j)\right)$$

and the corner (1, 1)

$$\nabla^2 A(1,1) = \frac{1}{\Delta^2}\left(A(N,1) + A(1,N) + A(1,2) + A(2,1) - 4A(1,1)\right),$$

with similar expression for the remaining three sides and three corners.

Exercise 13.13 Diastolic and systolic phases of the cardiac cycle correspond to respective relaxed and contracted states of the heart. A model of cardiac activity by E. C. Zeeman is given by

$$\varepsilon \frac{dx}{dt} = -(x^3 + ax + b)$$

$$\frac{db}{dt} = x - x_a$$

where x correspond to muscle fiber length, x_a is the typical fiber length in the diastolic phase, b is a chemical control, a represents tension and ε is a small positive number. Find the equilibrium point in this system of equations. Can this model exhibit chaos? Why or why not?

Exercise 13.14 Viral activity in an infected host organism is modeled by the following system neglecting the immune system response:

$$\frac{dV}{dt} = aY - bV$$

$$\frac{dX}{dt} = c - dX - \beta XV$$

$$\frac{dY}{dt} = \beta XV - fY$$

where V is the number of viral entities called virions, X and Y are the numbers of uninfected target cells and infected cells, respectively. The death rates of virions, uninfected and infected cells, are respectively given by b, d, and f. Uninfected cells are produced at a rate c and become infected at a rate proportional to the number of virions βV. Find the equilibrium points of this system. Can this model exhibit chaos? Why or why not?

14 Fractals and Complexity in the Life Sciences

14.1 Fractal Geometry

Physicists often approximate complex physical objects with simple shapes that can be described mathematically. Examples are cells and organelles modeled as perfect spheres, bacteria as rods, and cell membranes approximated as smooth surfaces. Smooth Euclidian shapes clearly cannot represent the majority of biological structures.

Biological structures do not have perfect symmetry and can only be approximated by perfect geometrical forms in some instances. Pine tree cones, for example, are not perfectly conical but have surface texture. Tree trunks are cylindrical at first approximation but on closer inspection are surrounded by a bark with surface roughness, discontinuities, and fragmentation. The rough bark provides smaller organisms with a more extensive habitat than is available to larger organisms. A small insect will travel a longer distance around a tree trunk compared to the same route traveled by a squirrel.

In 1975 Benoit Mandelbrot introduced the term *fractal* to describe complex geometrical objects that have similar features over infinitely many length scales. Iterative algorithms or rules can be applied to mathematically model fractal curves, surfaces,

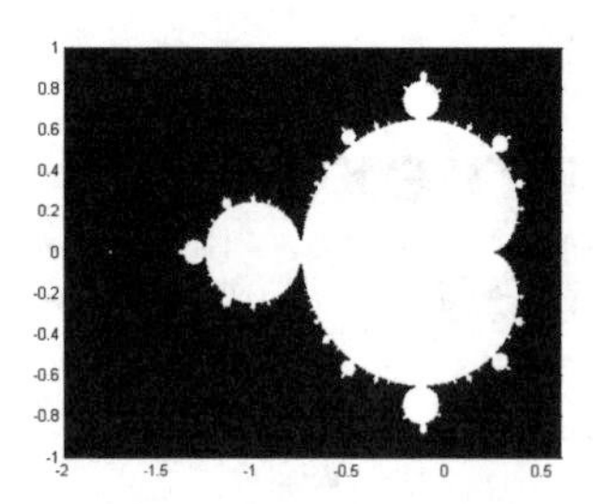

FIGURE 14.1 The Mandelbrot set.

and volumes. The first example that was investigated by iterated computation of complex numbers was

$$z_{n+1} = z_n^{\,2} + c \tag{14.1}$$

where z and c are complex numbers. Fractal curves are generated by picking initial points c in the complex plane and iterating the above formula until $|z| > |z|_{max}$, or until a specified maximum number of iterations are reached. The resulting pattern generated by repeated application of this formula shown in Figure 14.1 is known as the Mandelbrot set. For initial c values in the white region $z_n \rightarrow 0$, and for initial c values in the black region $z_n \rightarrow \infty$ upon repeated iteration of Equation 14.1. One can understand this behavior by squaring numbers on a pocket calculator. Zero is approached by repeatedly squaring a number x, where $|x| < 1$ while infinity is approached if $|x| > 1$. In the complex plane, the border between regions that either tend toward zero or infinity forms a fractal cure that exhibits self-similar structure on all scales of magnification. There is no lower limit where additional structure cannot be found. The following MATLAB code generates the Mandelbrot set in Figure 14.1:

```
x=linspace(-2,.6,1000);
y=linspace(-1,1,1000);

[X,Y]=meshgrid(x,y);
C= complex(X,Y);

Z_max=2;
it_max=50;
Z=C;

for k = 1:it_max
      Z=Z.^2+C;
end

contourf(x,y,double(abs(Z)<Z_max))
colormap bone

%adapted from Nicholas Higham “MATLAB guide”
```

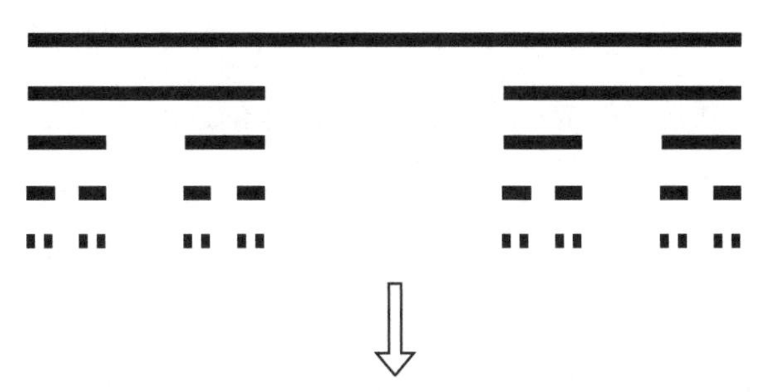

FIGURE 14.2 Cantor comb formed by successive removal of the center third of each subsequent line segment.

The simplest example of a fractal is the Cantor set or "Cantor comb." Removing successive thirds of a line segment ad infinitum as illustrated in Figure 14.2 forms the comb.

The final result is a set of infinitely many points with zero length. The dimensionality D of this set lies between the dimensionality of a single point and that of a

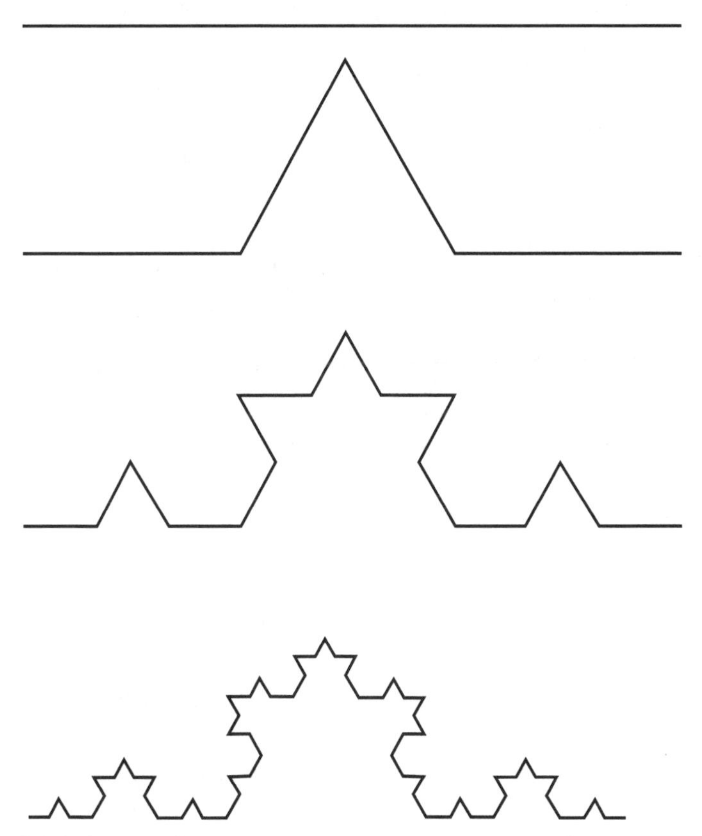

FIGURE 14.3 Successive iterations forming the Koch curve. The bottom curve is obtained after three iterations of the top line segment.

line where $0 < D < 1$. Noninteger dimensionality is characteristic of fractals. For the Cantor comb, it turns out that the fractal dimension is

$$D = \frac{\ln 2}{\ln 3}. \tag{14.2}$$

The Koch curve is formed by iteratively replacing the center third of each line segment by two equal line segments as shown in Figure 14.3.

The Koch curve is a continuous curve with dimensionality between that of a line and a plane or $1 < D < 2$ where

$$D = 2\frac{\ln 2}{\ln 3}. \tag{14.3}$$

The Koch snowflake shown in Figure 14.4 has the same dimension as the Koch curve. The Cantor comb, Koch curve, and the Koch snowflake each exhibit self-similar structure under increasing magnification.

The subject of fractals is usually associated with the study of chaos theory. It turns out that physical systems with chaotic temporal behavior are generally not accompanied by the formation of fractal structures in real space. However, chaotic systems can be described by trajectories in phase space whose cross sections are fractal curves without integer dimension. These curves are called strange attractors in the mathematical phase space. The connection between fractals and low-dimensional chaos is also evident in bifurcation diagrams and iterated maps used to model population dynamics as discussed in the previous chapter.

Fractal-like structures are often found in nature and in living organisms. Biological structures cannot form perfect mathematical fractals, however, because there will be a lower limit to the extent of scale invariance when viewed with increasing magnification. This is because a physical object cannot exhibit self-similar features on length scales smaller than the diameter of an atom.

14.1.1 Computing Fractal Dimension

Although there is no single length that describes a fractal, it is possible to characterize a fractal by its dimension D. The fractal dimension can be thought of as a measure of geometric complexity. Even though physical objects are not true mathematical fractals, we may determine the fractal dimension over some range of length scales.

If we set out to measure the length d of a one-dimensional curve by placing rulers end to end, then we will require a number of rulers N that is inversely proportional to the length of the rulers L since $d = N \cdot L$. If we halve the length of the rulers $L \rightarrow L/2$, then

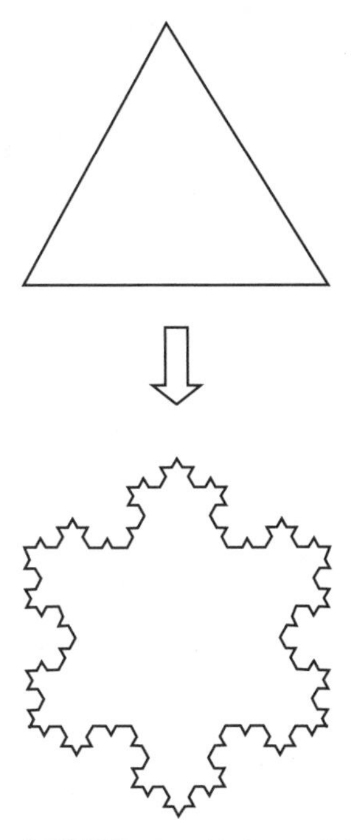

FIGURE 14.4 Koch snowflake formed after three iterations of the top triangle.

we must double their number $N \rightarrow 2N$ to cover the same d. A plot of N versus $1/L$ will be a straight line with slope equal to one, the same as the dimensionality of the curve.

Instead of rulers, imagine covering the curve with boxes. We will then count the number of boxes required to span the curve as a function of box size. The capacity dimension is calculated by covering the curve with d-dimensional boxes where the number of boxes of length ε is given by

$$N(\varepsilon) = L^d (1/\varepsilon)^d. \tag{14.4}$$

For a square of side L, the number of boxes of length $\varepsilon = L/2$ is given by $L^2/\varepsilon^2 = 4$ so that $d = 2$. A similar procedure gives $d = 1$ for a line and $d = 3$ for a cube. A formula for the capacity dimension is obtained by taking the logarithm of both sides of Equation 14.4:

$$\ln N(\varepsilon) = d \ln L + d \ln(1/\varepsilon). \tag{14.5}$$

We thus obtain

$$d = \frac{\ln N(\varepsilon)}{\ln L + \ln(1/\varepsilon)}. \tag{14.6}$$

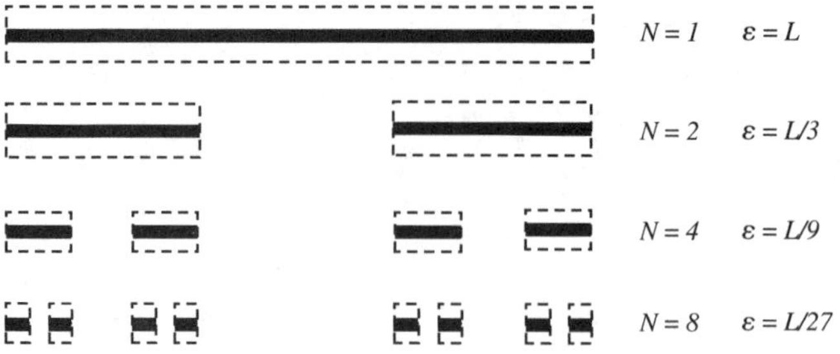

$N = 1$ $\varepsilon = L$

$N = 2$ $\varepsilon = L/3$

$N = 4$ $\varepsilon = L/9$

$N = 8$ $\varepsilon = L/27$

$$d_c = \lim_{\varepsilon \to 0} \frac{\ln N}{\ln(1/\varepsilon)} = \lim_{n \to \infty} \frac{\log 2^n}{\log L + \log 3^n} = \frac{\log 2}{\log 3}$$

FIGURE 14.5 Application of the box-counting method to calculate the fractal dimension of the Koch curve.

For small ε, $\ln L \ll \ln(1/\varepsilon)$, and we define the capacity dimension d_c in the limit $\varepsilon \to 0$ as

$$d_c = \frac{\ln N(\varepsilon)}{\ln(1/\varepsilon)}. \tag{14.7}$$

Figure 14.5 illustrates the application of the box-counting method to measure the fractal dimension of the Koch curve.

The box-counting method may be used to calculate the fractal dimension of extended objects such as the coast of Norway in Figure 14.6. The Norwegian coast exhibits scale invariance with numerous inlets called fjords with a hierarchy of fjords within fjords. The dimensionality of the coastline is about 1.52.

FIGURE 14.6 Fractal coast of Norway surrounded by phytoplankton blooms. (Satellite image by GeoEye, Copyright © 2009 GeoEye.)

Note that the capacity dimension does not take into account the distribution of points on the fractal. The information dimension d_I can be calculated by determining the probability p_i that a point on the fractal lies in a box of size ε. This method uses the Shannon entropy formula:

$$S(\varepsilon) = -\sum_{i=1}^{N} p_i \ln p_i . \tag{14.8}$$

The information dimension d_I becomes in the limit $\varepsilon \to 0$

$$d_I = \frac{\sum_{i=1}^{N} p_i(\varepsilon) \ln p_i(\varepsilon)}{\ln \varepsilon} . \tag{14.9}$$

14.2 Fractal Structures in Biology

Fractal structures provide an advantageous architecture for living organisms seeking to cover a large area while conserving the amount of building material. Examples are branching networks in circulatory systems, lungs, nerves, and plant structures. The fractal dimension of alveoli surfaces in the lung are ~2.97. Branching networks such as those found in biology can be modeled using simple algorithms and transformation rules as illustrated below.

14.2.1 Fractal Ferns

Barnsley's fern is an example of a mathematical fractal that resembles natural plant structures. The fern is generated by the iterative transformation

$$\begin{pmatrix} x_{n+1} \\ y_{n+1} \end{pmatrix} = \begin{pmatrix} a_j & b_j \\ c_j & d_j \end{pmatrix} \begin{pmatrix} x_n \\ y_n \end{pmatrix} + \begin{pmatrix} e_j \\ f_j \end{pmatrix} \tag{14.10}$$

where the coefficients

$$\begin{pmatrix} a_j & b_j \\ c_j & d_j \end{pmatrix} \text{ and } \begin{pmatrix} e_j \\ f_j \end{pmatrix} \tag{14.11}$$

are chosen with probability p_j, resulting in a rotation and translation of the vector (x_n, y_n) to (x_{n+1}, y_{n+1}). This affine transformation can be written in compact vector notation:

$$\mathbf{X}_{n+1} = A_j \mathbf{X}_n + B_j \tag{14.12}$$

where A_j are rotation matrices and B_j are translation vectors. The following MATLAB code simulates the fractal fern shown in Figure 14.7:

```
% Initialize Fern array

Jmax = 50000;

Fern = zeros(Jmax,2);

% Specify Parameters

A1=[.849 0.037;-0.037 0.849];
A2=[0.197 -.226; 0.226 0.197];
A3=[-0.150 0.283; 0.26 0.237];
A4=[0 0 ;0 0.16];

B1=[0.075; 0.18];
B2=[0.4; 0.049];
B3=[0.575; -0.084];
B4=[0.5;0];

% Iterate the matrix transformation X' = A*X+B

Imax= 40;

xn=[0; 0];

for j = 1:Jmax

x=[rand(1); rand(1)];

for i = 1:Imax

p = rand(1);

if p < .85

        xn = A1*x+B1;

elseif p < .92

        xn = A2*x+B2;

elseif p < .99

        xn = A3*x+B3;

else
        xn = A4*x+B4;

end

x=xn;

end

% build the fern array

Fern(j,1) = x(1);
Fern(j,2) = x(2);

end

plot(Fern(:,1),Fern(:,2),'.k','MarkerSize',1)
axis ([.3 .8 0 1.2])
axis equal
```

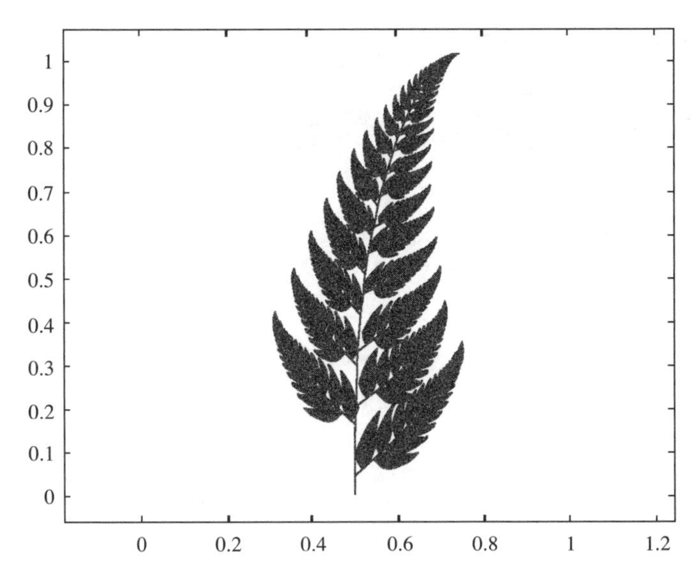

FIGURE 14.7 Fractal fern generated by iterations of the transformation Equation 14.10.

FIGURE 14.8 Branching network generated by the diffusion-limited aggregation (DLA) simulation. Similar networks are observed in bacteria colonies grown on agar plates.

14.2.2 Diffusion-Limited Aggregation

Diffusion-limited aggregate (DLA) structures results from the clumping of particles undergoing Brownian motion. Fractal branching networks are formed during the growth of crystals, coral formations, and colonies of bacteria.

DLA simulations are performed by random walks that are terminated when a boundary is reached. An initial seed particle is introduced, and then a second particle begins executing a random walk until it sticks to the seed particle. The aggregate begins to grow after more random walkers are terminated on the structure. The fractal dimensions of planar and volumetric DLA simulations are about 1.7 and 2.5, respectively. Comparable fractal dimensions are measured in planar bacteria colonies grown on agar. Figure 14.8 shows a DLA simulation performed on a 2-D lattice.

14.3 Power Laws in Biology

Many power laws in biology follow a simple form:

$$Y = Y_0 M^b \tag{14.13}$$

Table 14.1 Several Power Laws in Biology

Observable $Y \approx M^b$	Power Law Exponent b
Genome Length	$\frac{1}{4}$
Concentration of RNA	$-\frac{1}{4}$
Total Mitochondrial Mass Relative to Body Mass	$-\frac{1}{4}$
Basal Metabolic Rate	$\frac{3}{4}$
Heart Rate	$-\frac{1}{4}$
Life Span	$\frac{1}{4}$
Radii of Aorta and Tree Trunks	$\frac{3}{8}$

where b is a multiple of 1/4. For example, basal metabolic rate scales as $M^{3/4}$ over about four orders of magnitude of organism mass M. Other scaling laws related to mass include genome length, RNA concentration, the ratio of total mitochondrial mass to body mass, heart rate, life span, and radii of aorta and tree trunks. Power law exponents for these quantities are given in Table 14.1. This scaling could be related to the geometry and metabolism of living systems. For example, the total number of heartbeats given by the product of life span $\sim M^{1/4}$ times the heart rate $\sim M^{-1/4}$ is approximately a constant for mammals.

Evidently the biochemical reaction rate R in living organisms scales as $M^{3/4}$ divided by the total number of cells times a Boltzmann factor. Since the total number of cells is proportional to M we have

$$R \sim M^{-1/4} e^{-E_a/k_B T} \tag{14.14}$$

where $E_a \approx 0.65$ eV is the average activation energy for amphibians and several aquatic animals.

Zipf's Law

A power law is also found in the population of cities (colonies of humans). There are more small cities than large cities. Remarkably, the number of cities N of size s scales as $N(s) \sim s^{-1}$. This scaling was originally discovered by Zipf in 1949 and is verified over different geographical areas today. Zipf's law is an empirical finding and says nothing about why city populations are distributed according to a power law.

Power Laws in Brain Signals

Georgelin et al. (1999) performed an experimental study revealing power laws in α-wave electroencephalographic signals. The duration τ of α bursts is found to scale as $P(\tau) \propto \tau^{-\omega}$ with $\omega \approx 1.75$ over nearly two decades. The dynamic is similar to the behavior of a class of multitrait, self-organized, critical models with time-scale distributions $P(\tau) \approx \tau^{-7/4}$.

14.4 Self-Organized Criticality

Self-organized criticality (SOC) refers to a type of dynamical behavior exhibited by systems with many interacting degrees of freedom where minor fluctuations lead to system rearrangements without characteristic size or time scale. These rearrangements are referred to as avalanches with power laws describing their size, lifetime, and power spectrum. SOC-like behavior is often associated with the formation of fractals in nature but is distinctly different from low-dimensional chaos studied in the previous chapter. SOC systems are not described by equilibrium statistical mechanics or by continuous differential equations. Power laws are a hallmark of SOC systems with distributions that are straight lines when plotted on a log–log scale over some range, as illustrated in Figure 14.9.

The first algorithmic SOC model was developed in 1987 by Bak, Tang, and Wiesenfeld (BTW), originally proposed as a possible explanation of noise spectra with $1/f^{\alpha}$ frequency dependence. In the BTW model, sand grains are numerically added to a rectangular lattice building up the slope of the pile. When the sandpile is critically poised, the addition of a single sand grain can result in a domino effect where anywhere from a few grains to a large fraction of the sandpile may be shifted.

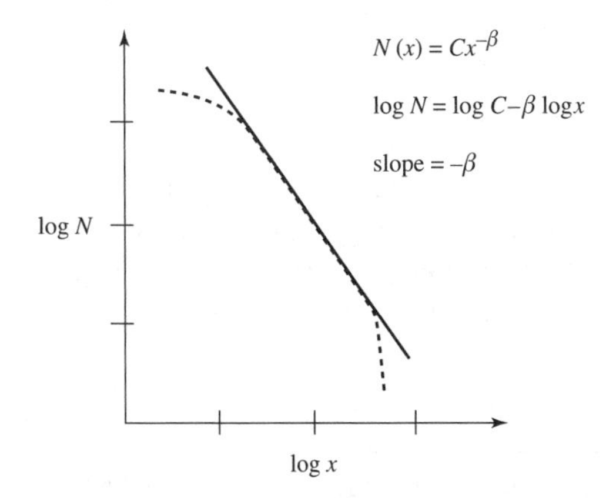

FIGURE 14.9 Graphical determination of the power law exponent from a distribution $N(x)$. Power laws in nature are finite ranged as indicated by deviations from the straight line fit at small and large values of x.

Experimentally, rice piles more precisely produce power law behavior in avalanche size and time duration compared with actual sandpiles.

SOC systems evolve into a critical state exhibiting avalanches of all sizes without the requirement of finely tuned parameters such as the interaction strength between particles. A system requiring a delicate adjustment of parameters to manifest power laws is not considered to be a robust SOC system. Furthermore, SOC systems will naturally evolve into a critical state with sufficient energy input. This behavior is more descriptively termed as *slowly driven interaction-dominated threshold dynamics* by Jensen (2004).

Characteristic of SOC systems are avalanches, or system rearrangements of size s, that obey power laws over some range of s

$$P(s) \sim s^{-\tau}. \tag{14.15}$$

For the probability distribution to be normalized

$$\int_0^\infty P(s)\,ds = 1 \tag{14.16}$$

we can approximate

$$P(s) = \frac{1}{\zeta(\tau)} s^{-\tau} \tag{14.17}$$

where $\zeta(\tau)$ is the Riemann zeta function

$$\zeta(\tau) = \sum_{s=1}^{\infty} s^{-\tau}. \tag{14.18}$$

In a sense, SOC models represent an idealization of natural processes that are often associated with the formation of spatial fractals. There will always be a finite range of power laws depending on the system size and energy input. Likewise, there will be a range of size scales where fractal-like structures have self-similarity, unlike true mathematical fractals, such as the Mandelbrot set, where additional complexity can be found zooming in without limit.

14.4.1 BTW Sandpile Model

The BTW model consists of a two-dimensional grid with an integer number $z_{i,j}$ specified at each grid location (i, j). The idealized sand is stacked randomly at each site until the slope at a given site exceeds a threshold value $z_{i,j} > z_{\text{threshold}}$ where

1	3	2	1	2
3	2	3	3	1
2	2	4	1	2
1	1	3	2	1
3	2	2	2	1

1	3	2	1	2
3	2	4	3	1
2	3	0	2	2
1	1	4	2	1
3	2	2	2	1

1	3	3	1	2
3	3	0	4	1
2	3	2	2	2
1	2	0	3	1
3	2	3	2	1

1	3	3	2	2
3	3	1	0	2
2	3	2	3	2
1	2	0	3	1
3	2	3	2	1

FIGURE 14.10 Avalanche propagating in the BTW sandpile model beginning from left to right. The avalanche size is equal to the number of sites toppled (here $N = 4$). The avalanche propagates for three time steps.

$z_{\text{threshold}} = 3$ so that $z_{i,j} = 4$. Sand is then redistributed to the nearest neighbors by first decreasing the slope of the site (i, j) by four

$$z_{i,j} \rightarrow z_{i,j} - 4. \tag{14.19}$$

The slope of the four nearest neighbors is then increased by one

$$\begin{aligned} z_{i\pm1,j} &\rightarrow z_{i\pm1,j} + 1 \\ z_{i,j\pm1} &\rightarrow z_{i,j\pm1} + 1. \end{aligned} \tag{14.20}$$

This rearrangement may cause one or more neighboring sites to exceed the threshold, resulting in further rearrangements until all the sites are less than $z_{\text{threshold}}$. The size of the avalanche is given by the total number N of sites toppled, while the avalanche duration is given by the total number of time steps the avalanche propagates. The resulting power law in avalanche size distribution is $N(s) \propto s^{-\tau}$ with $\tau \approx 1.1$. Power laws are also found in the lifetimes and power spectrum of avalanches. Figure 14.10 illustrates the propagation of an avalanche with a lifetime of three time steps and with a size $N = 4$.

When a site is toppled next to a boundary of the model, sand will fall off the edge as indicated by Figure 14.11.

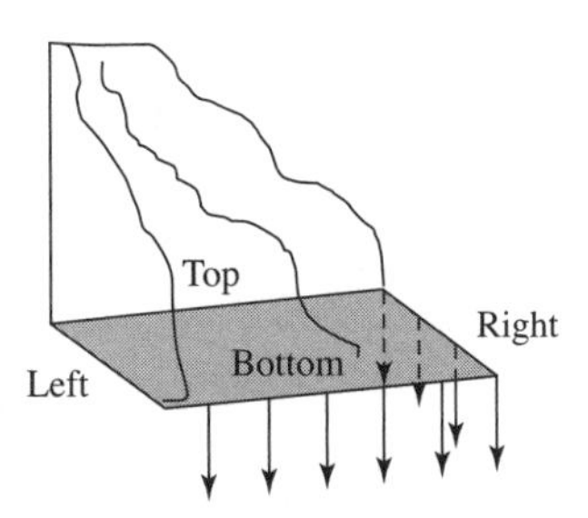

FIGURE 14.11 Boundary conditions of the BTW sandpile model. The left and top edges are constrained to be zero corresponding to a zero slope of the sandpile at its apex. If a site exceeds threshold on the bottom or the right boundary, then four is subtracted from every site on these two boundaries. This simulates sand falling off the edge as indicated by the arrows in the figure.

14.5 Extinction in the Bak–Sneppen Model

In the previous chapter we investigated predatory–prey-type interactions involving two or three species. Here we consider interactions of many species with varying fitness represented by a fitness landscape as shown in Figure 14.12.

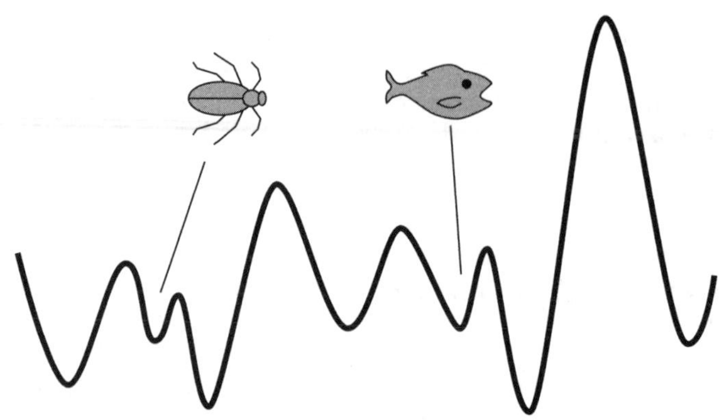

FIGURE 14.12 Schematic of an evolutionary fitness landscape. The fitness of each species is determined by its height on the curve.

The Bak–Sneppen extinction and evolution model proposed in 1993 includes nearest-neighbor interactions between species with fitnesses represented by random numbers on a one-dimensional lattice. A random number between 0 and 1 is initially assigned to each species representing its fitness, with 0 being the least fit. The species with the lowest fitness on the grid is then reassigned a fitness along with its two nearest neighbors. This operation assumes that the least fit species is most likely to evolve. Replacing the two neighboring sites with new random numbers models a coevolutionary interaction between species. For N interacting species, periodic boundary conditions are applied where the nearest neighbors of species 1 are species 2 and species N as shown in Figure 14.13.

The dynamics of the model shows comparatively rare large-scale rearrangements of the fitness landscape with smaller extinction avalanches occurring more frequently. This behavior is consistent with fossil records indicating that species survive for long

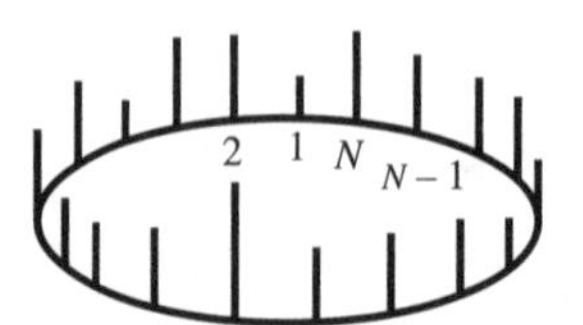

FIGURE 14.13 Fitness of N species represented by a random number between 0 and 1 in the Bak–Sneppen model of evolution. Periodic boundary conditions are applied where the nearest neighbors of species N are 1 and $N-1$.

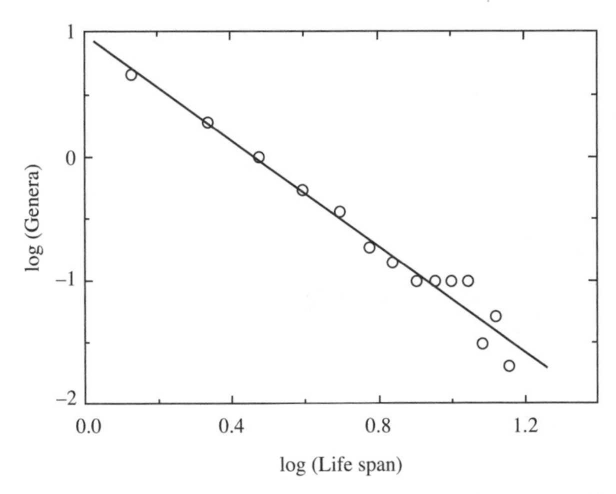

FIGURE 14.14 Log–log plot of the number of fossil genera vs. life span. The power law exponent is approximately equal to 2. (Data from D.M. Raup. 1991. A kill curve for Phanerozoic marine species. *Paleobiology* 17: 37–48.)

periods of time and then vanish on much shorter time scales. Also, the extinction of one species is often accompanied by the simultaneous extinction of other species. Eldredge and Gould proposed a punctuated equilibrium model in 1977 where there are long periods of tranquility interrupted by bursts of activity with many species becoming extinct. Evidence of punctuated equilibrium may be found in the fossil record. For example, Figure 14.14 plots the number of fossil genera vs. life span with power law ~2. This exponent is also reproduced by the Bak–Sneppen model.

14.6 Power Law Behavior in Chemical Reactions

Electrochemical reactions are probably the most experimentally assessable systems exhibiting power law behavior in terms of ease of setup and data acquisition. Parameters such as sample size and driving are easily varied. System size may be adjusted by changing the dimensions of the metal surface exposed to electrolyte, while driving is controlled by changing the concentration of electrolytic solutions. Power laws were observed during the reaction of metals in chloride solutions such as

$$\mathrm{Mg}(s)+\mathrm{CuCl_2}(aq)\rightarrow \mathrm{MgCl_2}(aq)+\mathrm{Cu}(s) \tag{14.21}$$

and

$$2\mathrm{Al}(s)+3\mathrm{CuCl_2}(aq)\rightarrow 2\mathrm{AlCl_3}(aq)+3\mathrm{Cu}(s). \tag{14.22}$$

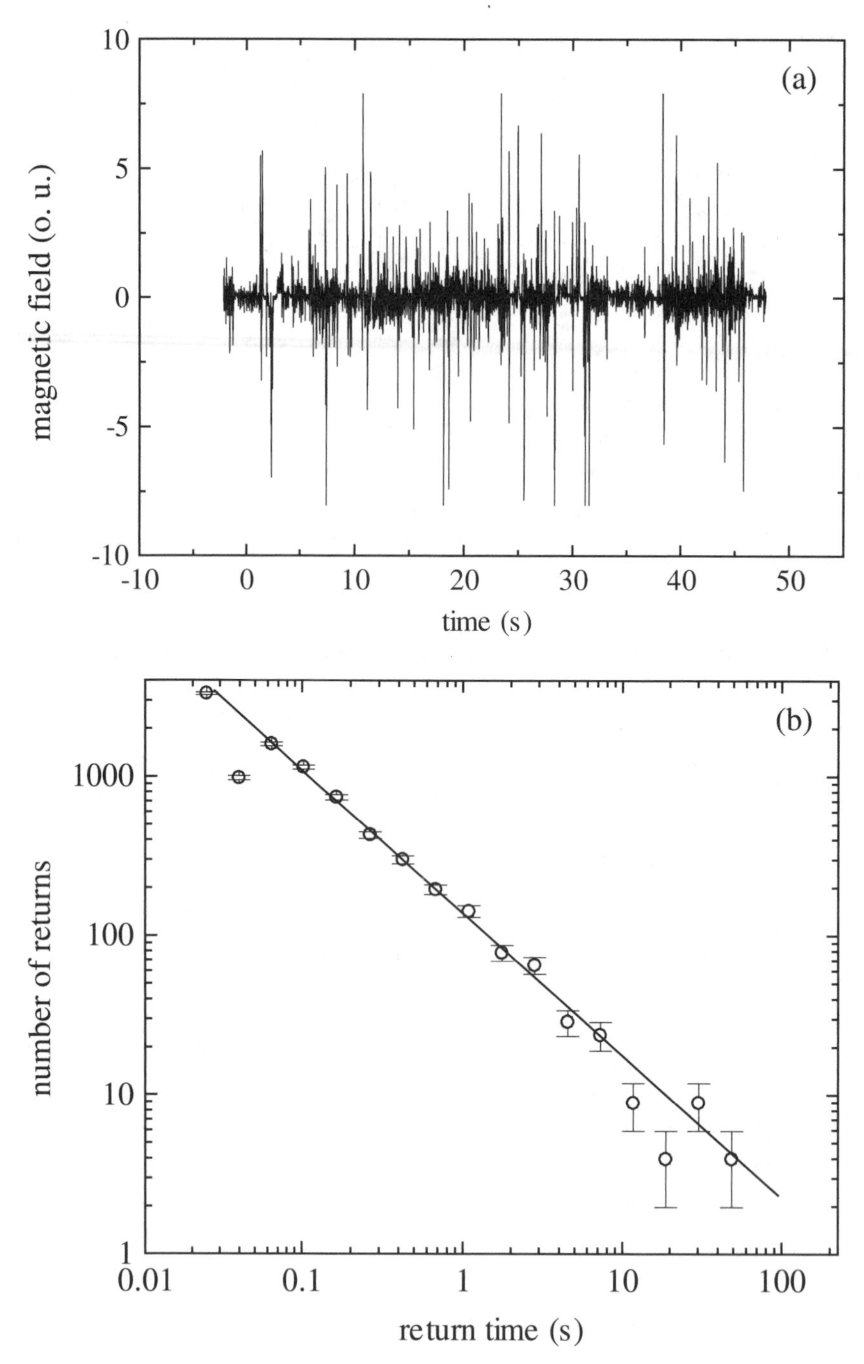

FIGURE 14.15 (a) Magnetic field recorded during the reaction of Mg with $CuCl_2$. (b) Distribution of return times between chemical avalanches of similar magnitude during the chemical reaction.

Using a SQUID magnetometer, $f^{-\alpha}$ noise was measured between two and three frequency decades with $\alpha \approx 1.8$ for the Mg reaction and $\alpha \approx 2.0$ for the Al reaction. Figure 14.15(a) shows the magnetic field generated by the Mg reaction. The distribution of peak magnetic field values scales such as $D(B_p) \propto |B_p|^{-\tau}$ with $\tau \approx 2.5$ for the Mg reaction and $\tau \approx 3.1$ for the Al reaction over about two decades. The distribution of times between chemical avalanches of similar magnitude $D(\Delta t) \propto \Delta t^{-a}$ is found with $a \approx 0.9$ for both Al and Mg reactions as shown in Figure 14.15(b). Similar power laws were measured using a two-pin electrode setup.

Power laws in electrochemical reactions are believed to originate from the dynamics of pitting and the passivation layer that can break off and re-form during the reaction. The passivation layer consists of fractal dendrite formations precipitated out of the electrolytic solution.

In these reactions, there is some transient time before power law behavior is observed after the electrolyte is introduced into the reaction cell. Similarly, in the BTW sandpile model, there is also a transient time before the slope of the pile reaches a critical value where avalanches are observed over all size scales.

Power laws have also been observed during the corrosion of stainless steel and during pitting of Fe–Cr alloys in chloride solutions. Note that it is possible to have $f^{-\alpha}$ noise in reactions where other power laws are not present. Not all systems that exhibit $f^{-\alpha}$ noise exhibit other SOC characteristics. Biochemical reactions such as glycolysis are more likely to exhibit nonlinear oscillations characteristic of low-dimensional systems rather than SOC, however.

14.7 The Game of Life

Cellular automata are numerical models that can produce complex patterns and temporal behavior observed in nature. In 1970 John Conway developed a cellular automaton model that mimics the birth, death, and survival of creatures represented as simple patterns on a rectangular lattice. Each nonboundary site on the 2-D grid is surrounded by eight nearest neighbors (including four sides and four corners). An occupation number of 1 or 0 is assigned to each grid point corresponding to occupied (alive = 1) or empty (dead = 0) states. The rules for iteratively updating a given site depend on the values of its nearest neighbors. The number of possible procedures for updating a site with eight nearest neighbors is $2^{512} \approx 10^{355}$. Conway's Game of Life represents one possible rule set out of these possibilities.

At each time step, all positions on the grid are updated according to the following rules:

1. An empty site surrounded by three occupied sites will become occupied, or give birth.

2. Any occupied site surrounded by fewer than two occupied sites will become unoccupied, or die of isolation.
3. Any occupied site with more than three neighbors will die of suffocation or overcrowding and become unoccupied.
4. Any occupied site with two or three neighbors will survive.

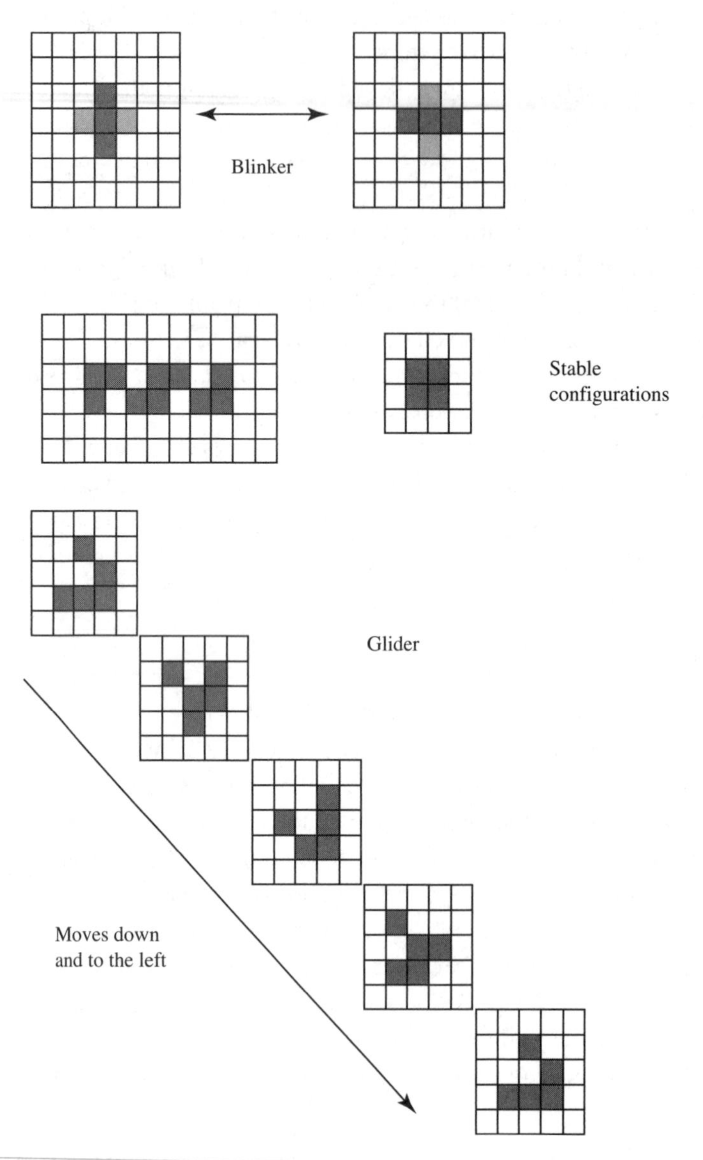

FIGURE 14.16 Patterns generated in the Game of Life simulation. In the top blinker pattern, dark gray squares are currently alive while light gray squares were alive during the previous time step.

From these simple rules, unpredictable and complex patterns arise. Creature-like stable, pulsating, and moving patterns arise including the following:

1. Crawling gliders that move diagonally along the grid by cyclic conformational changes.
2. Blinkers that alternate between three vertical and three horizontal cells.
3. Glider gun patterns that give birth to gliders.
4. Patterns that produce continuous trails of offspring while crawling.

Examples of blinking, static, and moving patterns are shown in Figure 14.16. The Game of Life cellular automaton generates patterns that are uniquely determined at later times from some initial pattern. Although the model is deterministic, it is not possible to predict future patterns because of the enormous number of possible configurations. For example, even for a 10×10 grid, there are $2^{100} \approx 10^{30}$ possible patterns, so that prediction becomes impossible after only a few time steps.

Consider a portion of the 2-D lattice shown below:

$$\left|\begin{array}{ccc} (i-1,j+1) & (i,j+1) & (i+1,j+1) \\ (i-1,j) & \boxed{(i,j)} & (i+1,j) \\ (i-1,j-1) & (i,j-1) & (i+1,j-1) \end{array}\right|. \tag{14.23}$$

The Game of Life rules may be implemented by calculating the sum Σ of occupation numbers A neighboring the site (i, j)

$$\Sigma = A_{i-1,j+1} + A_{i,j+1} + A_{i+1,j+1} + A_{i-1,j} + A_{i+1,j} + A_{i-1,j-1} + A_{i,j-1} + A_{i+1,j-1}. \tag{14.24}$$

A given site (i, j) survives or becomes occupied if Σ is equal to two or three. The site dies or becomes unoccupied if Σ is less than two or if Σ is greater than three. The following MATLAB program performs the Game of Life simulation:

```
N=75;       % grid size

% create an array of 1's and 0's randomly distributed

R =randn(N);
A =zeros(N);
Anew = zeros(N);
A(find(R>1)) =1;

for k=1:200
for i=2:N-1
for j=2:N-1

sum =A(i-1,j+1)+A(i,j+1)+A(i+1,j+1)+A(i+1,j)+...
        A(i-1,j)+ A(i-1,j-1)+A(i,j-1)+A(i+1,j-1);
```

```
            if sum < 2 && A(i,j)==1
                Anew(i,j)=0;
            end

            if sum >3 && A(i,j)==1
                Anew(i,j)=0;
            end

            if sum ==3 && A(i,j)==0
                Anew(i,j)=1;
            end
    end
    end

            imagesc(Anew);
            pause(.01);
            A = Anew;

    end
```

This program may be modified to include periodic boundary conditions as suggested in Exercise 14.7. Also, a Game of Life simulation is included with MATLAB that may be executed by typing "life" at the command prompt:

```
>> life
```

Intermediate and final results of this simulation are shown in Figures 14.17(a) and 14.17(b), respectively.

14.7.1 SOC in the Game of Life

Self-organized criticality (SOC) may be demonstrated in Conway's Game of Life. Given an initial configuration, the patterns will evolve and eventually reach a final stable configuration. Once the final configuration is reached, a single site is chosen at random and its occupation number is flipped. Redistributions of the grid then ensue with an avalanche size equal to the number of sites that are affected. The avalanche duration is equal to the number of time steps required for the next stable configuration to be reached. Power law distributions in size $D(s) \propto s^{-\tau}$ with $\tau \approx 1.3$ and time durations $D(t) \propto t^{-a}$ with $a \approx 1.4$ are obtained for the Game of Life for lattice sizes less than 1024×1024. Note that out of the 2^{512} possible updating algorithms, the Game of Life is the only automaton that has been found to exhibit SOC so far.

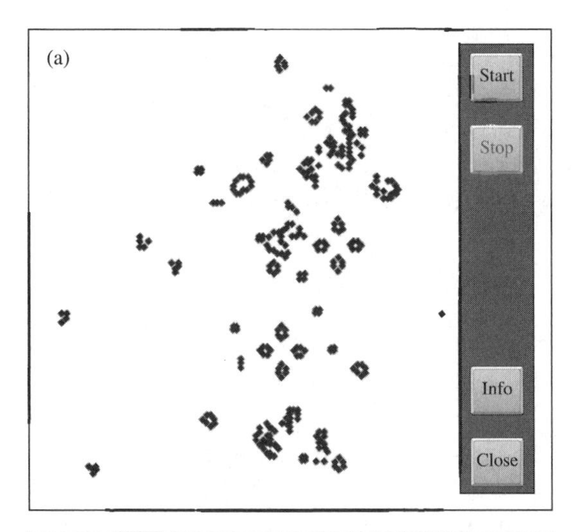

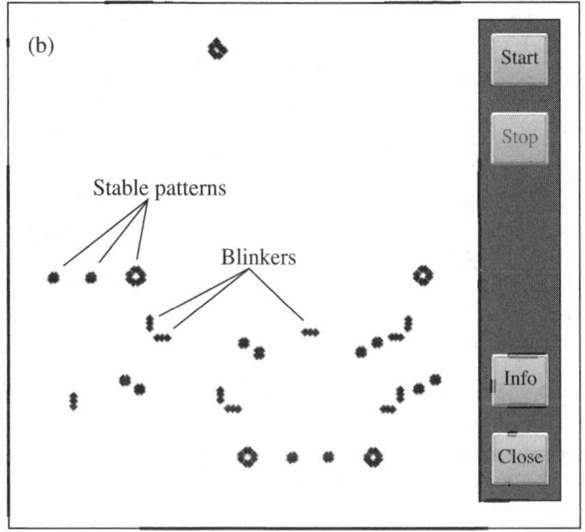

FIGURE 14.17 Game of Life simulations showing (a) a transient configuration and (b) a final configuration consisting of stable patterns and blinkers.

EXERCISES

Exercise 14.1 The Weierstrass function

$$f(x) = \sum_{n=1}^{\infty} \frac{\sin(\pi n^a x)}{\pi n^a}$$

describes a fractal curve that is only differentiable on the set of rational numbers $x = (2A+1)/(2B+1)$ where A and B are integers. Write a program to investigate the self-similarity of this function at different scales of magnification. Compare $f(x)$ over the unit interval with $a = 2$ and $a = 3$.

Exercise 14.2 Write programs to simulate diffusion-limited aggregation (DLA) in two and three dimensions. Also perform a DLA simulation where particles adhere to the boundary of the model instead of seed particles.

Exercise 14.3 Write a program to calculate the avalanche size distribution in the Bak, Tang, and Wiesenfeld (BTW) sandpile model. Also calculate the return time distribution and the power spectrum.

Exercise 14.4 Write a program to calculate the extinction size distribution in the Bak–Sneppen model. Also calculate the return time distribution and the power spectrum.

Exercise 14.5 Modify the Bak–Sneppen model to include meteoric impacts. Simulate meteoric impacts by "zeroing out" random fractions of the grid at random time intervals. Investigate the influence on the extinction event power laws and their ranges for different intensities of bombardment.

Exercise 14.6 Modify the Bak–Sneppen model to include climate change. Simulate climate change by "tilting the ring" or by continuously drifting the threshold around the ring. Investigate the influence on the extinction event power laws and their ranges for different tilts and frequencies of threshold variation.

Exercise 14.7 Modify the Game of Life simulation to include periodic boundary conditions. Compare final configurations of the model with and without periodic boundary conditions applied.

Exercise 14.8 Perform a 3-D Game of Life simulation on a cubic lattice where an occupied site survives if it has three, four, or five neighbors. An unoccupied site gives birth if it has exactly five neighbors. An occupied site with fewer than three or greater than five neighbors becomes unoccupied.

15 Life and the Universe

15.1 Astrobiology

Astrobiology is a multidisciplinary field that studies terrestrial biology while trying to answer larger questions concerning the possible distribution of life in the universe. Another focus of astrobiology is the study of how physical processes in the universe affect life on Earth. The field brings together the disciplines of astronomy, astrophysics, Earth sciences, and microbiology while providing a motivation for further space exploration. A key area of astrobiological research investigates terrestrial organisms that survive at the very limits of habitable temperature, pressure, and pH ranges to better understand conditions that might be supportive for life elsewhere in the solar system and in the universe.

15.2 Extremophiles

Extremophilic organisms on Earth are highly tenacious, existing in a wide range of seemingly inhospitable physical conditions. Microorganisms have been found living in rocks and ice core samples. Hypoliths can live inside of rocks while psychrophiles can survive at low temperatures in ice. Thermoacidophiles can survive both at high temperatures and in very acidic surroundings. Piezophiles can survive at high pressures in oceanic trenches up to 7 miles deep. Methanogens are methane-producing organisms that use hydrogen to anaerobically reduce carbon dioxide. The presence of methane in the Martian atmosphere suggests the possible presence of methanogens on Mars.

In 1977 the deep sea submersible Alvin discovered an ecosystem driven solely by energy from hydrothermal vents in regions completely devoid of sunlight as shown in Figure 15.1. This remote ecosystem supports giant tubeworms and other life forms near black-smoker vent features at a depth of over 2 km and at pressures well over 200 atmospheres. At the base of the black-smoker food chain is a bacterium that obtains its nutrients from hydrogen or hydrogen sulfide that wells up from the interior of the planet. These hyperthermophiles live at temperatures up to 120°C near the hydrothermal vents.

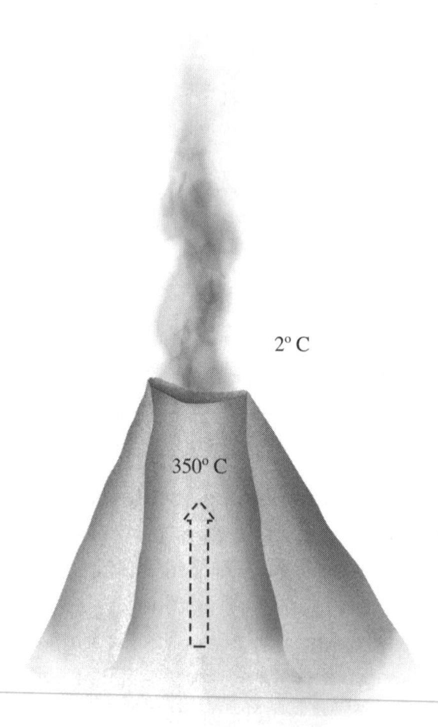

FIGURE 15.1 Deep sea hydrothermal vent 2.6 km below the surface of the ocean.

15.3 Primordial Soup, Interstellar Gas, and Dust

15.3.1 The Miller Experiment

Organic mixtures in Earth's early oceans are referred to as primordial soup. Working at the University of Chicago in 1952, Stanley Miller and Harold Urey developed an experiment that simulated the chemical evolution of complex molecules in Earth's early atmosphere and oceans. Their apparatus illustrated in Figure 15.2 consists of a closed-loop glass container initially filled with water, hydrogen, methane, and ammonia.

The lower flask contained liquid water representing primeval oceans. The ocean flask was heated to 100°C. The resulting vapors from the boiling mixture collected in the upper flask representing Earth's atmosphere. Electrical discharges supplied through a spark gap in the upper sphere simulated lightning in the atmosphere. The vapor in the upper sphere, rich in complex molecules, was passed through a condenser and returned to the ocean sphere.

A continuous exchange between the two glass spheres simulated the atmosphere–ocean cycle. Complex carbon compounds such as amino acids including glycine and alanine were soon formed from the NH_3, CH_4, and H_2 precursors.

The precursor compounds of the Miller experiment differed somewhat from compounds that made up Earth's early atmosphere, which consisted mostly of carbon dioxide, nitrogen, and water vapor. Variations of Miller's experiment with initial chemical compositions more representative of the young Earth's atmosphere and

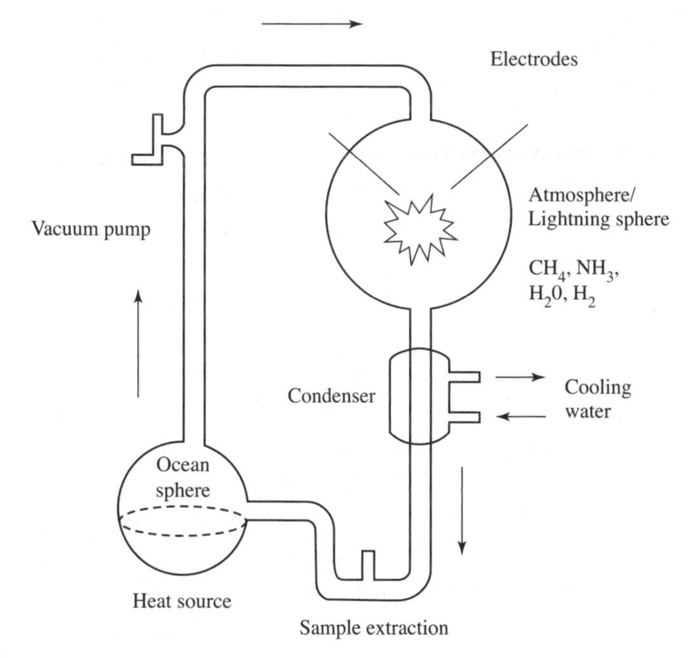

FIGURE 15.2 Stanley Miller's experimental apparatus for conducting primordial soup experiments.

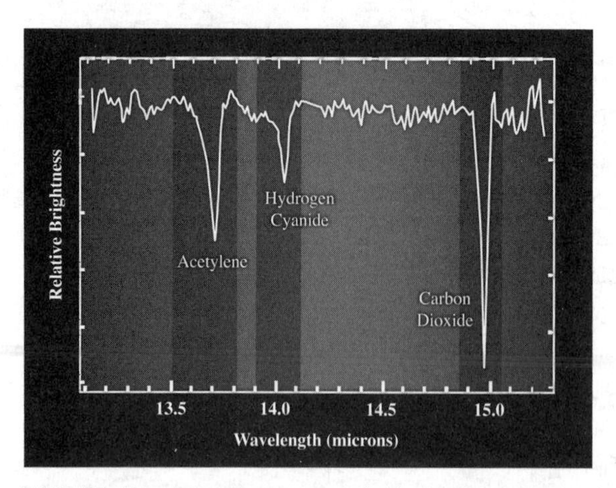

FIGURE 15.3 Infrared spectra recorded by the Spitzer space telescope revealing the presence of the organic compounds hydrogen cyanide (HCN), CO_2, and acetylene in the protostellar dust disk IRS 46. (Courtesy of Planetary Photo Journal Collection/Courtesy of nasaimages.org.)

ocean demonstrated that the same complex molecules could form in the terrestrial ocean atmosphere cycle. However, fewer complex molecules were formed using these precursors compared to Miller's original experiment.

So far, primordial soup experiments have not been able to produce proteins from chains of amino acids or simple sugars. Sidney Fox demonstrated in about 1960 that proteins, once formed, can self-assemble into micelle microspheres. Other sources that might promote the synthesis of complex molecules include solar ultraviolet radiation, molten lava, and ice. Complex ecosystems found near deep ocean hydrothermal vents support the hypothesis that complex molecules may form near these structures.

15.3.2 Chemistry of the Interstellar Medium

Research at NASA Ames Research Center demonstrated the formation of complex molecules using ultraviolet (UV) photochemistry including adenine, hydrogen cyanide, formaldehyde, and acetylene. Large molecules are formed in the interstellar medium (ISM) by photochemical processes that involve the interaction of UV photons with smaller molecules.

Spectroscopic studies of star-forming regions in the Orion Nebula by the Hubble space telescope have revealed molecules such as water, CO_2, CO, NH_3, amino acids, PAH, HCN, methane, methanol, formic acid, ethanol, and acidic acid. These molecules are also detected using radio astronomy and infrared (IR) spectroscopy. The Spitzer space telescope has detected the IR spectra of complex molecules surrounding a protostellar dust disk (Figure 15.3), planet-forming disks of sun-like and cool red stars (Figure 15.4), and a protostellar Herbig–Haro object (Figure 15.5). The discovery of complex molecules in the ISM suggests that many of these molecules may have been present in Earth's early oceans.

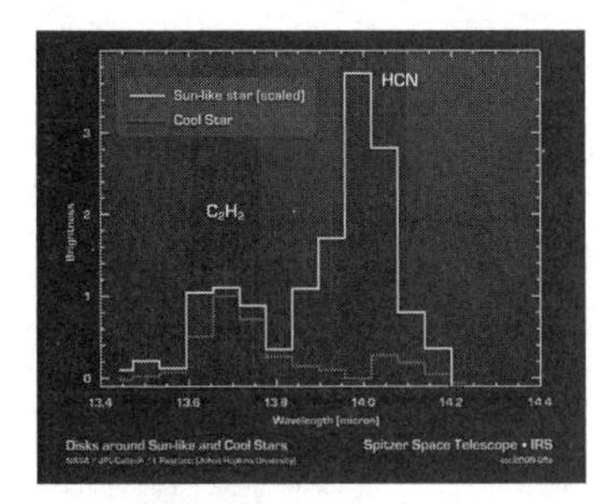

FIGURE 15.4 Infrared spectra recorded from the Spitzer space telescope showing hydrogen cyanide (HCN) and C_2H_2 in planet-forming disks surrounding sun-like and cool red stars. Less HCN is present in the planet-forming region surrounding the cool red star. (Courtesy of NASA/JPL-Caltech.)

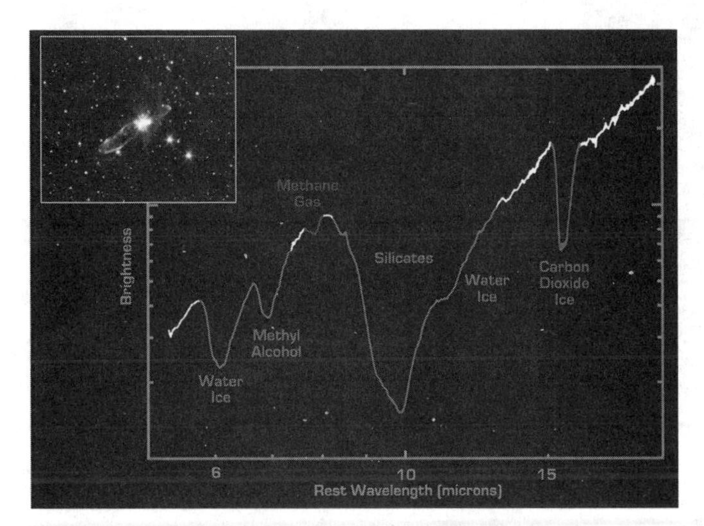

FIGURE 15.5 Infrared spectra recorded by the Spitzer space telescope indicating the presence of water ice, methane gas, methyl alcohol, and carbon dioxide gas in the protostellar Herbig–Haro object HH 46/47. A visual wavelength image of the young protostar is shown in the upper-left inset. (Courtesy of NASA Planetary Photo Journal Collection/Courtesy of nasaimages.org.)

15.4 Searches for Life in the Solar System

15.4.1 Mars

Space travelers touring the early solar system would have been dazzled by two water worlds. Earth, with plate tectonics later forming a single continent Pangaea, that broke up only 200 million years ago, and Mars, with water covering large portions of its Northern hemisphere. An artist rendition of Mars's early oceans is shown in Figure 15.6. Mars is currently a dry planet with a tenuous atmosphere composed mostly of CO_2. Because liquid water once flowed freely on its surface, Mars is believed to have been more hospitable for life in its past. The Martian atmosphere was much denser in

FIGURE 15.6 Artist reconstruction of Mars's oceans from data obtained from the FUSE spacecraft with the Mars Global Surveyor–MOLA instrument. (Courtesy of NASA Scientific Visualization Studio Collection/ Courtesy of nasaimages.org.)

its early history than it is today. Mars's weak gravitational field was insufficient to hold on to its lighter gases. Without its protective atmosphere, solar ultraviolet photons were able to break apart many of the water molecules into hydrogen and oxygen, which then escaped into space.

In 2002 the Mars Odyssey orbiter detected large quantities of hydrogen evidently locked into water ice in the top meter of the Martian regolith. Very minute traces of methane in the Martian atmosphere have been detected at infrared wavelengths by the planetary Fourier spectrometer (PFS) onboard the Mars Express orbiter. Since methane is broken down by solar radiation, it must be continually replenished by some unknown mechanism such as cometary impacts, geologic processes, or methane-producing life such as extremophile bacteria found on Earth that metabolize CO_2 and H_2 to form methane.

David McKay and coworkers at NASA Johnson Space Center performed studies on the Martian Meteorite ALH84001 depicted in Figure 15.7.

FIGURE 15.7 Mars meteorite ALH84001 contains micro fossil-like features formed over 3.6 billion years ago resembling terrestrial bacteria. The rock was dislodged from the surface of Mars by a meteoric impact roughly 16 million years ago and impacted the Earth 13,000 years ago. The specimen was found in Antarctica in 1984. (Courtesy of NASA/JPL-Caltech.)

These studies, originally reported in 1996 in the journal *Science*, revealed possible evidence for early microbial life on Mars. The meteorite was found in Antarctica in 1984 having been ejected from the surface of Mars as a result of a meteor impact about 15 million years ago and finally landing in Antarctica some 13 thousand years ago. Scanning electron microscopy images of the meteorite revealed structures (Figure 15.8) that resemble nanobacteria found on Earth but some 100 times smaller.

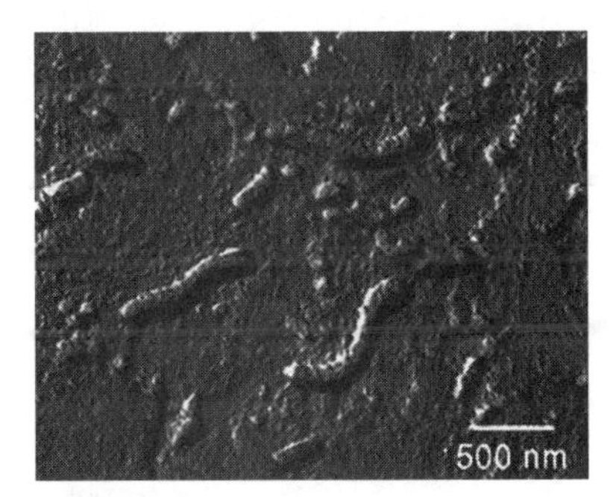

FIGURE 15.8 Transmission electron microscope image of Martian meteorite ALH84001 formed from a casting of a chip from the meteorite. The longest bacteria-like features are approximately 1 μm. (Courtesy of NASA Ames Research Center (NASA–ARC).)

The microfossil candidates included tiny carbonate globules (Figure 15.9), egg-shaped mineral deposits (Figure 15.10), and tube-like features (Figure 15.11).

The meteorite also contains hydrocarbon remnants similar to those produced by deceased microorganisms on Earth. Currently there is no unified scientific consensus supporting the hypothesis that ALH84001 contains evidence of past life on Mars. Table 15.1 lists evidence for and against the possibility of fossilized life in the Martian meteorite.

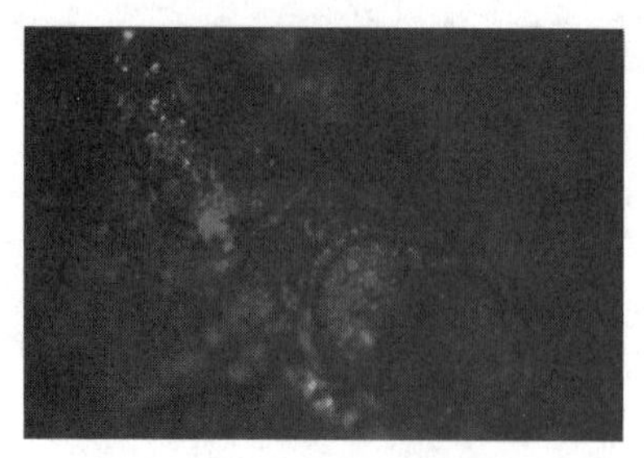

FIGURE 15.9 Carbonate mineral globules found in Martian meteorite ALH84001. The carbonate globules are similar to those formed by terrestrial bacteria. (Courtesy of NASA/JPL-Caltech.)

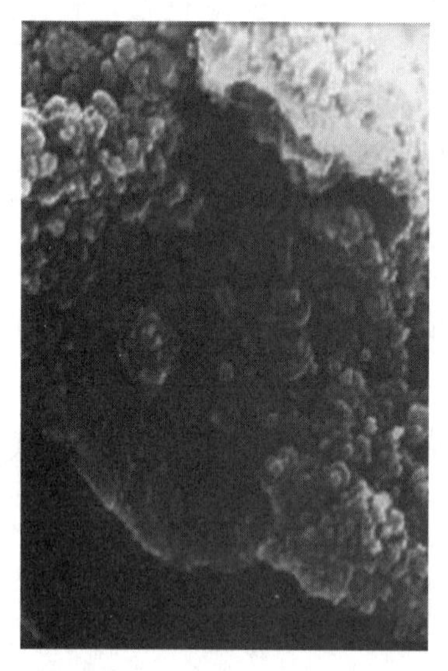

FIGURE 15.10 Egg-shaped mineral deposits found in Martian meteorite ALH84001. (Courtesy of NASA/JPL-Caltech.)

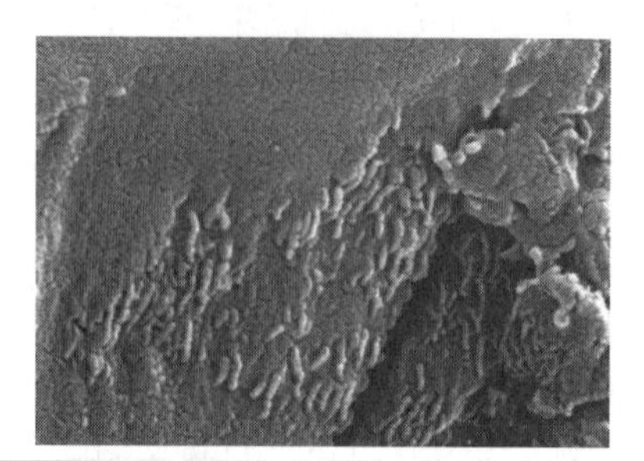

FIGURE 15.11 Tube-like features found in Martian meteorite ALH84001 imaged with an electron microscope. (Courtesy of NASA/JPL-Caltech.)

Table 15.1 Evidence For and Against Past Life in the Mars Meteorite

Past Life in Mars Meteorite ALH84001?	
Evidence for Life	**Evidence to the Contrary**
Complex organic (Polycyclic Aromatic Hydrocarbon) PAH molecules are found in the meteorite similar to those produced by the decomposition of biological material on Earth. PAH concentrations increase toward the center of the meteorite, suggesting they do not result from terrestrial contamination.	PAH molecules are also found in the interstellar medium and in carbonaceous meteorites formed in space. Biology is not a requirement for the presence of these organics. There is also the possibility that these organics have been introduced by terrestrial contamination.
The meteorite contains spherical carbonate globules with cores of manganese and iron carbonate with iron sulfide rings. Similar globules are associated with terrestrial bacteria in water. SQUID images of the magnetization distribution inside the meteorite suggest that the carbonate globules formed at temperatures below 40°C.	These globules can also from in the absence of bacteria on Earth. The temperatures at which the globules formed are estimated between 80° and 600°C). Carbonate globules are more likely to form from abiotic processes at higher temperatures.
The structures have a striking resemblance to terrestrial bacteria.	Fossil-like structures are only 20–500 nm in length and are roughly 10–100 times smaller than Earth bacteria. It is questionable whether genetic material could fit in such small volumes.
The structures contain magnetite crystals similar to magnetite crystals used by some terrestrial bacteria to navigate. On Earth similar crystals are left behind as magnetofossils after the bacteria die.	Magnetite crystals can also form near volcanic vents at temperatures between 500° and 800°C.
The relative abundance of C-12 compared to C-13 is suggestive of past life in the meteorite.	Repeated thawing and freezing of the meteorite in Antarctica and terrestrial contamination may be responsible for these carbon abundances.

Most of Earth's surface has been modified due to plate tectonics, volcanism, and erosion so that no geologic record exists of life that may have arisen during the first half-billion years of Earth's history. Any traces of chemical evolution in the prebiotic era have been obliterated. Mars is a one-plate planet and does not have plate tectonics like Earth. Evidence of transitioning biochemical evolution on Mars may one day reveal clues as to the emergence of self-replicating structures on Earth, perhaps similar to those found in ALH84001.

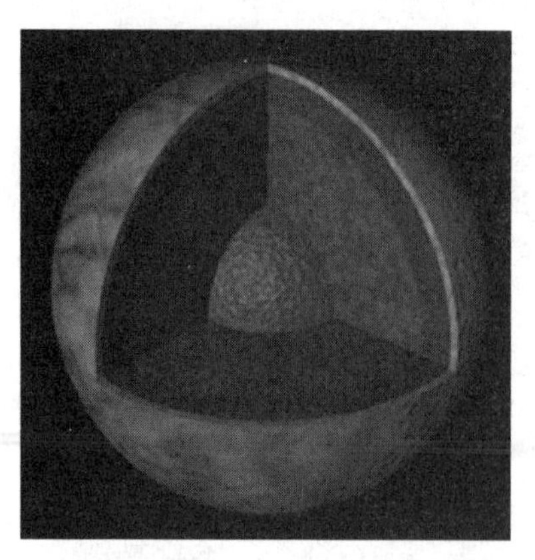

FIGURE 15.12 Artist rendition of the possible internal structure of Europa superimposed on surface images taken by the Voyager spacecraft. The central iron core is surrounded by a rocky shell covered by a liquid water ocean capped by the surface ice sheet. (Courtesy of NASA/JPL-Caltech.)

15.4.2 Europa

Europa is an icy moon of Jupiter about the size of Earth's Moon. Europa likely has a subsurface liquid water ocean beneath its ice crust that is believed to be roughly 10–30 km thick as illustrated in Figure 15.12.

Because of the presence of liquid water, Europa is considered the most likely candidate for extraterrestrial life in the solar system. The maximum surface temperature of the Jovian moon is about 110 K. The ocean is warmed by tidal interactions with Jupiter that flex the moon generating heat by internal friction. Models of tidal heating of the moon are shown in Figure 15.13.

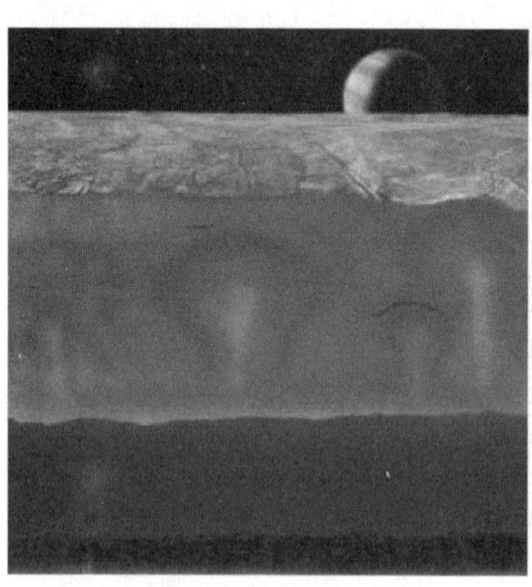

Tidal interaction with Jupiter heats interior

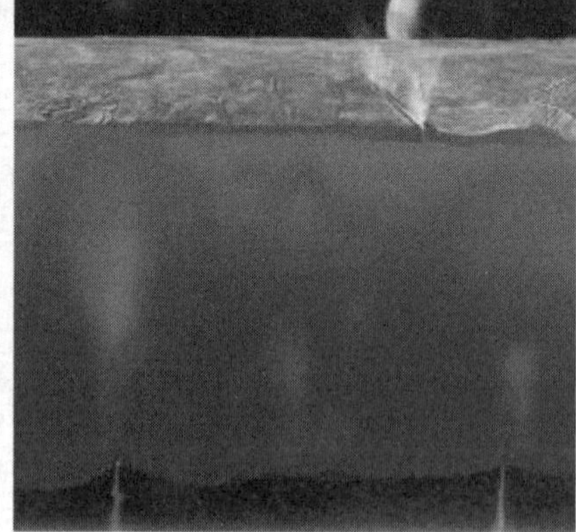

More tidal energy is coupled into the ice sheet

FIGURE 15.13 Rendition of two models of Europa's ice shell: the thin ice model (left) with a larger fraction of tidal energy heating Europa's interior, and the thick ice model (right) with more tidal energy coupling into the ice sheet. Melting and refreezing (thin ice model) or convective ice flows (thick ice model) results in a tumbled and broken "chaos" ice terrain observed on Europa's surface. (Courtesy of NASA/JPL-Caltech; artist: Michael Carroll.)

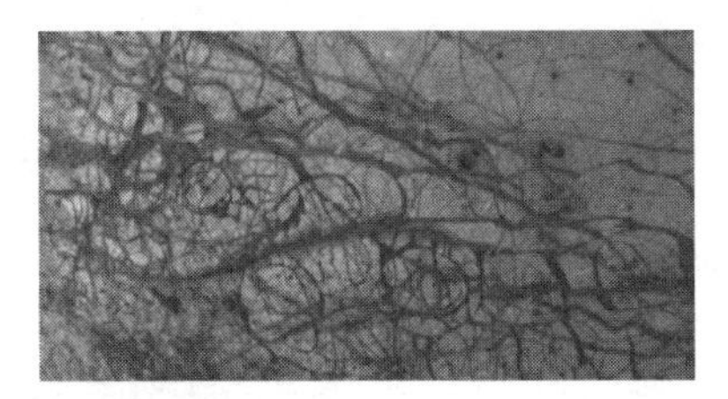

FIGURE 15.14 Polar ice features on Europa imaged by the Galileo spacecraft. Larger ice-plate fractures are roughly 30 km across. The dark areas are believed to be slushy ice contaminated by rocky material welling from under the ice sheet. (Courtesy of NASA/JPL-Caltech.)

Europa's surface is relatively young with few impact craters compared to other airless rocky bodies in the solar system. The Voyager and Galileo spacecrafts have imaged surface cracks and patches of darker ice on the surface shown in Figure 15.14. Other regions of the jumbled icy terrain known as Conamara Chaos resemble a scrabbled jigsaw puzzle.

Europa's magnetic field, as measured by the Galileo space probe, is believed to result from the motion of Europa's electrically conducting ocean in Jupiter's magnetic field. According to Faraday's law, eddy currents will be induced in a conductor moving in a magnetic field. Europa's ~100 nT magnetic field results from induced currents in its salt-rich ocean. By comparison, Earth's magnetic field ~50 μT results from the dynamo effect set up by its molten interior. Europa, however, is too small to have retained a molten interior.

15.4.3 Lake Vostok

Lake Vostok is a subglacial freshwater lake located roughly 3 km under the Antarctic ice sheet. The lake appears as a flat area when imaged by space radar as shown in Figure 15.15. The lake depth ranges between 300 and 800 meters at its deepest point. Ice core samples taken above the surface of the water date the lake to over a half-million years old. These ice cores are also found to support microbial life.

The discovery of life in Lake Vostok would increase the hopes of finding life under the ice sheet of Europa. Any life forms found in the lake would have to survive high pressures and oxygen levels roughly 50 times greater than found in ordinary fresh-water

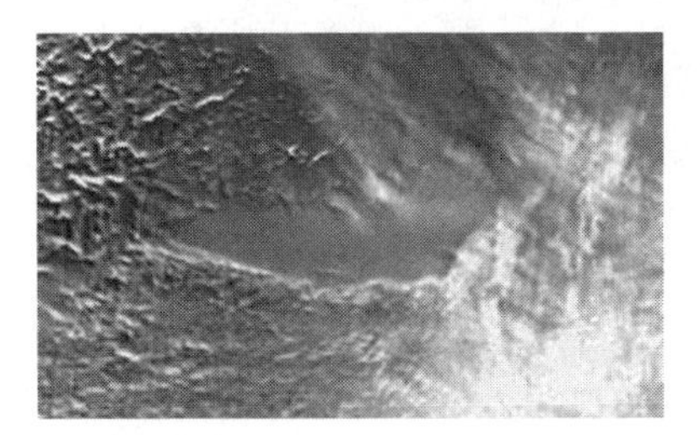

FIGURE 15.15 Radar image of Lake Vostok taken from space. (Courtesy of NASA Scientific Visualization Studio Collection/Courtesy of nasaimages.org.)

lakes. Even though geothermal heat and tidal forces warm the lake, the temperature remains slightly below the freezing point of water at atmospheric pressure.

15.5 Search for Life Outside the Solar System

15.5.1 Extrasolar Planets

Nearly 400 extrasolar planets have been detected to date. Extrasolar planets and their stars orbit about a common center of mass. This causes a slight wobble in the position of a star that can be detected as a Doppler shift in the stellar spectrum. In eclipsing systems, a star's spectrum may be subtracted to reveal the planetary spectrum as illustrated in Figure 15.16.

15.5.2 SETI Initiatives

The first searches for signals of extraterrestrial origin were conducted by Nicholi Tesla and later by Frank Marconi, credited with the invention of the radio. Philip Morrison and Giuseppe Coccni originally conceived the Search for Extraterrestrial Intelligence (SETI) program at Cornell University. The first SETI search was conducted in 1960 by Frank Drake using the 26-meter radio telescope in Green Bank, Maryland. The NASA SETI program was initiated in 1988 and began conducting searches in 1992 before the U.S. Congress canceled the program in 1993. Other SETI Initiatives include the SETI institute in California that has continued a portion of the NASA research using private funding as well as the SETI Australia Center in Sydney.

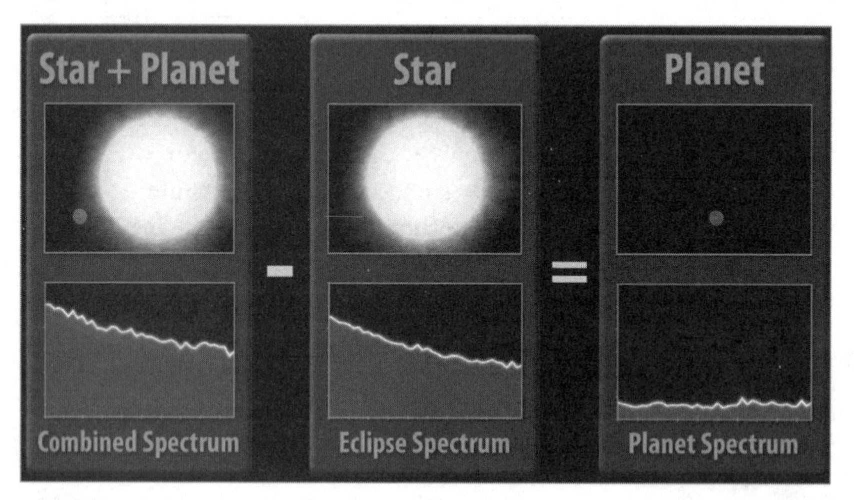

FIGURE 15.16 Schematic illustrating the technique for separating a planet's spectrum from that of its host star. The eclipse spectrum of the star is subtracted from the combined spectrum of the planet + star. In this way, the Spitzer space telescope was able to obtain the first two spectra ever recorded of planets outside our solar system. These spectra revealed the presence of dry clouds high in the stratospheres of the hot Jupiters HD 209458b and HD 189733b. (Courtesy of NASA Spitzer Space Telescope Collection/ Courtesy of nasaimages.org.)

15.5.3 Anticoded Signals

Anticoded signals can be arranged into a two-dimensional array forming a picture. Consider the signal S consisting of a string of 1's and 0's that might represent long and short radio bursts, respectively,

$$S = \{01110011100010011111001000111001010\}. \tag{15.1}$$

Two pictures can be created if the total signal length $n(S)$ is the product of two prime numbers. Here $n(S) = 35$ corresponds to either 5 rows and 7 columns or 7 rows and 5 columns. Figure 15.17 shows the picture formed by arranging S into 7 rows and 5 columns. Arranging the signal into 5 rows and 7 columns generates no recognizable pattern.

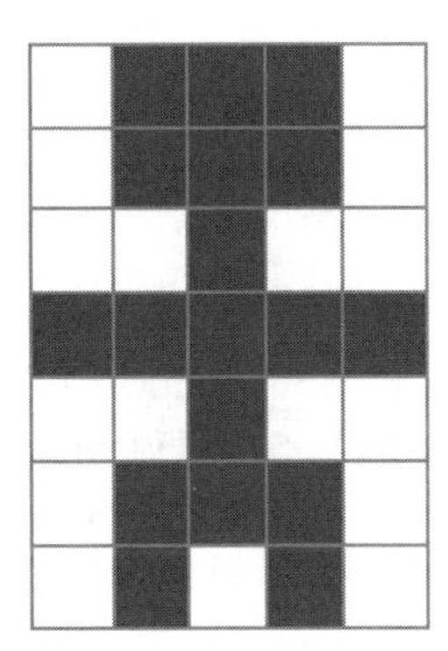

FIGURE 15.17 Picture formed by the anticoded signal S in Equation 15.1 arranged in 7 rows and 5 columns.

A famous anticoded signal was directed at the globular cluster M13 in 1974. Among other things, the message displayed pictorial representations of our solar system, the human figure, and the DNA double helix. This message will reach M13 in approximately 25 thousand years.

15.5.4 Frequency Domain Searches

To record an anticoded message forming a picture, it is necessary to receive pulses in the time domain. Instead of listening for signals broadcast in time, frequency-domain searches can scan a wider segment of the radio spectrum. Artificial signals at specific wavelengths will appear as peaks on the frequency axis. For example, a radio power spectrum of Earth would reveal peaks at the broadcast frequencies of different radio stations.

The program SETI@home employs multiple computers to process data from the Arecibo radio telescope. Data collected from the radio telescope is parceled into

FIGURE 15.18 SETI@home screen saver displaying fast Fourier transform (FFT) operations on data from the Arecibo radio telescope in Puerto Rico.

107-second segments that can be analyzed by desktop PCs. Data are downloaded from the university of California (UC)-Berkeley computer servers and processed on the user's PC. Radio spectra are displayed in the screen-saver format illustrated in Figure 15.18.

Post-processing results are uploaded to the UC-Berkeley servers, and another segment of data is downloaded. Any prominent peaks in the fast Fourier transform (FFT) spectrum are analyzed for Doppler spreading due to the Earth's rotation. Such Doppler spreading would not be present in a signal originating from Earth. The Arecibo radio telescope is not a steerable telescope so that extraterrestrial signal amplitudes would also rise to a maximum and then attenuate as the Earth rotates.

The SETI@home software performs a FFT algorithm with 15 different frequency resolutions ranging from 0.075 Hz to 1200 Hz. Smaller frequency resolutions require longer time series and analyses times. To perform a FFT with a resolution of 0.075 Hz will require 13.42 seconds of data.

15.5.5 Radio Interferometry

Interferometers employ multiple telescopes for greater sensitivity and angular resolution. Two radio telescopes have the equivalent resolution as a single radio telescope with a dish size equal to the baseline separation between the telescopes. The Very Large Array (VLA) operated by the National Radio Astronomy Observatory in New Mexico consists of 27 radio telescopes that can be extended 36 km. Space Very Long Baseline Interferometry (SVLBI) employs combinations of space- and ground-based telescopes with a baseline equal to the distance between Earth and space-based telescopes. Future SVLBI telescopes in geosynchronous orbits, such as those depicted in Figure 15.19, could have a baseline of 12.2 Earth radii.

Figure 15.20 shows a possible configuration of two telescopes forming an interferometer in orbit around the Sun with a baseline of two astronomical units.

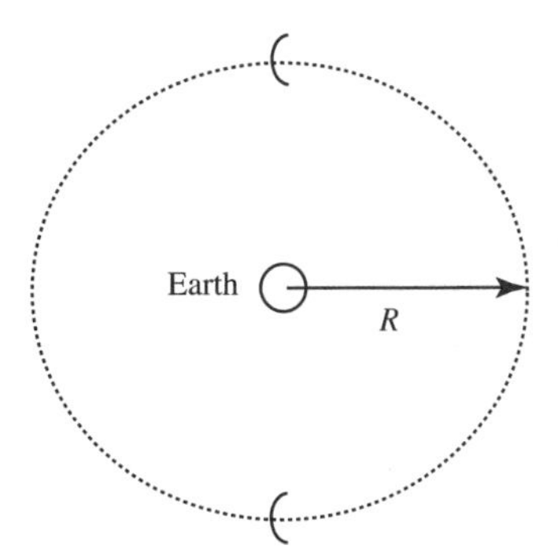

FIGURE 15.19 Depiction of a possible space-based radio interferometer in geosynchronous orbit.

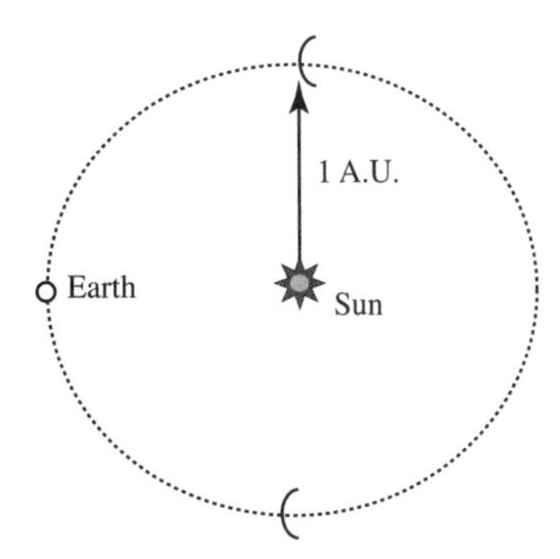

FIGURE 15.20 Depiction of a possible space-based radio interferometer in orbit about the Sun.

15.5.6 The Drake Equation

In 1960 Frank Drake, then at Cornell University, Ithaca, N.Y. developed an estimate for the number of communicable civilization N in a galaxy:

$$N = R^* \cdot f_p \cdot n_e \cdot f_\ell \cdot f_i \cdot f_e \cdot L. \tag{15.2}$$

The terms in the Drake equation are defined in Table 15.2, with estimates ranging from pessimistic to most optimistic.

An alternate formulation of the Drake equation replaces $R^* \rightarrow N^*$ where N^* is the current number of stars in a galaxy. Also, $L \rightarrow L/T_g$ where the term L/T_g is the fraction of the age of a galaxy that an average intelligent civilization can broadcast. Thus we have

$$N = N^* \cdot f_p \cdot n_e \cdot f_\ell \cdot f_i \cdot f_e \cdot \frac{L}{T_g}. \tag{15.3}$$

Table 15.2 Terms in the Drake Equation with Estimated Ranges

R^*	Average rate of star formation in the galaxy (in stars/year)	~10
f_p	Fraction of the above stars with planetary systems	$10^{-2} - 0.5$
n_e	Average number of planets belonging to planetary systems above that could potentially support life	$10^{-2} - 1$
f_ℓ	Fraction of above planets on which life ultimately arises	$10^{-2} - 1$
f_i	Fraction of planets supporting life where intelligent life emerges	$10^{-2} - 1$
f_e	Fraction of above planets with intelligent life forms that go on to emit detectable radiation into space	$10^{-2} - 1$
L	Average lifetime of the above communicating civilizations (in years)	$10^2 - 10^8$

The total number of stars in a galaxy is obtained by integrating

$$N^* = \int_0^{T_g} R^*(t)\,dt = \langle R^* \rangle_{\text{avg}} T_g. \tag{15.4}$$

Since the rate of star formation is maximal roughly 10^9 years after the big bang, as estimated from gamma ray bursters, it is reasonable to expect the number of habitable zones is maximal at a characteristic time as suggested in Figure 15.21. Using spectroscopic techniques with hyperfine resolution and ultra-long baseline interferometry, astronomers may one day estimate the average number of communicable civilizations per galaxy at different look-back times, up to the present age.

The Drake equation may be modified to include a recurrence factor $1 + n_r$ where n_r is the average number of times a new civilization arises on a planet where a previous civilization has already lived. If only one communicable civilization arises on a planet, then $n_r = 0$. We might also multiply the Drake equation by a realization factor

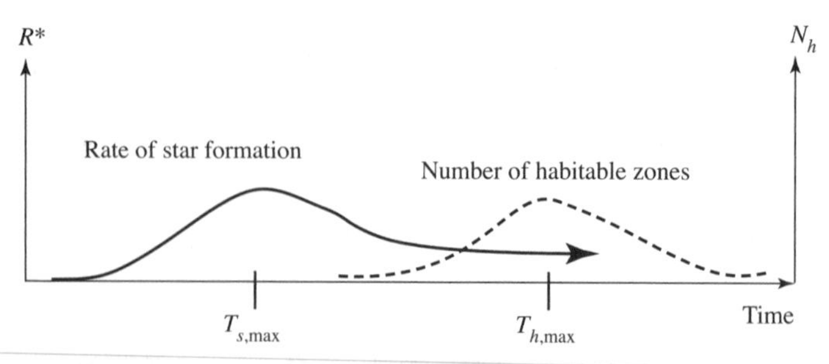

FIGURE 15.21 Schematic of the star formation rate R^* and the possible number of habitable zones in the universe. The peak in the star formation rate at the time $T_{s,\max}$ is followed by a peak in the number of habitable zones at $T_{h,\max}$.

proportional to the average time it takes a civilization to realize that it might be unwise to advertise its presence in a highly competitive galactic community.

15.6 Implications for Life in the Multiverse Picture

15.6.1 The Multiverse

In this last section we discuss more recent ideas in modern cosmology and the implications for life in these models. We are interested in the distribution of life in our universe as well as the distribution of life in an ensemble of possible universes. Multiverse models are motivated, in part, by ideas of string theory and inflationary cosmology. According to string theorists, the universe in which we find ourselves is described by a vacuum state that is one among approximately 10^{500} possible vacua. Each vacuum is associated with a specific set of physical constants. Figure 15.22 shows a Venn diagram of the set of all physical constants with subsets of constants where universes evolve, stars form, and life is possible.

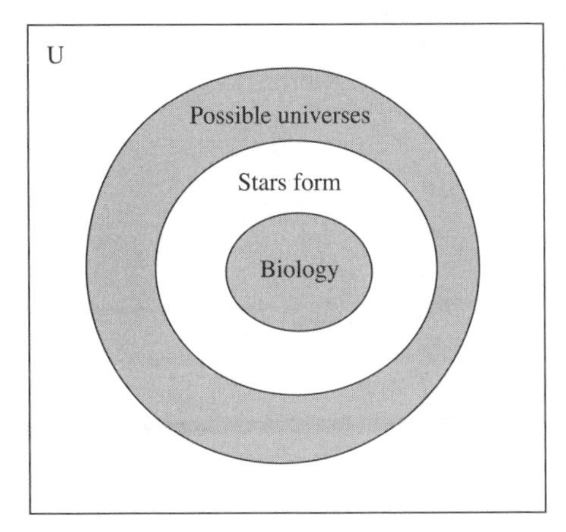

FIGURE 15.22 Venn diagram of the fundamental constants (G, h, c, ...). The square U represents the universal set of all possible values and combinations of constants. The outer circle represents those constant combinations resulting in universes that evolve (including universes with just supermassive black holes or expanding gas). The innermost circle represents constant combinations compatible with life.

Figure 15.23 depicts a reduced space of constants (G, h, c) corresponding to possible universes with a subset of universes capable of sustaining life.

If the number of string theory vacua is finite, then the parameter surface in N-dimensional hyperspace surrounding the life zone is not smooth and may be multiply connected or fractal. The position of our universe in a dimensionless space of cosmological constant Λ and amplitude of density fluctuations Q is shown in Figure 15.24. Stars cannot form for values of Q less than 10^{-6}, while only supermassive black holes are formed for Q greater than 10^{-3}. Gravitationally bound objects cannot form for values of Λ above the diagonal line in this figure.

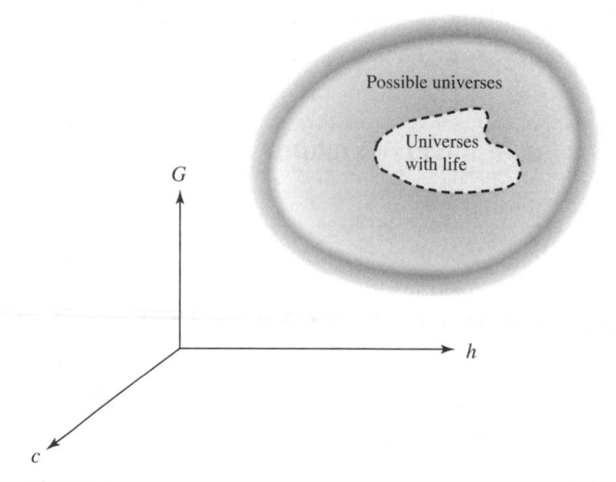

FIGURE 15.23 Reduced parameter space of fundamental constants (*G, h, c*). The outer shaded region represents those constant combinations resulting in universes that evolve. The smaller subregion contains those sets of constants compatible with life.

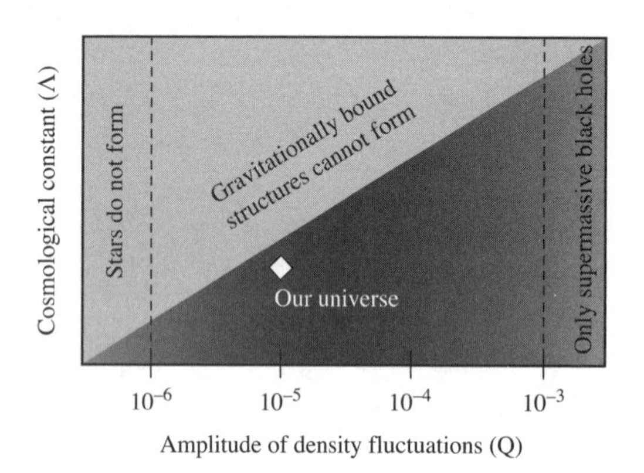

FIGURE 15.24 Q–Λ parameter space. Universes favorable to life may form in the darker shaded region below the diagonal. Our universe is denoted by the diamond with Q~10^{-5}. (Adapted from Livio and Rees. 2005. Anthropic reasoning. *Science* 309: 1022–1023.)

15.6.2 The Anthropic Principle

In its simplest form, the anthropic principle states that our universe must be compatible with life simply because life is present. The principle has also been employed as a selection effect to explain how we might find ourselves in a particular universe among an ensemble of universes with different physical properties. There is some debate in the scientific community on whether making predictions based on the anthropic principle is consistent with the scientific method. In order for a theory to be scientific, there must be some prediction of the theory that could, at least in principle, be falsified by experimental measurement.

15.6.3 Multiverse Cosmological Models

Below we review a few models discussed by Rodger Penrose and Paul Davies. None of these models relies on the ensemble of physical constants discussed above. In fact, Einstein believed that the physical constants are truly fundamental in nature and therefore form a unique set.

Andrei Linde: This model proposes a cosmology with eternal inflation. Pocket universes are nucleated in a larger expanding space with higher vacuum energy, or false vacuum. This cosmology leads to a Cantor-like fractal structure of the multiverse as universes (U) nucleate as bubbles of lower vacuum energy in the expanding false vacuum (FV) as illustrated in Figure 15.25. This process is known as chaotic inflation. According to Alan Guth, the originator of the inflationary theory, the inflation must have an origination and is only eternal in the forward time direction. In this model, the number of universes produced each second is increasing. The separation between universes is truly immense with inflationary domains on the order of $\exp(10^{11})$ cm. This is $\sim 10^{10}$ orders of magnitude greater than the Hubble radius of our universe $\sim 10^{28}$ cm.

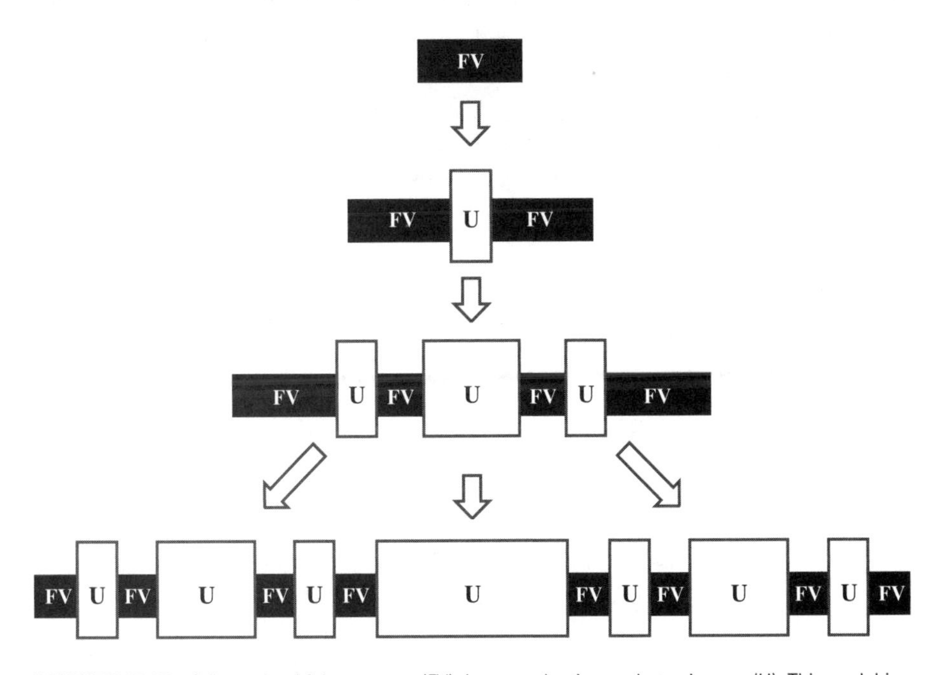

FIGURE 15.25 Schematic of false vacuum (FV) decay nucleating pocket universes (U). This model is referred to as chaotic inflation because the geometry resembles the Cantor comb discussed in Chapter 14.

Lee Smolin: Universes are generated by the gravitational collapse of stars forming black holes. Upon gravitational collapse of a star, a separate region of space–time begins expanding to form a new universe connected to the parent universe by an Einstein–Rosen bridge (or wormhole) as shown in Figure 15.26. The wormhole is

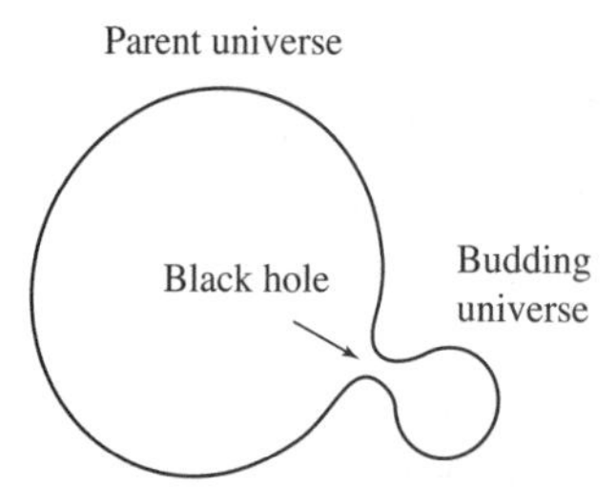

FIGURE 15.26 Smolin's model of universe production by gravitational collapse. Here a single universe sprouts from the gravitational collapse of a star forming a black hole (resembling budding yeast). The model places a restriction on the parameter space population density where only universes capable of producing black holes will reproduce (cosmological natural selection). The resulting fantastic web of universes is connected by Einstein–Rosen bridges (wormholes). The network is similar to nanofiber-connected bacteria colonies or neural networks.

eventually severed as the black hole evaporates through quantum mechanical processes at the event horizon. The radiative evaporation process predicted by Steven Hawking results in a black hole lifetime proportional to the cube of its mass. For example, a one-solar-mass black hole has a lifetime on the order of 10^{67} years. This is roughly 57 orders of magnitude greater than the present age of the universe.

Smolin's model of the multiverse would seem to be friendlier for life because only universes capable of forming black holes could reproduce as illustrated in Figure 15.27. Lifeless universes lacking stars ending as black holes could not reproduce except, perhaps, through a cascade of primordial black hole production.

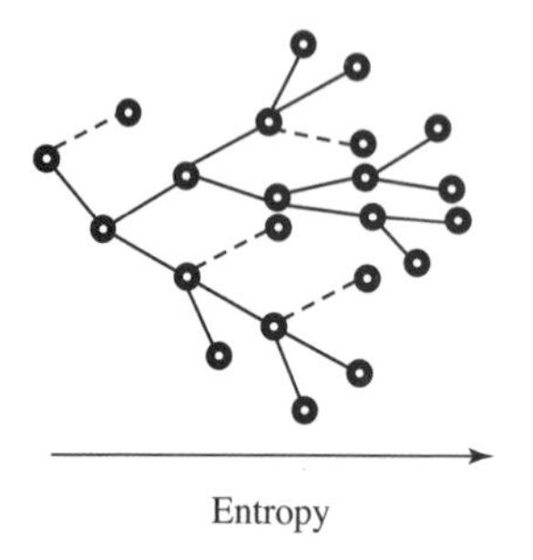

FIGURE 15.27 Smolin leaf diagram. Only universes capable of forming black holes can reproduce. Sterile universes are connected by broken branches in this figure.

SOC in the Multiverse?

Smolin's model (as well as Linde's) has a feature that the universe production rate is apparently increasing with each generation as shown schematically in Figure 15.28. We consider the possibility that a dynamical equilibrium could be established where production is balanced by extinction in a self-organized critical multiverse. One characteristic of SOC dynamics is the formation of fractal structure that is evident in

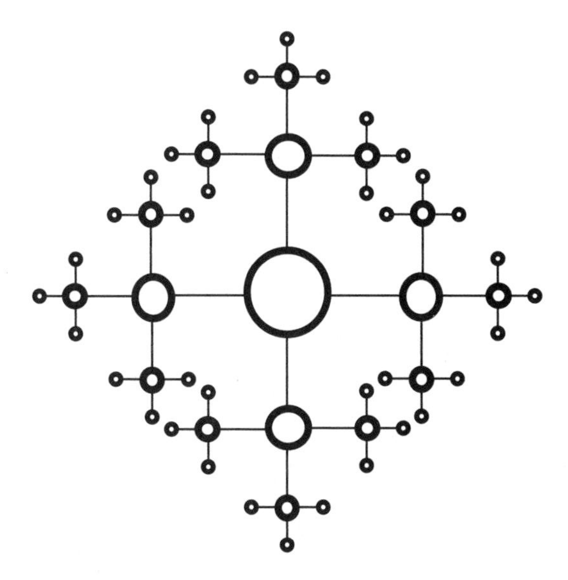

FIGURE 15.28 Smolin branching network. The web of universes (small circles) is connected by wormholes (line segments). This idealized schematic shows three generations of baby universes spawned from the central universe. Fractal-like structure is also evident in Smolin's model with entropy increasing in the radial direction. Self-similarity in this model increases indefinitely in the direction of increasing entropy.

Smolin's model. Wormhole entanglement could perhaps play a role in the dynamics of the multiverse by dampening cascade production of universes by primordial black holes.

Some attractive features of Smolin's model are these: (1) it is connected by the "standard Legos" of theoretical physics including black holes and Einstein–Rosen bridges; (2) it has a high degree of connectivity and fractal structure similar to the Mandelbrot set in Chapter 14; (3) it is possibly the most biophilic model spawning fewer universes incapable of forming stellar black holes; and (4) the model is subject to falsification since properties of stars forming black holes can be observed and simulated. It is therefore a scientific theory. A 3-D rendering of Smolin's model resembling a prickly pear is shown in Figure 15.29.

Roger Penrose: There is only one universe that undergoes cycles where the final stage of the last cycle transitions to the initial stage of the first cycle. Evolution of the universe can be viewed as a natural succession of subsequent phase transitions. Energy released in the big bang is a latent heat of vacuum phase transition. An analogous process would be liquid water transitioning to ice with latent heat generation.

Steinhardt and Turok: Their model is based on the interaction of topological features described by string theory, or membranes, that are sometimes called branes. Universes are membranes imbedded in a higher dimensional space. The separation between membranes is incredibly small. Big bang events occur when two of these membranes

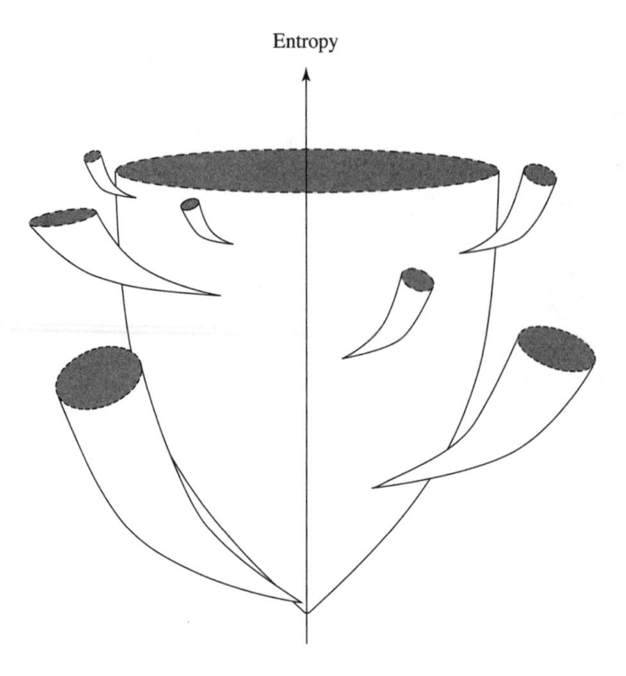

FIGURE 15.29 3-D rendering of Smolin's cosmological model (prickly pear) with baby universes spawned from gravitational collapse of stars forming black holes. Figure 15.28 can be thought of as a horizontal slice of this diagram.

collide releasing an enormous amount of energy. Steinhardt and Turok object to employing the anthropic principle as a selection effect.

Tegmark: All universes are possible, not only with different physical constants, but also with different physical laws. The central idea is that a multiverse consisting of all universes is a simpler description of nature than a cosmology consisting of a subset of universes. According to the principle of Occam's Razor, the simpler description is usually the more favorable one. An analogy would be a straight line in Euclidean space that requires less information to specify compared to a line with a set of points removed. Tegmark's model is an interesting application of philosophical reasoning to physical theory.

EXERCISES

Exercise 15.1 QuickField Thermal Analysis: Model the temperature distribution surrounding a ~1-m length, 0.1-m diameter hydrothermal vent black smoker with a temperature of 300° to 400°C neglecting convection. Take the surrounding ocean temperature between 2° to 4°C. Perform the calculation in axial symmetry.

Exercise 15.2 QuickField Stress Analysis: Calculate the stress distribution in the ice sheet of Europa due to tidal forces exerted by Jupiter. Take the diameter of Europa as 3130 km with an ice sheet thickness ~20 km with a Young's modulus of 6 GPa. Apply the tidal force components

$$F_z = \frac{2GMmr}{R^3}\cos\theta$$

$$F_r = -\frac{GMmr}{R^3}\sin\theta$$

along the spherical surface of the icy body where $M = 1.9 \cdot 10^{27}$ kg is the mass of Jupiter, $m = 4.8 \cdot 10^{22}$ kg is the mass of Europa, $R = 6.71 \cdot 10^5$ km is the average center-to-center separation between the planet and the moon, and $r = 1565$ km is the radius of Europa. The gravitational constant is $G = 6.673 \cdot 10^{-11}$ Nm^2/kg^2. Compare stress distributions for apoapsis $R_{max} = 676{,}938$ km and periapsis $R_{min} = 664{,}862$ km. Note that it is only necessary to mesh the ice sheet and not the ocean or planet interior. Perform the calculation in axial symmetry with the z-axis passing through the centers of each body. You may wish to exploit planar symmetry about the $\theta = \pi/2$ plane to simplify the model.

FIGURE 15.30 Jovian magnetic field imaged by NASA's Cassini spacecraft. (Courtesy of NASA/JPL-Caltech.)

Exercise 15.3 QuickField AC Magnetic Simulation: Model the induced currents in Europa's electrically conducting ocean due to its motion through Jupiter's magnetic field. Take the thickness of the water layer to be 100 km. Take the amplitude of the Jovian magnetic field (Figure 15.30) to be 150 nT along the z direction with a period equal to Europa's orbital period of 3.55 days. Take the electrical conductivity of the salt-rich ocean to be $\sigma = 0.5$ S/m. Plot the out-of-phase component of Europa's induced magnetic field.

Exercise 15.4 Figure 15.31 shows a rendition of hydrobot and cryobot probes that are proposed to explore the subsurface ocean of Europa. The cryobot probe melts its way through the ice sheet by liquefying water with an on-board nuclear power supply. Water refreezes above cryobot as it proceeds through the ice layer. When the water layer is reached, the cryobot releases the hydrobot probe which then explores the ocean below. How much energy is required to penetrate 20 km of ice if the diameter of the cryobot is 0.5 m? Assume that the temperature of the ice shell varies linearly between 100 K at its surface to 270 K at its base. Take the latent heat of melting of water to be 334 kJ/kg with a specific heat of 4.186 kJ/kg and a density of 1000 kg/m^3.

FIGURE 15.31 Artist's rendition of the proposed cryobot and submersible hydrobot exploring the subsurface ocean of Jupiter's moon Europa. (Courtesy of NASA Cassini-Huygens Collection/Courtesy of nasaimages.org.)

Exercise 15.5 Given that the total mass of all living matter on Earth is $\sim 10^{14}$ kg, consider leading terms in the Drake equation to estimate the total mass of living state material in the universe. According to your estimate (not including dark matter and dark energy), what mass fraction of the observable universe is alive? Take the mass of the baryonic matter to be 10^{52} kg.

Exercise 15.6 Calculate the entropy of a one-solar-mass black hole

$$S_{\text{b.h.}} = 4\pi M^2 \frac{Gk_B}{\hbar c}$$

where $M = 2 \cdot 10^{30}$ kg. Calculate the temperature of the black hole from the relationship

$$\frac{1}{T} = \frac{\partial S}{\partial E}$$

with $E = Mc^2$.

Lastly, calculate the heat capacity of the black hole

$$C = \frac{\partial E}{\partial T}.$$

Exercise 15.7 Calculate the angular resolution θ of space-based radio interferometers ($\lambda = 30$ cm) with baselines B of 12.2 Earth radii (radius of Earth = 6378 km) and 2 astronomical units (1 A. U = 150 million km). Angular resolution is given by

$$\theta = \frac{\lambda}{B}$$

where u is in radians. In general, multiple telescopes with different baselines are required to form high-quality images. What is the smallest feature S that could be resolved in the Alpha Centauri system approximately $r = 4.2$ light years away (1 light year 5 9.46 trillion km) by interferometer arrays with the effective baselines above? Use $S = r\theta$.

1 Mathematical Formulas

Exponential Functions

$$\sin x = \frac{e^{ix} - e^{-ix}}{2i}$$

$$\cos x = \frac{e^{ix} + e^{-ix}}{2}$$

$$\sinh x = \frac{e^{x} - e^{-x}}{2}$$

$$\cosh x = \frac{e^{x} + e^{-x}}{2}$$

Factorials

$$N! = N \cdot (N-1) \cdot (N-2) \cdots 3 \cdot 2 \cdot 1$$

$$0! = 1! = 1$$

Logarithms

$$x = a^{y}$$

$$y = \log_{a} x$$

The logarithm and exponential functions are inverse functions

$$f(f^{-1}(x)) = f^{-1}(f(x)) = x$$

so that

$$\ln(e^{x}) = e^{\ln x} = x$$

$$\log(10^{x}) = 10^{\log x} = x$$

$$\log_{a}(a^{x}) = a^{\log_{a} x} = x.$$

Properties of Logarithm Functions

$$\ln(a \cdot b) = \ln a + \ln b$$

$$\ln\left(\prod_i a_i\right) = \sum_i \ln a_i$$

$$\ln\left(\frac{a}{b}\right) = \ln a - \ln b$$

$$\ln a^n = n \ln a$$

$$\log 10 = 1$$

$$\ln e = 1$$

$$\log_a a = 1$$

$$\ln\frac{1}{a} = -\ln a$$

Derivatives of Logarithm Functions

$$\frac{d}{dx}\ln x = \frac{1}{x}$$

$$\frac{d}{dx}\ln f(x) = \frac{f'(x)}{f(x)}$$

Integrals of Logarithm Functions

$$\int \frac{dx}{x} = \ln|x| + \text{const.}$$

$$\int \ln x dx = x \ln|x| - x + \text{const.}$$

$$\int \frac{f'(x)}{f(x)} dx = \ln|f(x)| + \text{const.}$$

Gaussian Integrals

$$\int_0^\infty x^n e^{-\beta x} dx = \frac{n!}{\beta^{n+1}}$$

$$\int_0^\infty e^{-\alpha x^2}dx = \frac{1}{2}\sqrt{\frac{\pi}{\alpha}}$$

$$\int_0^\infty x^2 e^{-\alpha x^2}dx = -\frac{\partial}{\partial\alpha}\int_0^\infty e^{-\alpha x^2}dx = -\frac{\partial}{\partial\alpha}\frac{1}{2}\sqrt{\frac{\pi}{\alpha}} = \frac{1}{4}\sqrt{\frac{\pi}{\alpha^3}}$$

$$\int_0^\infty x^4 e^{-\alpha x^2}dx = \left(-\frac{\partial}{\partial\alpha}\right)^2\int_0^\infty e^{-\alpha x^2}dx = \left(-\frac{\partial}{\partial\alpha}\right)^2\frac{1}{2}\sqrt{\frac{\pi}{\alpha}} = \frac{3}{8}\sqrt{\frac{\pi}{\alpha^5}}$$

$$\int_0^\infty x^{2n} e^{-\alpha x^2}dx = \left(-\frac{\partial}{\partial\alpha}\right)^n\frac{1}{2}\sqrt{\frac{\pi}{\alpha}}$$

$$\int_0^\infty x e^{-\alpha x^2}dx = \frac{1}{2\alpha}$$

$$\int_0^\infty x^3 e^{-\alpha x^2}dx = \left(-\frac{\partial}{\partial\alpha}\right)\int_0^\infty x e^{-\alpha x^2}dx = \left(-\frac{\partial}{\partial\alpha}\right)\frac{1}{2\alpha} = \frac{1}{2\alpha^2}$$

$$\int_0^\infty x^5 e^{-\alpha x^2}dx = \left(-\frac{\partial}{\partial\alpha}\right)^2\int_0^\infty x e^{-\alpha x^2}dx = \left(-\frac{\partial}{\partial\alpha}\right)^2\frac{1}{2\alpha} = \frac{1}{\alpha^3}$$

$$\int_0^\infty x^{2n+1} e^{-\alpha x^2}dx = \left(-\frac{\partial}{\partial\alpha}\right)^n\frac{1}{2\alpha}$$

Chain Rule for Derivatives

$$\frac{d}{dx}(f\cdot g) = f'g + g'f$$

$$\frac{d}{dx}(f(g(x))) = g'f'(g(x))$$

$$\frac{d}{dx}f(ax) = af'(x)$$

Useful Approximations

$(1+x)^n \approx 1+ nx \qquad x \ll 1$

$e^x \approx 1+ x \qquad x \ll 1$

$\sinh x \approx x \qquad x \ll 1$

Stirling's formula

$\ln N! \approx N \ln N - N$

Error Functions

The error function is similar to a Gaussian integral but with a finite upper limit

$$\operatorname{erf}(x) = \frac{2}{\sqrt{\pi}} \int_0^x \exp(-t^2)\,dt.$$

The complementary error function is

$$\operatorname{erfc}(x) = \frac{2}{\sqrt{\pi}} \int_x^\infty \exp(-t^2)\,dt.$$

The sum of the error function and complementary error is equal to one:

$$\operatorname{erf}(x) + \operatorname{erfc}(x) = \frac{2}{\sqrt{\pi}} \int_0^\infty \exp(-t^2)\,dt = 1.$$

Thus, we can immediately calculate the complementary error function given the error function

$$\operatorname{erfc}(x) = 1 - \operatorname{erf}(x).$$

The derivative of the error function is

$$\frac{d}{dx}\operatorname{erf}(x) = \frac{2}{\sqrt{\pi}} \exp(-x^2).$$

The error function is odd so that

$$\operatorname{erf}(-x) = -\operatorname{erf}(x).$$

For small values of its argument the error function has the limiting form

$$\operatorname{erf}(x) \approx \frac{2}{\sqrt{\pi}}x \qquad x \ll 1.$$

Scalar Vector Product

The scalar or dot product between two vectors **A** and **B** is

$$\mathbf{A}\cdot\mathbf{B} = AB\cos(\theta)$$

where θ is the angle between the vectors. Because the result is a scalar quantity, the dot product is also known as the scalar product.

Vector Cross Product

The magnitude of the cross product between two vectors **A** and **B** is

$$|\mathbf{A}\times\mathbf{B}| = AB\sin(\theta)$$

where θ is the angle between the vectors. The cross product between two vectors gives a vector that is perpendicular to both **A** and **B** so that if $\mathbf{A}\times\mathbf{B} = \mathbf{C}$, then $\mathbf{A}\cdot\mathbf{C} = \mathbf{B}\cdot\mathbf{C} = 0$.

The cross product is also known as the vector product.

Divergence of a Vector Field

Given the vector field $\mathbf{E}(x, y, z) = E_x(x, y, z)\mathbf{i} + E_y(x, y, z)\mathbf{j} + E_z(x, y, z)\mathbf{k}$, the divergence of **E** is defined as the dot product of the del operator with **E** or

$$\nabla\cdot\mathbf{E} = \left(\frac{\partial}{\partial x}\mathbf{i} + \frac{\partial}{\partial y}\mathbf{j} + \frac{\partial}{\partial z}\mathbf{k}\right)\cdot(E_x\mathbf{i} + E_y\mathbf{j} + E_z\mathbf{k}),$$

which gives

$$\nabla\cdot\mathbf{E} = \left(\frac{\partial E_x}{\partial x} + \frac{\partial E_y}{\partial y} + \frac{\partial E_z}{\partial z}\right).$$

Curl of a Vector Field

The curl of a vector field is defined as the cross product of the del operator with **E**

$$\nabla \times \mathbf{E} = \left(\frac{\partial}{\partial x}\mathbf{i} + \frac{\partial}{\partial y}\mathbf{j} + \frac{\partial}{\partial z}\mathbf{k}\right) \times (E_x\mathbf{i} + E_y\mathbf{j} + E_z\mathbf{k}).$$

This may be expressed as the determinate

$$\nabla \times \mathbf{E} = \begin{vmatrix} \mathbf{i} & \mathbf{j} & \mathbf{k} \\ \frac{\partial}{\partial x} & \frac{\partial}{\partial y} & \frac{\partial}{\partial z} \\ E_x & E_y & E_z \end{vmatrix} = \left(\frac{\partial E_z}{\partial y} - \frac{\partial E_y}{\partial z}\right)\mathbf{i} - \left(\frac{\partial E_z}{\partial x} - \frac{\partial E_x}{\partial z}\right)\mathbf{j} + \left(\frac{\partial E_y}{\partial x} - \frac{\partial E_x}{\partial y}\right)\mathbf{k}.$$

Laplacian of a Scalar Field

The Laplacian of a scalar field is defined as the dot product of the del operator with the gradient of the scalar field $\nabla \cdot \nabla T(x, y, z)$. In Cartesian coordinates the scalar Laplacian has the simple form

$$\nabla^2 T = \left(\frac{\partial}{\partial x}\mathbf{i} + \frac{\partial}{\partial y}\mathbf{j} + \frac{\partial}{\partial z}\mathbf{k}\right) \cdot \left(\frac{\partial T}{\partial x}\mathbf{i} + \frac{\partial T}{\partial y}\mathbf{j} + \frac{\partial T}{\partial z}\mathbf{k}\right) = \frac{\partial^2 T}{\partial x^2} + \frac{\partial^2 T}{\partial y^2} + \frac{\partial^2 T}{\partial z^2}.$$

Discrete Laplacian

A central difference numerical approximation to $\nabla^2 T(x, y)$ is obtained from

$$\frac{\partial^2 T}{\partial x^2} = \frac{T_{i+1,j} - 2T_{i,j} + T_{i-1,j}}{\Delta x^2} + O(\Delta x^2)$$

and

$$\frac{\partial^2 T}{\partial y^2} = \frac{T_{i,j+1} - 2T_{i,j} + T_{i,j-1}}{\Delta y^2} + O(\Delta y^2)$$

where (i, j) are indices corresponding to the x- and y-axes, respectively. If we let $\Delta x = \Delta y = \Delta$, we can write the discrete form of the Laplacian in two dimensions

$$\frac{\partial^2 T}{\partial x^2} + \frac{\partial^2 T}{\partial y^2} = \frac{T_{i+1,j} + T_{i-1,j} + T_{i,j+1} + T_{i,j-1} - 4T_{i,j}}{\Delta^2}.$$

Divergence Theorem

Gauss's divergence theorem relates the volume integral of the divergence of a vector field **E** to the integral of the vector **E** dotted into the surface normal and evaluated over the surface bounding the volume v

$$\iiint\limits_{\text{vol}} \nabla \cdot \mathbf{E}\, dv = \iint\limits_{\text{surf}} \mathbf{E} \cdot \mathbf{n}\, da.$$

Stokes's Theorem

Stokes's theorem relates the line integral of a vector field **E** around a closed path Γ to the surface integral evaluated over any capping surface bound by the contour

$$\iint\limits_{\text{surf}} \nabla \times \mathbf{E} \cdot \mathbf{n}\, da = \oint\limits_{\Gamma} \mathbf{E} \cdot d\ell.$$

Gradient Theorem

The line integral of the gradient of a scalar function may be evaluated

$$\int_{\mathbf{a}}^{\mathbf{b}} \nabla f \cdot d\ell = f(\mathbf{b}) - f(\mathbf{a})$$

where the integral only depends only on the value of the function at the end points.

2 Overview of MATLAB®

M-Files

In many cases it is better to create an M-file when writing code blocks more than a few lines in length. The M-file is saved under a filename such as "predator_prey.m." M-files are much easier to edit than code entered in the Command window. *After code is executed in the Command window, it is necessary to use the Up Arrow on the keyboard to bring up a given line to modify/correct it.*

To create an M-file select (File → New → M-file) and then enter the code and save the file (File → Save as . . . → filename.m).

As an example, consider the M-file consisting of the code block

```
1+1
```

saved as "sample.m." The M-file is executed in the Command window by typing "sample" at the Command prompt

```
>> sample
```

resulting in the output

```
ans =

     2
```

Basic Math and Syntax

```
% comment
; output suppression
5+3-4*2-2/3    %  addition, subtraction, multiplication, division
3^2            %  exponentiation
sqrt(pi)       %  square root
```

Logical Operators

```
~ % not
|| % logical or
&& % logical and
& % logical and for arrays
| % logical or for arrays
```

Conditionals

```
if condition
    operations;
end;

if condition
    operations1;
else
    operations2;
end;

if condition1
    operations1;
elseif condition2
    operations2;
else
    operations3;
end;
```

Loops

```
for i=1:n              % i=1, 2, 3,…,n
    operations;
end;

for i=1:2:n            % i=1, 3, 5, 7, 9,…,n
    operations;
end;

for i=1:-3:-10         % i=1, -2, -5, -8
    operations;
end;

while condition
    operations;
end;
```

Vectors and Matrices

`v=1:5`	% vector v=(1, 2, 3, 4, 5)
`u=-1:2:7`	% vector u= (-1, 1, 3, 5, 7)
`w=u.\1`	% vector w = (-1, 1, 1/3, 1/5 ,1/7)
`A = [1, 2, 3; 4, 5, 6; 7, 8, 9]`	% 3 by 3 matrix
`A'`	% transpose of matrix A
`rand(5)`	% creates a 5x5 matrix of random numbers (uniform)
`rand(5,5)`	% also creates a 5x5 matrix of random numbers
`zeros(5)`	% creates a 5x5 matrix of zeros
`randn(3)`	% creates a 3x3 matrix % of normally distributed random numbers

Complex Numbers

`z= 3 + i`	% complex number in form a + bi
`real(z)`	% real part of z (returns 3)
`imag(z)`	% imaginary part of z (returns 1)
`abs(z)`	% magnitude of complex number (returns sqrt(10))

Plotting Graphs

Simple Graphs

`t=0:.01:5;`	%define time variable and increment
`plot(t,cos(3*t).*exp(-t))`	%plot a single curve

Log–Log and Semilog Graphs

`semilogy(x,y)`	% plots x-axis linear and y-axis log
`semilogx(x,y)`	% plots y-axis linear and x-axis log
`loglog(x,y)`	% plots both axes log scale

Multiple Curves on One Graph

```
%plot two curves on one graph
plot(t,1./(t-3),t,exp(t)),legend('1/(t-3)','exp(t)') % two curves
```

```
%plot the first six Bessel functions on the same graph
x=0:.1:15;

plot(x,besselj(0,x))

hold on
for n= 1:5
    y=besselj(n,x);
    plot(x,y)
end
hold off
```

Contour Plots of Complex Functions

```
[X,Y]=meshgrid(-3:0.01:3,-3:0.01:3);
H=besselh(0,2,X+i*Y);
contour(X,Y,real(H),-1:.1:1),hold on
contour(X,Y,imag(H),-1:.1:1); hold off
% also try besselj, bessely, besseli, besselk
```

Plots Equipotentials and Vector Field on Same Graph

```
v=-2:0.2:2;
[x,y]=meshgrid(v);
z=y.*exp(-x.^2-y.^2);
[px,py]=gradient(z,.2,.2);
contour(v,v,z),hold on,quiver(v,v,px,py),hold off
```

Calculating Sums

```
% calculate the sum of the first five natural numbers
sum=0;

for n=1:5
    sum=sum+n;
end
sum            % outputs 15

% vectorized sums eliminates loops and execute faster

n=1:5;
sum(n)         % outputs 15
```

Fast Fourier Transform

The discrete Fourier transform (DFT) may be evaluated using the fast Fourier transform (FFT) algorithm in MATLAB. The FFT of a time series given by $x = f(t)$ may be expressed with default variables x and w in MATLAB as

```
y= fft(x) % returns the Fast Fourier transform
of time series x
y= fft(x,n) % returns the n-point FFT.
If n is greater that the
          % if n is greater than the number of points in x
          % then x is padded with zeros to length n
```

Clearing

```
>> clf    % clear current figure window
>> clear % removes items from the workspace
```

Importing and Exporting Data

MATLAB's Import Wizard can be used to import Excel® worksheets (.xls), text files (.csv, .txt, .dat), and MAT-file array (.mat):

File → Create Time Series from File

For MATLAB workspace arrays:

File → Import from Workspace File → Array Data

This example creates and reads an ASCII data file from the command line

```
>> B= [1 2 3];
>> save c:\docs\jim.dat B- Ascii

>> P=dlmread('c:\docs\jim.dat');
>> P
```

resulting in the output

```
P =
      1    2    3
```

Many more read and write options are available. See also "Importing and Exporting Data" from the MATLAB Help feature.

Symbolic ToolBox

The following examples require the Symbolic Toolbox. These can be executed at the Command line. Remember to use the Up Arrow key to modify/correct previous operations.

Simple Plots

```
syms x
ezplot(x*sin(1/x), [-.1,.1])
ezplot(besselj(0,x),[0,10])
```

Surface Plots

```
syms x y
ezmesh(besselj(0,sqrt(x^2+y^2)), [-10,10])
```

Combination Surface and Contour Plots

```
syms x y
ezmeshc(sin(x)*exp(-x^2-y^2))
% also use the Rotate 3D tool to view different orientations
```

3-D Parametric Curve

```
syms t
ezplot3(sin(cos(t)),cos(sin(t)),t,[0,15])
% also use the Rotate 3D tool to view different orientations
```

Symbolic Integration

```
syms x a
a= 3;
int(dirac(x-a)*besselj(0,x),0,inf)

syms a b positive
syms x
f=exp(-a*x^2+b*x)
int(f,x,0,inf)
pretty(ans)     %use to make the answer more readable
```

Symbolic Differentiation

```
syms x
f=besselj(2,x);
diff(f)
```

Symbolic Solution of Differential Equations

```
dsolve('D2I + 5*DI +6*I = 9*sin(t)', 'I(0)=0', 'DI(0)=0')

r = dsolve('DX+Y=Z','DY+3*Z=Y','DZ+X=2*Z','X(0)=1','Y(0)=0','Z(0)=1')
r.X          % returns X(t) where dsolve is evaluated first
r.Y          % returns Y(t)
r.Z          % returns Z(t)
```

Symbolic Summation

```
syms x k
symsum(exp(-k*x),0,inf)

syms n
symsum((-1)^n/n^2,1,inf)

syms n
symsum(1/2^n,1,inf)
```

Taylor Expansion

```
syms x
taylor(besselj(1,x),8)  % terms up to but not including order 8
```

Symbolic Solution to Systems of Equations

```
syms a b c x y z
S=solve('a*x+b*y=3', 'c*y+b*z=2', 'x+z =1', x,y,z)
S.x       %returns x value
S.y       %returns y value
S.z       %returns z value
```

Factoring Expressions

```
syms x y
factor(x^6-y^6)
```

Simplifying Expressions

```
syms a b positive
f=(a^3+b^3)/(a+b)
simplify(f)
```

3 Derivation of the Heat Equation

The partial differential equation describing local heat flow in a material body with a one-dimensional temperature gradient is

$$\frac{\partial Q}{\partial t} = \lambda A \frac{\partial T}{\partial x}. \tag{A3.1}$$

We seek to write equation A3.1 as a partial differential equation over a single scalar field $T(x, t)$. The thermal energy stored in a body of volume V with constant temperature is given by $Q = mcT$, or with variable temperature and mass density

$$Q = c \iiint_{\text{vol}} T \rho \, dV \tag{A3.2}$$

where c is the specific heat in $J/kg \cdot K$, T is the absolute temperature in Kelvins, and ρ is the mass density in kg/m^3. The power lost due to the dissipation of thermal energy through a surface bounding the volume V is given by

$$\frac{\partial Q}{\partial t} = \lambda \iint_{\text{surf}} \nabla T \cdot d\mathbf{a}. \tag{A3.3}$$

Substituting Equation A3.2 and applying Gauss's divergence theorem gives

$$c \iiint_{\text{vol}} \frac{\partial T}{\partial t} \rho \, dV = \lambda \iiint_{\text{vol}} \nabla^2 T \, dV. \tag{A3.4}$$

Note that Equation A3.4 holds over arbitrary volumes so that the integrands can be equated giving the heat equation

$$\frac{\partial T}{\partial t} = \alpha \nabla^2 T \tag{A3.5}$$

where $\alpha = \lambda / c\rho$.

4 Derivation of Shannon's Entropy Formula

In the Microcanonical Ensemble, the number of microstates is given by

$$\Omega = \frac{N!}{N_1!N_2!\ldots N_m!}. \tag{A4.1}$$

Boltzmann's statistical formula for the entropy is

$$S = k_B \ln \Omega = k_B \ln \frac{N!}{N_1!N_2!\ldots N_m!}. \tag{A4.2}$$

Using the fact that the log of a ratio is a difference of logs

$$S = k_B(\ln N! - \ln N_1!N_2!\ldots N_m!). \tag{A4.3}$$

Next, applying Stirling's approximation to the first term and using the fact that the log of a product is a sum of logs to transform the second term we have

$$S = k_B\left(N \ln N - N - \sum_{i=1}^{m} \ln N_i! \right). \tag{A4.4}$$

Now applying Stirling's approximation to the last term in the previous expression

$$S = k_B\left(N \ln N - N - \sum_{i=1}^{m} (N_i \ln N_i - N_i)\right) \tag{A4.5}$$

and since $\sum_{i=1}^{m} N_i = N$ we have that

$$S = k_B\left(N \ln N - \sum_{i=1}^{m} N_i \ln N_i \right). \tag{A4.6}$$

Upon factoring N,

$$S = Nk_B\left(\ln N - \sum_{i=1}^{m}\frac{N_i}{N}\ln N_i\right), \tag{A4.7}$$

which can be written as

$$S = Nk_B\left(-\sum_{i=1}^{m}\frac{N_i}{N}\ln\frac{N_i}{N}\right). \tag{A4.8}$$

Thus we have the Shannon entropy formula

$$S = Nk_B\left(-\sum_{i=1}^{m}P_i\ln P_i\right) \tag{A4.9}$$

where $P_i = N_i/N$. Note that $S \geqslant 0$ since $P_i \leqslant 1$ and $\ln P_i \leqslant 0$.

Entropy in the Canonical Ensemble

We can now calculate the entropy in the Canonical Ensemble at temperature T using the Shannon entropy formula and the statistical Boltzmann factor where the probability of the ith state is

$$P_i = \frac{\exp\left(-\frac{E_i}{k_BT}\right)}{Z}. \tag{A4.10}$$

Substituting Equation A4.10 into the Shannon formula

$$S = Nk_B\left(-\sum_{i=1}^{m}\frac{\exp\left(-\frac{E_i}{k_BT}\right)}{Z}\ln\frac{\exp\left(-\frac{E_i}{k_BT}\right)}{Z}\right) \tag{A4.11}$$

and expanding the logarithm

$$S = Nk_B\sum_{i=1}^{m}\frac{\exp\left(-\frac{E_i}{k_BT}\right)}{Z}\left(\frac{E_i}{k_BT} + \ln Z\right) \tag{A4.12}$$

then distributing the sum

$$S = \sum_{i=1}^{m} \frac{N}{T} \underbrace{\frac{E_i \exp\left(-\frac{E_i}{k_B T}\right)}{Z}}_{E_i P_i} + N k_B \ln Z \sum_{i=1}^{m} \underbrace{\frac{\exp\left(-\frac{E_i}{k_B T}\right)}{Z}}_{P_i} \tag{A4.13}$$

gives us

$$S = \frac{N}{T}\langle E_i \rangle + N k_B \ln Z \tag{A4.14}$$

where $\langle E_i \rangle = \sum_i P_i E_i$ and $\sum_i P_i = 1$.

We can now write the entropy

$$S = \frac{E}{T} - \frac{F}{T} \tag{A4.15}$$

where $E = N\langle E_i \rangle$ and $F = -N k_B T \ln Z$. The familiar expression of the free energy is then retrieved

$$F = E - TS. \tag{A4.16}$$

Note also that the entropy can be expressed in terms of the free energy

$$S = -\frac{\partial F}{\partial T} \tag{A4.17}$$

so that

$$S = \frac{\partial}{\partial T} N k_B T \ln Z = N k_B \left(\ln Z + T \frac{\partial}{\partial T} \ln Z \right) \tag{A4.18}$$

and

$$S = N k_B \ln Z - N \frac{1}{T} \frac{\partial}{\partial \beta} \ln Z \tag{A4.19}$$

where

$$\frac{\partial}{\partial \beta} = -k_B T^2 \frac{\partial}{\partial T} \tag{A4.20}$$

and finally

$$S = N k_B \left(\ln Z + \beta \langle E \rangle \right). \tag{A4.21}$$

EXERCISES

Exercise A4.1 Obtain expressions similar to Equations A4.11–A4.21 for the Grand Canonical Ensemble using the Gibbs factor Equation 5.83.

5 Thermodynamic Identities

Internal Energy

$$dE = TdS - PdV + \mu dN$$

$$\left(\frac{\partial S}{\partial E}\right)_{N,V} = \frac{1}{T}, \left(\frac{\partial S}{\partial V}\right)_{E,N} = \frac{P}{T}, \left(\frac{\partial S}{\partial N}\right)_{E,V} = -\frac{\mu}{T}$$

Helmholtz Free Energy

$$dF = -SdT - PdV + \mu dN$$

$$\left(\frac{\partial F}{\partial N}\right)_{T,V} = \mu, \left(\frac{\partial F}{\partial V}\right)_{T,N} = -P, \left(\frac{\partial F}{\partial T}\right)_{V,N} = -S$$

Gibb's Free Energy

$$dG = dE - TdS - SdT + PdV + VdP$$

$$dG = -SdT + VdP + \mu dN$$

$$\left(\frac{\partial G}{\partial N}\right)_{T,P} = \mu, \left(\frac{\partial G}{\partial P}\right)_{T,N} = V, \left(\frac{\partial G}{\partial T}\right)_{P,N} = -S$$

6 Kramers–Kronig Transformations

The Kramers–Kronig transforms serve as a useful check for the consistency of experimental impedance data and enable the computation of complementary ac components. The principal value P is taken in the following integral transformations.

Impedance transformations where $Z = R + iX$, R is the resistance, and X is the reactance:

$$R(\omega) - R(\infty) = \frac{2}{\pi} P \int_0^\infty \frac{\omega' X(\omega') - \omega X(\omega)}{\omega'^2 - \omega^2} d\omega'. \tag{A6.1}$$

The DC resistance increment is then

$$R(0) - R(\infty) = \frac{2}{\pi} P \int_0^\infty \frac{X(\omega')}{\omega'} d\omega'. \tag{A6.2}$$

The frequency dependent reactance is

$$X(\omega) = -\frac{2\omega}{\pi} P \int_0^\infty \frac{R(\omega') - R(\omega)}{\omega'^2 - \omega^2} d\omega'. \tag{A6.3}$$

The phase angle between resistance and reactance is

$$\phi(\omega) = \frac{2\omega}{\pi} P \int_0^\infty \frac{\ln|Z(\omega')|}{\omega'^2 - \omega^2} d\omega' \text{ where } \phi = \tan^{-1}\left(\frac{X}{R}\right). \tag{A6.4}$$

Admittance transformations $Y = G + iB$ where G is the conductance and B is the susceptance:

$$G(\omega) - G(\infty) = \frac{2}{\pi} P \int_0^\infty \frac{\omega' B(\omega') - \omega B(\omega)}{\omega'^2 - \omega^2} d\omega' \quad \text{(A6.5)}$$

$$G(0) - G(\infty) = \frac{2}{\pi} P \int_0^\infty \frac{B(\omega')}{\omega'} d\omega'. \quad \text{(A6.6)}$$

Conductivity and permitivity transformations for $\varepsilon'' = \frac{\sigma}{\omega}$ and $\varepsilon' = \varepsilon$ where $\varepsilon^* = \varepsilon' - i\varepsilon''$:

$$\varepsilon'(\omega) - \varepsilon'(\infty) = \frac{2}{\pi} P \int_0^\infty \frac{\omega' \varepsilon''(\omega')}{\omega'^2 - \omega^2} d\omega' \quad \text{(A6.7)}$$

$$\varepsilon''(\omega) = \frac{2\omega}{\pi} P \int_0^\infty \frac{\varepsilon'(\omega') - \varepsilon'(\infty)}{\omega'^2 - \omega^2} d\omega' \quad \text{(A6.8)}$$

$$\varepsilon'(0) - \varepsilon'(\infty) = \frac{2}{\pi} P \int_0^\infty \frac{\varepsilon''(\omega')}{\omega'} d\omega' \quad \text{(A6.9)}$$

From these transformations, the AC conductivity can be calculated from the permitivity and vice versa.

Solution to the One-Dimensional Schrödinger Equation

The Schrödinger equation is a wave equation that describes the probability of locating a particle as a function of position and time. A particle such as an electron is described by a wavefunction Ψ called the amplitude. Ψ is a complex number. The probability of locating the electron in a volume dv is $\Psi^* \Psi dv$ where Ψ^* is the complex conjugate obtained by replacing $i \rightarrow -i$ in the expression for Ψ. The probability of locating an electron somewhere in space is one as expressed by the normalization condition

$$\iiint_{\text{vol}} \Psi^* \Psi \, dv = 1.$$

The time-dependent Schrödinger equation in three dimensions is a diffusion equation similar to Fick's second law as mentioned in Chapter 6

$$-\frac{\hbar^2}{2m}\nabla^2\Psi(\mathbf{r}, t) + V(x, t)\Psi(\mathbf{r}, t) = i\hbar\frac{\partial}{\partial t}\Psi(\mathbf{r}, t)$$

where V is a potential energy function. Certain quantum systems are in states where the probability of locating the electron is independent of time. If we can express the wave function as

$$\Psi(\mathbf{r}, t) = \psi(\mathbf{r})e^{i\omega t}$$

where $\omega = E/\hbar$ and E is the energy of the electron, then we see that

$$\Psi^*\Psi = \psi(\mathbf{r})^2 e^{i\omega t} e^{-i\omega t} = \psi(\mathbf{r})^2.$$

Ψ is then referred to as a *stationary state* satisfying the time-independent Schrödinger equation

$$-\frac{\hbar^2}{2m}\nabla^2\Psi + V(x)\Psi = E\Psi$$

where V is not changing with time. In Chapter 8 we used the particle-in-a-box model to describe light absorption in biomolecules and the vibrational spectra of atoms. The one-dimensional box is described by the potential

$$V(x) = \begin{cases} 0 & 0 < x < L \\ \infty & x \leq 0 \text{ or } x \geq L \end{cases}.$$

In one dimension the Schrödinger equation is

$$\frac{\partial^2 \psi}{\partial x^2} = -\frac{2mE}{\hbar^2}\psi$$

inside the box where $V(x) = 0$. Making the substitution

$$k^2 = \frac{2mE}{\hbar^2}$$

the general solution to this differential equation is

$$\psi(x) = A\sin(kx) + B\cos(kx).$$

Outside the box the potential is infinite so we require that the wave function vanish at the walls $x = 0$ and $x = L$. At $x = 0$, $\psi(0) = A\sin(0) + B\cos(0) = B$ and we have that $B = 0$. At the wall $x = \mathrm{L}$, $\psi(L) = A\sin(kL)$ so we have that

$$kL = n\pi \qquad n = 1, 2, 3 \ldots$$

for $\psi(L) = 0$. For integer values of n, the above condition permits only discrete or quantized energy levels

$$E_n = \left(\frac{\hbar^2\pi^2}{2mL^2}\right)n^2.$$

Now that we have the energy levels, we return to the wave functions now expressed as

$$\psi_n(x) = A\sin\left(\frac{n\pi x}{L}\right).$$

The constant A is determined by requiring that the wave functions are normalized so that the probability of locating the electron somewhere along the x-axis is one expressed by

$$\int_{-\infty}^{\infty} \psi^* \psi dx = 1.$$

Since the electron is constrained between 0 and L, we have to only integrate over this interval, and our normalization condition becomes

$$A^2 \int_0^L \sin^2\left(\frac{n\pi x}{L}\right) dx = 1.$$

Making use of the trigonometric identity

$$\sin^2\left(\frac{n\pi x}{L}\right) = \frac{1}{2}\left(1 - \cos\frac{2n\pi x}{L}\right)$$

we have

$$\int_0^L \sin^2\left(\frac{n\pi x}{L}\right) dx = \frac{L}{2}$$

so that

$$A = \sqrt{\frac{2}{L}},$$

and the wave function is now

$$\psi_n(x) = \sqrt{\frac{2}{L}} \sin\left(\frac{n\pi x}{L}\right).$$

This wave function also describes a heavy molecule constrained to a one-dimensional box of length L. However, a heavier particle has a lower energy inversely proportioned to its mass. The partition function for a molecule constrained to a 1-D box can be obtained from the energy levels

$$Z = \sum_n \exp\left(-\frac{E_n}{k_B T}\right) = \sum_n \exp\left(-\frac{\hbar^2 \pi^2}{2mL^2 k_B T} n^2\right).$$

For sufficiently large L and T, the energy levels are closely spaced, and we approximate the sum by an integral

$$Z = \int_0^\infty \exp\left(-\frac{\hbar^2 \pi^2}{2mL^2 k_B T} n^2\right) dn.$$

This is a Gaussian integral, also tabulated in Appendix 1, with solution

$$Z = \frac{\sqrt{\pi}}{2}\sqrt{\frac{2mL^2k_BT}{\hbar^2\pi^2}} = \sqrt{\frac{2\pi mk_BT}{h^2}}L.$$

Remember that Z is a dimensionless quantity. We can express it in terms of a characteristic length $Z = L/L_Q$ where

$$L_Q = \frac{h}{\sqrt{2\pi mk_BT}}$$

represents a quantum length. L_Q is roughly 0.2 Å at room temperature for a nitrogen molecule.

EXERCISES

Exercise A7.1. Calculate L_Q for a carbon atom at room temperature.

Exercise A7.2. A molecule is constrained to a 3-D box of side L. Show that $Z = V/V_Q$ where $V_Q = L_Q^3$. Write an expression for Z if the particle is constrained in a box with sides of unequal lengths a, b, and c.

8 Biophysical Applications of QuickField®

What Is QuickField?

QuickField is a Finite Element Method CAD modeling software that supports a variety of electromagnetic, thermal, and stress-analysis problems that have either two-dimensional x–y symmetry or axial symmetry. Problem types supported include Electrostatics, Magnetostatics, AC Magnetics, Transient Magnetics, DC Conduction, AC Conduction, Transient and Steady State Thermal Analysis, Stress Analysis, and Electrical Circuits. Multiphysics analysis capabilities include various couplings between Thermal, Stress, Electrostatic, Magnetostatic, Time Harmonic Magnetic, and AC and DC Conduction. Table A8.1 gives an overview of analysis types supported by QuickField.

Table A8.1 Capabilities of QuickField FEM Software

Analysis and Symmetry	Material Properties	Sources	Boundary Conditions	Post-Processing Results
Electrostatics • *X–Y* Symmetry • Axial Symmetry	• Isotropic Permittivity • Orthotropic Permittivity • Air $\varepsilon = \varepsilon_0$	• Voltages • Point Charges • Surface Charges • Volume Charges	• Dirichlet Condition (Specified Voltages) • Neumann Condition (Specified Charge Density) • Constant Potential with Specified Total Charge	• Voltages • Electric Fields • Field Gradients • Flux Density • Surface Charges • Capacitances • Electric Forces • Torques • Electric Energy
Magnetostatics • *X–Y* Symmetry • Axial Symmetry	• Isotropic Permeability • Orthotropic Permeability • Air $\mu = \mu_0$ • **B–H** curves: Ferromagnets and Superconductors	• Line Currents • Current Density • Uniform Field • Permanent Magnets	• Dirichlet Condition (Vector Potential **A**) • Neumann Condition (Tangential **B** Field) • Superconductor (Zero Normal **B**)	• Vector Potential **A** • Magnetic Fields (Flux Density **B**) • Field Intensity **H** • Magnetic Forces • Torques • Magnetic Energy • Flux Linkages • Inductances (Self and Mutual)
AC Magnetics • *X–Y* Symmetry • Axial Symmetry	• Isotropic Permeability • Orthotropic Permeability • Air $\mu = \mu_0$ • Conductivity	• Line Currents (Time Harmonic) • Current Density (Time Harmonic) • Uniform AC Magnetic Field • AC Voltage	• Dirichlet Condition (Vector Potential **A**) • Neumann Condition (Tangential **B** Field) • Superconductor (Zero Normal **B**)	• Vector Potential **A** • Magnetic Fields (Flux Density **B**) • Field Intensity **H** • Maxwell and Lorentz Forces (Peak and Time Averages) • Maxwell and Lorentz Torques (Peak and Time Averages) • Magnetic Energy • Flux Linkages • Inductance (Self and Mutual) • Eddy Current • Joule Heat • Impedances

Analysis and Symmetry	Material Properties	Sources	Boundary Conditions	Post-Processing Results
Transient Magnetics • X–Y Symmetry • Axial Symmetry	• Isotropic Permeability • Orthotropic Permeability • Air $\mu = \mu_0$ • **B–H** curves: Ferromagnets and Superconductors	• Line Currents (Time Dependent) • Current Density (Time Dependent) • Uniform Field • Permanent Magnets	• Dirichlet Condition (Vector Potential **A**) • Neumann Condition (Tangential **B** Field) • Superconductor (Zero Normal **B**)	• Vector Potential **A** • Magnetic Fields (Flux Density **B**) • Field Intensity **H** • Magnetic Forces • Magnetic Torques • Magnetic Energy • Flux Linkages • Inductances (Self and Mutual) • Eddy Currents
DC Conduction • X–Y Symmetry • Axial Symmetry	• Isotropic Resistivity • Orthotropic Resistivity	• Voltage • Current Density	• Dirichlet Condition (Specified Voltage) • Normal Derivative (Surface Current Density) • Constant Potential (Given Constraint)	• Voltages • Current Density • Electric Fields • Total Currents • Power Loss
AC Conduction • X–Y Symmetry • Axial Symmetry	• Isotropic Resistivity • Orthotropic Resistivity • Isotropic Permittivity • Orthotropic Permittivity	• AC Voltages • Current Density (Time Harmonic)	• Dirichlet Condition (Specified Voltages) • Normal Derivative (Surface Current Densities) • Constant Potential (Given Constraint)	• Voltages • Current Density • Electric Fields • Total Currents • Power Loss • Voltages • Electric Forces (Peak and Time Averages) • Electric Torques (Peak and Time Averages)
Thermal Analysis • X–Y Symmetry • Axial Symmetry	• Thermal Conductivity (Isotropic for Temperature Dependent) • Thermal Conductivity (Orthotropic Temperature Independent)	• Volume Heat Densities (Constant and Temperature Dependent) • Convective Sources • Radiative Sources	• Specified Temperatures • Boundary Heat Flows • Convection • Radiation	• Temperatures • Thermal Gradients • Heat Flux Density • Total Heat Loss or Gain Over Subregion of Model

(continued)

Table A8.1 (*continued*)

Analysis and Symmetry	Material Properties	Sources	Boundary Conditions	Post-Processing Results
	• Specific Heat (Temperature Dependent)	• Joule Heat Sources (Imported from DC or AC Conduction or AC or Transient Magnetic Analysis)	• Specified Constraints for Constant Temperature Boundaries	• Time Dependence of Quantities (Transient Analysis)
Stress Analysis • *X–Y* Symmetry • Axial Symmetry	• Isotropic Materials • Orthotropic Materials	• Body Forces • Pressure • Thermal Strain • Imported Electric or Magnetic Force from Electric or Magnetic Analysis	• Prescribed Displacements • Elastic Spring Supports	• Displacement • Stress Components • Principle Stress • von Mises Stress • Tresca • Mohr–Coulomb • Drucker–Prager • Hill Criteria
Electrical Circuits (AC Magnetics and Transient Magnetics)	• Resistors • Capacitors • Inductors • Conductivity and Permittivity of Connected Model Blocks	• AC Voltage and Current • DC Voltage and Current • Transient Voltage and Current	• Electrical Wires	• Voltage • Current • Frequency Response

Basic Organization of QuickField

The basic steps in setting up, solving, and analyzing problem results in QuickField are outlined in Figure A8.1. These steps include specifying the problem type, designing the geometrical model, and assigning material properties and boundary conditions. Finally the problem is solved, and the results are analyzed.

Biophysical Applications of QuickField

QuickField can be applied to a wide range of biophysical problems involving current flow, thermal and stress analysis, bioimpedance, and equivalent circuit models.

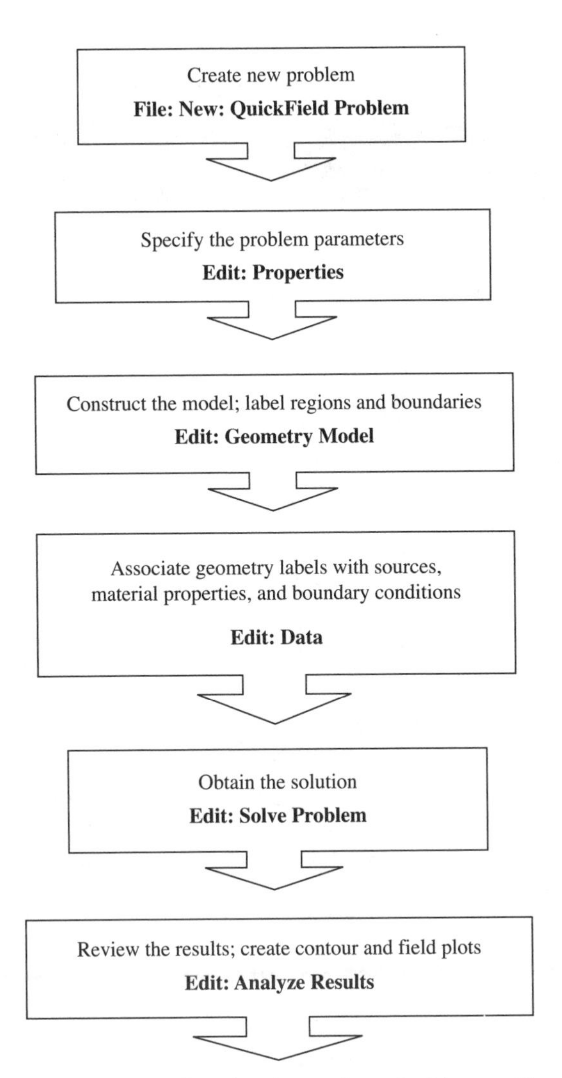

FIGURE A8.1 Procedure for creating and solving a problem in QuickField.

Biophysical problems that can be simulated in QuickField are outlined in Table A8.2. Biological membranes can be modeled with finite-element mesh-grading spanning several orders of magnitude. A hyperfine mesh is required to simulate a lipid bilayer with a thickness of 7 nm compared to the cell size of ~10 μm.

Table A8.2 Biophysical Applications in QuickField

Analysis and Symmetry	Biophysical Calculations in QuickField
Electrostatics • *X–Y* Symmetry • Axial Symmetry	• Field distributions of molecular ion and dipole sources • Plasma and organelle membrane potentials and field distributions • Forces and torques on membrane-bound protein dipole sources • Dielectrophoretic force acting on cells and biomolecules in nonuniform electric fields • Electrorotation forces on cells of various shapes
Magnetostatics • *X–Y* Symmetry • Axial Symmetry	• Magnetic fields generated by bioelectric currents • Magnetotactic bacteria • Current dipole model of action potentials • Forces and torques acting between bioelectric current sources • Forces and torques acting on bioelectric sources in external B fields • Magnetostatic forces acting on weakly diamagnetic biomaterials in high magnetic fields
AC Magnetics • *X–Y* Symmetry • Axial Symmetry	• AC susceptometry of tissue • AC currents induced in tissue and cells by alternating magnetic fields • Joule heating of tissue by alternating magnetic fields • AC currents and induced potentials across plasma and organelle membranes • Magnetic response of ferrite particles in tissue (Liver susceptometry)
Transient Magnetics • *X–Y* Symmetry • Axial Symmetry	• Transient currents induced in biomaterials by magnetic field pulses and periodic waveforms (including square wave, triangle, sawtooth, rectified sine wave, delta train) • Transient currents and induced potentials across plasma and organelle membranes
DC Conduction • *X–Y* Symmetry • Axial Symmetry	• DC current flow through biomaterials • Bioelectric return currents due to action potentials • DC currents through gap junctions, plasma, and organelle membranes
AC Conduction • *X–Y* Symmetry • Axial Symmetry	• Bioimpedance calculations of tissue and cells • AC currents induced in tissue and cells by alternating magnetic fields
Thermal Analysis • *X–Y* Symmetry • Axial Symmetry	• Heat transfer between biomaterials via radiation, convection, and conduction
Stress Analysis • *X–Y* Symmetry • Axial Symmetry	• Stresses and strains in biomaterials with specified Young's modulus and Poisson's ratio • Coupled electric stresses in biomaterials • Coupled thermal stresses in biomaterials
Electrical Circuits (AC Magnetics and Transient Magnetics)	• Equivalent circuit models of tissue and cells under electrical stimulation • Dendrite passive signal propagation • Gap junction circuit models • Circuit models of electrode interfaces • Electrical circuits linked to tissue elements constructed in AC and Transient Magnetics modules

Biological Material Properties

Table A9.1 Young's Moduli of Biomaterials

Material	Young's Modulus (GPa)
Soft Biological Tissue	0.0002
Human Cartilage	0.024
Collagen	0.6–1.2
Human Tendon	0.6
Compact Bone in Tension	17
Compact Bone in Compression	10
Cortical Bone (longitudinal)	9.6
Cortical Bone (transverse)	17.4
Elastin	0.0006–0.0011
Blood Vessels	0.0002
Lipid Bilayer (membrane)	0.012
Erythrocytes Membranes	0.056
Lymphocyte Membranes	0.08

Table A9.2 Thermal Conductivities of Biomaterials

Material	Thermal Conductivity (W/m K)
Skin	0.34
Muscle	0.46
Fat (pig)	0.17

Table A9.3 Geometric and Electrical Properties of Cellular Components

Mitochondrion (spherical model)	Dimension	Conductivity (S/m)	Relative Permittivity
Outer Membrane	7 nm (thickness)	$0.95\ 10^{-6}$	12.1
Inner Membrane Space	0.92 μm	0.4	54.0
Inner Membrane	7 nm (thickness)	$0.95\ 10^{-6}$	3.4
Matrix	0.832 μm	0.121	54
Cellular Plasma Membrane	5 nm (thickness)	$1.0\ 10^{-5}$–$1.0\ 10^{-7}$	8–11.3
Cytoplasm	1~10 μm	0.5	80
Extracellular Medium	1~10 μm	1.0	80

Table A9.4 DC Resistivities of Cardiac Tissues

Tissue During Cardiac Phase	DC Resistivity (Ohm-m)
Blood	1.38–1.54 Ohm-m (between end-systole and end-diastole phases)
Myocardium	4.2 Ohm-m
Lungs	14.2–15 Ohm-m (between end-systole and end-diastole phases)
Thorax (average)	40 Ohm-m

Cellular and Organelle Dimensions

Cell/Organelle	Dimension (microns)
Mitochondrion	0.1–5 μm
E. coli	2 μm
Liver Cell	20 μm
Nerve Cell	5 μm (diameter) (1 m length)
Erythrocyte (red blood cell)	2 μm (thickness) 8 μm (diameter)
Bacteria	0.2–2 μm (diameter) 1–400 μm (length)
Cell Membrane	5–8 nm (thickness)

Table A9.5 AC Electrical Properties of Tissue

Tissue	Conductivity (S/m)	Relative Permittivity
Skin	10^{-3} (kHz) 0.1 (MHz)	10^6 (kHz) 10^3 (MHz)
Bone	0.01 (DC)	3,000–10,000 (DC)
Fat	0.03 (10^2–10^6 Hz)	10^6 (100 Hz) 10 (MHz)
Fat	0.1–0.3 (0.4–5 GHz)	5 (0.4–5 GHz)
Breast Fat	0.2–1 (0.4–5 GHz)	5–50 (0.4–5 GHz)
Breast Tumor	0.7–3 (0.4–5 GHz)	47–67 (0.4–5 GHz)
Brain	0.1 (10^2–10^6 Hz)	10^7 (100 Hz) 10^3 (MHz)
Muscle	0.4 (10^2–10^6 Hz)	10^6 (kHz) 10^3 (MHz)
Blood	0.7 (DC)	10^5 (kHz)

Table A9.6 Physical Constants

Physical Quantity	Symbol	Numerical Value
Avogadro's number	N_A	6.022×10^{23} mol^{-1}
Boltzmann's constant	k_B	1.381×10^{-23} J/K
Gas constant	$R = N_A k_B$	8.314 J/mol·K
Electron mass	m_e	9.109×10^{-31} kg
Electron volt	eV	1.602×10^{-19} J
Electron charge	e	1.602×10^{-19} C
Permeability of free space	μ_0	$4\pi \times 10^{-7}$ T·m/A
Permittivity of free space	ε_0	8.854×10^{-12} C^2/N·m^2
Planck's constant	h	6.626×10^{-34} J·s
Plank's constant/2π	$\hbar$	1.055×10^{-34} J·s
Proton mass	m_p	1.673×10^{-27} kg
Speed of light	$c = \frac{1}{\sqrt{\mu_0 \varepsilon_0}}$	2.998×10^8 m/s
Gravitational constant	G	$6.673 \cdot 10^{-11}$ Nm^2/kg^2
Acceleration due to gravity	g	9.81 m/s^2
Fine structure constant	$\alpha = \frac{1}{4\pi\varepsilon_0}\frac{e^2}{\hbar c}$	$\frac{1}{137.04}$ (Dimensionless)
Bjerrum length	$\ell_B = \frac{e^2}{4\pi\varepsilon k_B T}$	~0.7 nm ($\varepsilon = 78\varepsilon_0$ $T = 273$ K) (In water)
Thermal energy	$k_B T_{\text{room}}$	$\sim\frac{1}{40}$ eV ~ 4.1 pN·nm

(continued)

Table A9.6 (*continued*)

Physical Quantity	Symbol	Numerical Value
Stefan-Boltzmann constant	σ	$5.67 \cdot 10^{-8} \frac{\mathrm{W}}{\mathrm{m^2K^4}}$
Faraday constant	$F = N_A e$	$9.6485 \cdot 10^4\ \mathrm{C \cdot mol^{-1}}$
Bohr magneton	$\mu_B = \frac{e\hbar}{2m_e}$	$9.274 \cdot 10^{-24}\ \mathrm{J/T} = 5.788 \cdot 10^{-5}\ \mathrm{eV/T}$
Flux quantum	Φ_0	$2.068 \cdot 10^{-15}\ \mathrm{Wb}$
Energy Conversion	1 eV/molecule	23 kcal/mole

10 Solutions of the Linearized Poisson–Boltzmann Equation

The linearized Poisson–Boltzmann (P–B) equation $\nabla^2 U_e = \kappa^2 U_e$ has the same mathematical form as the Helmholtz equation and may be solved using the technique of separation of variables in various orthogonal coordinate systems as shown in Table A10.1. The expressions below can be discretized for finite difference or Monte Carlo methods in instances where the linearized P–B equation is not separable, or for numerically solving the nonlinear P–B equation.

Tabulated solutions of the Helmholtz equation in the literature usually correspond to the form $\nabla^2 y = -\chi^2 y$ so some care must be taken in adapting solutions to the linearized P–B equation. Also, there is considerable variance in functional forms and notations used by different authors. The forms presented below are similar to Moon and Spencer with the replacements $\eta \rightarrow \xi$, $\theta \rightarrow \eta$ in prolate, oblate and bispherical coordinate systems, while α and β are used for planar coordinates in elliptic and bicylindrical coordinates.

Table A10 Separability of the linearized P–B equation in several orthogonal coordinate systems. Other separable combinations are possible in 2-D Cartesian and Cylindrical systems. In some cases, analytical forms and approximations may be found when the P–B equation is not separable

Coordinate System	3D	2D	1D
Cartesian	$U_e(x, y, z) = X(x)Y(y)Z(z)$	$U_e(x, y) = X(x)Y(y)$	$U_e(x) = X(x)$
Cylindrical	$U_e(r, z, \phi) = R(r)Z(z)\Phi(\phi)$	$U_e(r, z) = R(r)Z(z)$	$U_e(r) = R(r)$
Elliptic Cylindrical	Not Separable	$U_e(\alpha, \beta) = \Lambda(\alpha)\Omega(\beta)$	No Solution
Bicylindrical	Not Separable	Not Separable	No Solution
Spherical	$U_e(r, \theta, \phi) = R(r)\Theta(\theta)\Phi(\phi)$	$U_e(r, \theta) = R(r)\Theta(\theta)$	$U_e(r) = R(r)$
Prolate Spheroidal	$U_e(\xi, \eta, \phi) = \Upsilon(\xi)\Theta(\eta)\Phi(\phi)$	$U_e(\xi, \eta) = \Upsilon(\xi)\Theta(\eta)$	No Solution
Oblate Spheroidal	$U_e(\xi, \eta, \phi) = \Upsilon(\xi)\Theta(\eta)\Phi(\phi)$	$U_e(\xi, \eta) = \Upsilon(\xi)\Theta(\eta)$	No Solution
Bispherical	Not Separable	Not Separable	No Solution

Boundary Conditions

The boundary conditions for electric potential U_e are Dirichlet (constant potential) or Neumann (constant surface charge density). A fixed layer of cations and anions will develop at electropositive and electronegative electrode/electrolyte interfaces. This is known as the Helmholtz layer resulting in a modified surface potential. It is the modified potential just above the Helmholtz layer that is taken as the boundary condition for the P–B equation.

Cartesian Coordinates

In Cartesian coordinates, the linearized P–B equation $\nabla^2 U_e(x, y, z) = \kappa^2 U_e$ is written as

$$\frac{\partial^2 U_e}{\partial x^2} + \frac{\partial^2 U_e}{\partial y^2} + \frac{\partial^2 U_e}{\partial z^2} = \kappa^2 U_e \tag{A10.1}$$

where $-\infty < x < \infty$, $-\infty < y < \infty$, $-\infty < z < \infty$. Applying separation of variables we assume a product solution

$$U_e(x, y, z) = X(x)\,Y(y)\,Z(z). \tag{A10.2}$$

Substituting this form of U_e into the P–B equation and dividing by U_e gives

$$\frac{1}{X}\frac{\partial^2 X}{\partial x^2} + \frac{1}{Y}\frac{\partial^2 Y}{\partial y^2} + \frac{1}{Z}\frac{\partial^2 Z}{\partial z^2} = \kappa^2. \tag{A10.2}$$

Each term on the left–hand side must be equal to a constant if the sum is κ^2 for all x, y and z, since these variables may vary independently

$$\frac{1}{X}\frac{\partial^2 X}{\partial x^2} = -\alpha^2 \quad \frac{1}{Y}\frac{\partial^2 Y}{\partial y^2} = -\beta^2 \quad \frac{1}{Z}\frac{\partial^2 Z}{\partial z^2} = \alpha^2 + \beta^2 + \kappa^2 \tag{A10.3}$$

where here α^2 and β^2 are separation constants and

$$U_e(x, y, z) = \begin{Bmatrix} \sin \alpha x \\ \cos \alpha x \end{Bmatrix} \begin{Bmatrix} \sin \beta y \\ \cos \beta y \end{Bmatrix} \begin{Bmatrix} \sinh \sqrt{\alpha^2 + \beta^2 + \kappa^2} z \\ \cosh \sqrt{\alpha^2 + \beta^2 + \kappa^2} z \end{Bmatrix} \tag{A10.4}$$

where $\begin{Bmatrix} \sin \alpha x \\ \cos \alpha x \end{Bmatrix}$ denotes $c_1 \sin \alpha x + c_2 \cos \alpha x$, etc. We may also express the circular functions as a linear combination of complex exponentials

$$\begin{Bmatrix} \sin \alpha x \\ \cos \alpha x \end{Bmatrix} \rightarrow \begin{Bmatrix} e^{i\alpha x} \\ e^{-i\alpha x} \end{Bmatrix} \tag{A10.5}$$

with different values for c_1 and c_2 determined by the boundary conditions. Also, we may equivalently write the hyperbolic functions as a linear combination of real exponentials

$$\begin{Bmatrix} \sinh\sqrt{\alpha^2+\beta^2+\kappa^2}z \\ \cosh\sqrt{\alpha^2+\beta^2+\kappa^2}z \end{Bmatrix} \rightarrow \begin{Bmatrix} e^{\sqrt{\alpha^2+\beta^2+\kappa^2}z} \\ e^{-\sqrt{\alpha^2+\beta^2+\kappa^2}z} \end{Bmatrix}. \tag{A10.6}$$

Note that the separation constants could also have been chosen so that the hyperbolic functions corresponded to either the x- or the y-coordinates. If an electrolytic solution is contained in a rectangular conduit where there is no potential variation in the z-direction, we may choose

$$\frac{1}{X}\frac{\partial^2 X}{\partial x^2} = -\alpha^2 \quad \frac{1}{Y}\frac{\partial^2 Y}{\partial y^2} = \alpha^2 + \kappa^2 \tag{A10.7}$$

so that

$$U_e(x,y) = \begin{Bmatrix} \sin\alpha x \\ \cos\alpha x \end{Bmatrix} \begin{Bmatrix} \sinh\sqrt{\alpha^2+\kappa^2}\,y \\ \cosh\sqrt{\alpha^2+\kappa^2}\,y \end{Bmatrix}. \tag{A10.8}$$

If the conduit is sufficiently long so that there is also no potential variation in the y-direction we have

$$\frac{1}{X}\frac{\partial^2 X}{\partial x^2} = \kappa^2 \tag{A10.9}$$

so that $U_e(x) = \begin{Bmatrix} \sinh\kappa x \\ \cosh\kappa x \end{Bmatrix}$, or $U_e(x) = c_1\sinh\kappa x + c_2\cosh\kappa x$. Applying the Dirichlet boundary conditions $U_e(0) = V_0$ and $U_e(L) = -V_0$ gives $c_2 = V_0$ and $-V_0 = c_1\sinh\kappa L + V_0\cosh\kappa L$ so that

$$U_e(x) = -V_0\frac{(1+\cosh\kappa L)}{\sinh\kappa L}\sinh\kappa x + V_0\cosh\kappa x. \tag{A10.10}$$

Consider the equivalent problem expressing $U_e(x) = \begin{Bmatrix} e^{\kappa x} \\ e^{-\kappa x} \end{Bmatrix}$ with $U_e(x) = c_1e^{\kappa x}+c_2e^{-\kappa x}$ where we now take the $x = 0$ plane between the two electrodes. Applying the boundary conditions $U_e\left(-\frac{L}{2}\right) = V_0$ and $U_e\left(\frac{L}{2}\right) = -V_0$ gives

$V_0 = c_1 e^{-\kappa\frac{L}{2}} + c_2 e^{\kappa\frac{L}{2}}$ and $-V_0 = c_1 e^{\kappa\frac{L}{2}} + c_2 e^{-\kappa\frac{L}{2}}$ so that $c_1 = -c_2$ and $V_0 = c_1\left(e^{-\kappa\frac{L}{2}} - e^{\kappa\frac{L}{2}}\right)$ which gives $c_1 = \dfrac{V_0}{2\sinh\kappa\frac{L}{2}}$ and we can write the potential in the more compact form

$$U_e(x) = -V_0\frac{\sinh\kappa x}{\sinh\kappa\frac{L}{2}}. \tag{A10.11}$$

The surface charge density at $x = \pm L/2$ is then

$$\sigma = -\varepsilon\frac{\partial U_e}{\partial n} = \pm\varepsilon\left.\frac{\partial U_e}{\partial x}\right|_{x=\pm L/2} = \mp\varepsilon V_0\kappa\frac{\cosh\kappa\frac{L}{2}}{\sinh\kappa\frac{L}{2}} = \mp\varepsilon V_0\kappa\coth\kappa\frac{L}{2} \tag{A10.12}$$

where the unit normal is $-\hat{\mathbf{i}}$ on the right plate and $\hat{\mathbf{i}}$ on the left plate where σ is respectively negative and positive. In general, the analytical form of potentials varying in two or three dimensions will be much more complicated often involving infinite sums.

Cylindrical Coordinates

The cylindrical coordinates (r, z, ϕ) are related to the Cartesian coordinates

$$\begin{aligned} x &= r\cos\phi \\ y &= r\sin\phi \\ z &= z \end{aligned} \tag{A10.13}$$

The linearized P–B equation is written

$$\frac{1}{r}\frac{\partial}{\partial r}\left(r\frac{\partial U_e}{\partial r}\right) + \frac{\partial^2 U_e}{\partial z^2} + \frac{1}{r^2}\frac{\partial^2 U_e}{\partial \phi^2} = \kappa^2 U_e \tag{A10.14}$$

where the separation constants α^2 and n^2 may be chosen so that

$$U_e(r, z, \phi) = \begin{Bmatrix} J_n(\alpha r) \\ Y_n(\alpha r) \end{Bmatrix}\begin{Bmatrix} \sinh\sqrt{\alpha^2+\kappa^2}z \\ \cosh\sqrt{\alpha^2+\kappa^2}z \end{Bmatrix}\begin{Bmatrix} \sin(n\phi) \\ \cos(n\phi) \end{Bmatrix} \tag{A10.15}$$

where $J_n(\alpha r)$ and $Y_n(\alpha r)$ are Bessel functions. In axial symmetry without ϕ dependence,

$$\frac{1}{r}\frac{\partial}{\partial r}\left(r\frac{\partial U_e}{\partial r}\right) + \frac{\partial^2 U_e}{\partial z^2} = \kappa^2 U_e \tag{A10.16}$$

with solutions

$$U_e(r, z, \phi) = \begin{Bmatrix} J_0(\alpha r) \\ Y_0(\alpha r) \end{Bmatrix} \begin{Bmatrix} \sinh\sqrt{\alpha^2 + \kappa^2}z \\ \cosh\sqrt{\alpha^2 + \kappa^2}z \end{Bmatrix}. \tag{A10.17}$$

If U_e is only a function of r then

$$\frac{\partial}{\partial r}\left(r\frac{\partial U_e}{\partial r}\right) = \kappa^2 r U_e \tag{A10.18}$$

with solutions

$$U_e(r) = \begin{Bmatrix} I_0(\kappa r) \\ K_0(\kappa r) \end{Bmatrix}. \tag{A10.19}$$

The modified Bessel function $K_0(\kappa r)$ diverges at $r = 0$ and exponentially decays at large values of r while $I_0(\kappa r)$ is finite at $r = 0$ and diverges as $r \to \infty$. As an example consider the potential outside of a long cylinder of radius a with surface charge density σ in an electrolytic solution. Since the origin is excluded from the problem, we choose

$$U_e(r) = c_1 K_0(\kappa r). \tag{A10.20}$$

The constant c_1 is determined from the Neumann boundary condition

$$\sigma = -\varepsilon\frac{\partial U_e}{\partial n} = -\varepsilon\frac{\partial U_e}{\partial r}\bigg|_{r=a} = \varepsilon\kappa c_1 K_1(\kappa a) \tag{A10.21}$$

so that

$$U_e(r) = \frac{\sigma K_0(\kappa r)}{\varepsilon\kappa K_1(\kappa a)}. \tag{A10.22}$$

Note that the axially symmetric potential between two concentric cylinders will be expressed in term of both K_0 and I_0 Bessel functions.

Bicylindrical Coordinates

The bicylindrical coordinates (α, β, z) are related to the Cartesian coordinates

$$x = \frac{a \sinh \alpha}{\cosh \alpha - \cos \beta}$$
$$y = \frac{a \sin \beta}{\cosh \alpha - \cos \beta} \qquad \text{(A10.23)}$$
$$z = z$$

where $-\infty < \alpha < \infty$, $0 \leqslant \beta \leqslant 2\pi$, $-\infty < z < \infty$. The coordinate surfaces are circular cylinders parallel to the z-axis and planes perpendicular to the z-axis.

The linearized P–B equation in bicylindrical coordinates is written

$$\frac{(\cosh \alpha - \cos \beta)^2}{a^2}\left[\frac{\partial^2 U_e}{\partial \alpha^2} + \frac{\partial^2 U_e}{\partial \beta^2}\right] + \frac{\partial^2 U_e}{\partial z^2} = \kappa^2 U_e. \qquad \text{(A10.24)}$$

If the variation in the z-direction can be neglected, we may write

$$\frac{\partial^2 U_e}{\partial \alpha^2} + \frac{\partial^2 U_e}{\partial \beta^2} = \frac{\kappa^2 a^2}{(\cosh \alpha - \cos \beta)^2} U_e. \qquad \text{(A10.25)}$$

Unfortunately the linearized P–B equation is not separable in bicylindrical coordinates so that it must be solved numerically. The bicylindrical coordinate system is ideal for modeling the potential distribution in an electrolyte between two cylindrical electrodes. The interaction between two double stranded DNA molecules can also be modeled solving the nonlinear bicylindrical P–B equation.

Elliptic Cylindrical Coordinates

The elliptic cylindrical coordinates (α, β, z) are related to the Cartesian coordinates

$$x = a \cosh \alpha \cos \beta$$
$$y = a \sinh \alpha \sin \beta \qquad \text{(A10.26)}$$
$$z = z$$

where $0 \leqslant \alpha < \infty$, $0 \leqslant \beta \leqslant 2\pi$, $-\infty < z < \infty$. The coordinate surfaces are elliptic (α = const.) and hyperbolic (β = const.) cylinders sharing the same focal line (confocal) with planes perpendicular to the z-axis. The elliptic cylinders have semi major axis a.

The linearized P–B equation in elliptic cylindrical coordinates is written

$$\frac{1}{a^2(\sinh^2 \alpha + \sin^2 \beta)}\left[\frac{\partial^2 U_e}{\partial \alpha^2} + \frac{\partial^2 U_e}{\alpha \beta^2}\right] + \frac{\partial^2 U_e}{\partial z^2} = \kappa^2 U_e \qquad \text{(A10.27)}$$

Substituting $U_e(\alpha, \beta, z) = \Lambda(\alpha)\Omega(\beta)Z(z)$ a product solution may be found where Λ and Ω are expressed in terms of Mathieu functions. Without variation in the z-direction we have

$$\frac{\partial^2 U_e}{\partial \alpha^2} + \frac{\partial^2 U_e}{\partial \beta^2} = \kappa^2 a^2 \left(\sinh^2\alpha + \sin^2\beta\right) U_e. \tag{A10.28}$$

Elliptic coordinates are useful for modeling the potential distribution in an electrolyte surrounding a flat electrode of finite width.

Spherical Coordinates

The spherical coordinates (r, θ, ϕ) are related to the Cartesian coordinates

$$\begin{aligned} x &= r\sin\theta\cos\phi \\ y &= r\sin\theta\sin\phi \\ z &= r\cos\theta \end{aligned} \tag{A10.29}$$

where $r \geq 0$, $0 \leq \theta \leq \pi$, $0 \leq \phi < 2\pi$. The coordinate surfaces are spheres of constant r, cones of constant θ and half planes of constant ϕ parallel to the z-axis.

The linearized P–B equation is written

$$\frac{1}{r^2}\frac{\partial}{\partial r}\left(r^2\frac{\partial U_e}{\partial r}\right) + \frac{1}{r^2\sin\theta}\frac{\partial}{\partial \theta}\left(\sin\theta\frac{\partial U_e}{\partial \theta}\right) + \frac{1}{r^2\sin^2\theta}\frac{\partial^2 U_e}{\partial \phi^2} = \kappa^2 U_e. \tag{A10.30}$$

Substituting $U_e(r, \theta, \phi) = R(r)\Theta(\theta)\Phi(\phi)$ with separation constants $\ell(\ell+1)$ and m^2, we obtain

$$U_e(r,\theta,\phi) = \begin{Bmatrix} i_\ell(\kappa r) \\ k_\ell(\kappa r) \end{Bmatrix} \begin{Bmatrix} P_\ell^m(\cos\theta) \\ Q_\ell^m(\cos\theta) \end{Bmatrix} \begin{Bmatrix} \sin(m\phi) \\ \cos(m\phi) \end{Bmatrix}. \tag{A10.31}$$

In axial symmetry,

$$\frac{\partial}{\partial r}\left(r^2\frac{\partial U_e}{\partial r}\right) + \frac{1}{\sin\theta}\frac{\partial}{\partial \theta}\left(\sin\theta\frac{\partial U_e}{\partial \theta}\right) = \kappa^2 r^2 U_e \tag{A10.32}$$

with solutions

$$U_e(r,\theta) = \begin{Bmatrix} i_\ell(\kappa r) \\ k_\ell(\kappa r) \end{Bmatrix} \begin{Bmatrix} P_\ell(\cos\theta) \\ Q_\ell(\cos\theta) \end{Bmatrix} \tag{A10.33}$$

where $i_\ell(\kappa r)$ and $k_\ell(\kappa r)$ are modified spherical Bessel functions and P_ℓ and Q_ℓ are Legendre functions of the first and second kind, respectively. If the potential only varies in the radial direction,

$$\frac{\partial}{\partial r}\left(r^2 \frac{\partial U_e}{\partial r}\right) = \kappa^2 r^2 U_e \tag{A10.34}$$

with solutions

$$U_e(r,\theta) = \begin{Bmatrix} \frac{1}{r}\exp(\kappa r) \\ \frac{1}{r}\exp(-\kappa r) \end{Bmatrix}. \tag{A10.35}$$

Here the positive exponential will have to be discarded in unbounded regions outside a conducting sphere. The potential in an electrolyte between two concentric spheres will consist of a linear combination of both positive and negative exponentials.

Oblate Spheroidal Coordinates

The oblate spheroidal coordinates (ξ, η, ϕ) are related to the Cartesian coordinates

$$\begin{aligned} x &= a\cosh\xi\cos\eta\cos\phi \\ y &= a\cosh\xi\cos\eta\sin\phi \\ z &= a\sinh\xi\sin\eta \end{aligned} \tag{A10.36}$$

where $\xi \geqslant 0$, $0 \leqslant \eta \leqslant \pi$, $0 \leqslant \phi < 2\pi$. The coordinate surfaces are oblate spheroids (ξ = const.) with semi major axis a, and hyperboloids of one sheet (η = const.) revolved about the z-axis, and half planes (ϕ = const.) parallel to the z-axis.

The linearized P–B equation in oblate spheroidal coordinates is written

$$\begin{aligned} &\frac{1}{a^2\left(\sinh^2\xi + \sin^2\eta\right)}\left[\frac{1}{\cosh\xi}\frac{\partial}{\partial\xi}\left(\cosh\xi\frac{\partial U_e}{\partial\xi}\right) + \frac{1}{\cos\eta}\frac{\partial}{\partial\eta}\left(\cos\eta\frac{\partial U_e}{\partial\eta}\right)\right] \\ &+ \frac{1}{a^2\cosh^2\xi\cos^2\eta}\frac{\partial^2 U_e}{\partial\phi^2} = \kappa^2 U_e. \end{aligned} \tag{A10.37}$$

Substituting $U_e(\xi,\eta,\phi) = \Upsilon(\xi)\Theta(\eta)\Phi(\phi)$ with separation constants $\ell(\ell+1)$ and m^2 gives

$$U_e(\xi,\eta,\phi) = \begin{Bmatrix} P_\ell^m(\kappa a, i\sinh\xi) \\ Q_\ell^m(\kappa a, i\sinh\xi) \end{Bmatrix} \begin{Bmatrix} P_i^m(\kappa a, \cos\eta) \\ Q_\ell^m(\kappa a, \cos\eta) \end{Bmatrix} \begin{Bmatrix} \sin(m\phi) \\ \cos(m\phi) \end{Bmatrix} \tag{A10.38}$$

in terms of the oblate spheroidal wave functions. In axial symmetry without ϕ dependence,

$$\begin{aligned}&\frac{1}{\cosh\xi}\frac{\partial}{\partial\xi}\left(\cosh\xi\frac{\partial U_e}{\partial\xi}\right)+\frac{1}{\cos\eta}\frac{\partial}{\partial\eta}\left(\cos\eta\frac{\partial U_e}{\partial\eta}\right)\\&=\kappa^2a^2(\sinh^2\xi+\sin^2\eta)U_e\end{aligned} \tag{A10.39}$$

with solutions

$$U_e(\xi,\eta)=\begin{Bmatrix}P_\ell(\kappa a, i\sinh\xi)\\Q_\ell(\kappa a, i\sinh\xi)\end{Bmatrix}\begin{Bmatrix}P_\ell(\kappa a,\cos\eta)\\Q_\ell(\kappa a,\cos\eta)\end{Bmatrix}. \tag{A10.40}$$

The oblate spheroidal P–B equation is useful for modeling the potential distribution surrounding disk-shaped flats electrodes in electrolytic solutions. This form of the P–B equation may be used to model the counter-ion distribution surrounding living cells where many cell shapes are more accurately modeled as oblate spheroids than perfect spheres.

The general spheroidal wave functions $P_\ell^m(c, z)$ and $Q_\ell^m(c, z)$ wave functions and are not the same as the more familiar associated Legendre functions $P_\ell^m(z)$ and $Q_\ell^m(z)$. The $P_\ell^m(c, z)$ are always odd with $P_\ell^m(c, 0) = 0$ while the $Q_\ell^m(c, z)$ are always even with $Q_\ell^m(c, 0) = 1$.

See also
http://mathworld.wolfram.com/SpheroidalWaveFunction.html
http://mathworld.wolfram.com/OblateSpheroidalWaveFunction.html

Prolate Spheroidal Coordinates

Prolate spheroidal coordinates are useful for modeling the potential distribution surrounding needle-shaped electrodes as well as counterion distributions surrounding rod-shaped bacteria. The prolate spheroidal coordinates (ξ, η, ϕ) are related to the Cartesian coordinates

$$\begin{aligned}x&=a\sinh\xi\sin\eta\cos\phi\\y&=a\sinh\xi\sin\eta\sin\phi\\z&=a\cosh\xi\cos\eta\end{aligned} \tag{A10.41}$$

where $\xi\geqslant 0$, $0\leqslant\eta\leqslant\pi$, $0\leqslant\phi<2\pi$. The coordinate surfaces are prolate spheroids (ξ = const.) with semi major axis a, hyperboloids of two sheets (η = const.) revolved about the z-axis and half planes (ϕ = const.) parallel to the z-axis.

The linearized P–B equation in prolate spheroidal coordinates is written

$$\frac{1}{a^2(\sinh^2\xi+\sin^2\eta)}\left[\frac{1}{\sinh\xi}\frac{\partial}{\partial\xi}\left(\sinh\xi\frac{\partial U_e}{\partial\xi}\right)+\frac{1}{\sin\eta}\frac{\partial}{\partial\eta}\left(\sin\eta\frac{\partial U_e}{\partial\eta}\right)\right]$$
$$+\frac{1}{a^2\sinh^2\xi\sin^2\eta}\frac{\partial^2 U_e}{\partial\phi^2}=\kappa^2 U_e. \tag{A10.42}$$

Substituting $U_e(\xi, \eta, f) = \Upsilon(\xi)\Theta(\eta)\Phi(\phi)$ with separation constants $\ell(\ell+1)$ and m^2 gives the prolate spheroidal wave functions

$$U_e(\xi,\eta,\phi)=\begin{Bmatrix}P_\ell^m(i\kappa a,\cosh\xi)\\ Q_\ell^m(i\kappa a,\cosh\xi)\end{Bmatrix}\begin{Bmatrix}P_\ell^m(i\kappa a,\cos\eta)\\ Q_\ell^m(i\kappa a,\cos\eta)\end{Bmatrix}\begin{Bmatrix}\sin(m\phi)\\ \cos(m\phi)\end{Bmatrix}. \tag{A10.43}$$

In axial symmetry we have

$$\frac{1}{\sinh\xi}\frac{\partial}{\partial\xi}\left(\sinh\xi\frac{\partial U_e}{\partial\xi}\right)+\frac{1}{\sin\eta}\frac{\partial}{\partial\eta}\left(\sin\eta\frac{\partial U_e}{\partial\eta}\right)$$
$$=\kappa^2a^2\left(\sinh^2\xi+\sin^2\eta\right)U_e \tag{A10.44}$$

with solutions

$$U_e(\xi,\eta)=\begin{Bmatrix}P_\ell(i\kappa a,\cosh\xi)\\ Q_\ell(i\kappa a,\cosh\xi)\end{Bmatrix}\begin{Bmatrix}P_\ell(i\kappa a,\cos\eta)\\ Q_\ell(i\kappa a,\cos\eta)\end{Bmatrix}. \tag{A10.45}$$

The $Q_\ell^m(c, z)$ and $P_\ell^m(c, z)$ obey orthogonality conditions useful for solving potential problems

$$\int_{-1}^{1}Q_\ell^m(c,z)Q_{\ell'}^{m'}(c,z)dz=\int_{-1}^{1}P_\ell^m(c,z)P_{\ell'}^{m'}(c,z)dz$$
$$=\frac{2}{2\ell+1}\frac{(\ell+m)!}{(\ell-m)!}\delta_{\ell\ell'}\delta_{mm'} \tag{A10.46}$$

where $\delta_{\ell\ell'}\delta_{mm'} = 1$ if $\ell = \ell'$ and $m = m'$ and zero otherwise so that

$$\int_{-\pi}^{\pi}[Q_\ell^m(c,\cos\eta)]^2d\eta=\int_{-\pi}^{\pi}[P_\ell^m(c,\cos\eta)]^2d\eta=\frac{2}{2\ell+1}\frac{(\ell+m)!}{(\ell-m)!}. \tag{A10.47}$$

See also http://mathworld.wolfram.com/ProlateSpheroidalWaveFunction.html

Bispherical Coordinates

The bispherical coordinates (ξ, η, ϕ) are related to the Cartesian coordinates

$$x = \frac{a \sin \eta \cos \phi}{\cosh \xi - \cos \eta}$$
$$y = \frac{a \sin \eta \sin \phi}{\cosh \xi - \cos \eta} \qquad \text{(A10.48)}$$
$$z = \frac{a \sinh \xi}{\cosh \xi - \cos \eta}$$

where $-\infty < \xi < \infty$, $0 \leq \eta \leq \pi$, $0 \leq \phi < 2\pi$. The coordinate surfaces are spheres (ξ = const.), apple-shaped surfaces revolved about the z-axis ($\eta < \pi/2$), spindle shapes ($\eta > \pi/2$), and half planes (ϕ = const.) parallel to the z-axis.

The linearized P–B equation in bispherical coordinates is written

$$\frac{(\cosh \xi - \cos \eta)^3}{a^2 \sin \eta}\left[\frac{\partial}{\partial \xi}\left(\frac{\sin \eta}{\cosh \xi - \cos \eta}\frac{\partial U_e}{\partial \xi}\right) + \frac{\partial}{\partial \eta}\left(\frac{\sin \eta}{\cosh \xi - \cos \eta}\frac{\partial U_e}{\partial \eta}\right)\right]$$
$$+ \frac{(\cosh \xi - \cos \eta)^2}{a^2 \sin^2 \eta}\frac{\partial^2 U_e}{\partial \phi^2} = \kappa^2 U_e. \qquad \text{(A10.49)}$$

In axial symmetry we have

$$\frac{\partial}{\partial \xi}\left(\frac{\sin \eta}{\cosh \xi - \cos \eta}\frac{\partial U_e}{\partial \xi}\right) + \frac{\partial}{\partial \eta}\left(\frac{\sin \eta}{\cosh \xi - \cos \eta}\frac{\partial U_e}{\partial \eta}\right) = \frac{\kappa^2 a^2 \sin \eta}{(\cosh \xi - \cos \eta)^3} U_e.$$
(A10.50)

The bispherical P–B equation can be used to model the interaction between charged particles in an electrolyte and hence can be used to determine the stability of colloidal suspensions.

EXERCISES

Exercise A10.1 Two planar electrodes located at $x = 0$ and $x = L$ are immersed in a one–one valent electrolyte. Analytically solve the nonlinear form of the P–B equation for the potential U_e between the electrodes with the boundary conditions $U_e(0) = V_0$ and $U_e(L) = -V_0$.

$$\frac{\partial^2 U_e}{\partial x^2} = -\frac{2C_0 q}{\varepsilon}\sinh\left(\frac{qU_e}{k_B T}\right)$$

Compare your solution to the linearized P–B equation where $\kappa^2 = \dfrac{2C_0 q^2}{\varepsilon k_B T}$ for two species with equal and opposite charges.

Exercise A10.2 Write finite difference formulae to numerically solve the linear and the nonlinear P–B equations in cylindrical and bicylindrical coordinates.

References and Further Reading

Chapter 1

1. Alberts, B., A. Johnson, J. Lewis, M. Raff, K. Roberts, and P. Walter. 2007. *Molecular biology of the cell.* 5th ed. Garland Science Taylor & Francis Group, New York.
2. Chopra, A., and C. H. Linewater. 2007. *The major elemental abundance differences between life, the oceans and the sun.* Planetary Science Institute, Canberra, Australia.
3. Mielczarek, E. V. 2006. Resource letter PFBi-1: Physical frontiers in biology. *American Journal of Physics* 74: 375–381.
4. Schrödinger, E. 1956. *What is life? and other scientific essays.* Doubleday, Garden City, NY.
5. Wingreen, N. S. 2006. A glossary of cellular components. *Physics Today* 59: 80–81.

Chapter 2

1. Griffiths, D. J. 1999. *Introduction to electrodynamics.* 3rd ed. Prentice-Hall, Upper Saddle River, NJ.
2. Lamoreaux, S. K. 1997. Demonstration of the Casimir force in the 0.6 to 6 μm range. *Physical Review Letters* 78: 5–8.
3. Milonni, P. W. 1994. *The quantum vacuum: An introduction to quantum electrodynamics.* Academic Press, London.
4. Parsegian, V. A. 2006. *Van der Waals forces: A handbook for biologists, chemists, engineers, and physicists.* Cambridge University Press, New York.
5. Philips, R., and S. R. Quake. 2006. The biological frontier of physics. *Physics Today* 59: 38–43.
6. Schlesener, F., A. Hanke, and S. Dietrich. 2003. Critical Casimir forces in colloidal suspensions. *Journal of Statistical Physics* 110: 981–1013.

Chapter 3

1. Alberts, B. 1989. *Molecular Biology of the Cell,* Vol. 1. Courier Corporation, North Chelmford, MA.
2. Claycomb, J. R. 2008. *Applied electromagnetics using QuickField™ and MATLAB®*. Jones and Bartlett, Sudbury, MA.
3. Glaser, R. 2001. *Biophysics.* Springer-Verlag, Heidelberg, Germany.
4. Serway, R. A., J. W. Jewett, Jr. 2006. *Principles of physics,* 4th ed., Brooks/Cole – Thomson Learning, Belmont, CA.

Chapter 4

1. Baker, G. L. 1976. DNA helix-to-coil transition: A simplified model. *American Journal of Physics* 44: 599–601.
2. Bradoujié, K., J. D. Swain, A. Widomn, and N. Y. Srivastava. 2009. The Casimir effect in biology: The role of molecular quantum electrodynamics in linear aggregates of red blood cells, 60 years of the Casimir effect. *Journal of Physics: Conference series* 161; 012035.
3. Chan, H. S., and K. A. Dill. 1993, February. The protein folding problem. *Physics Today* 24–32.
4. Cotterill, R. 2003. *Biophysics: An introduction.* England: John Wiley & Sons, West Sussex.
5. Dean, D. S., and R. R. Hogan. 2005. The thermal Casimir effect in lipid bilayer tubules. *Physical Review* E 71: 04i907–11 [arXiv: cond-mat/0410609v1].
6. Dill, K. A. 1985. Theory for the folding and stability of globular proteins. *Biochemistry* 24: 1501–1509.
7. http://folding.stanford.edu
8. Kittel, C. 1909. Phase transition of a molecular zipper. *American Journal of Physics* 37: 917–920.
9. Rodriguez, R., K. Thompson, and M. Stewart. 2009. *Lecture notes: Biophysics and bio-imaging,* Centenary College of Louisiana, Shreveport, LA, http://www.centenary.edu/biophysics/bphy304
10. Schrödinger, E. 1956. *What is Life? and other scientific essays.* Doubleday, Garden City, NY.
11. Sneppen, K., and G. Zocchi. 2005. *Physics in molecular biology.* Cambridge University Press, Cambridge, UK.
12. Socci, N. D., J. N. Onuchic, and P. G. Wolynes. 1999. Stretching lattice models of protein folding. *Proceedings of the National Academy of Sciences* 96: 2031–2035.

13. Styer, O. F. 2008. Entropy and evolution. *American Journal of Physics* 76: 1031–1033.
14. Udgaonkar, F. B. 2001. Entropy in biology. *Resonance* 61–66.
15. Wang, J., and G. M. Crippen. 2004. Statistical mechanics of protein folding with separable energy functions. *Biopolymers* 74: 214–220.

Chapter 5

1. Beard, D. A., and H. Qian. 2008. *Chemical biophysics: Quantitative analysis of cellular systems.* Cambridge University Press, Cambridge, U.K.
2. Cotterill, R. 2003. *Biophysics: An introduction.* John Wiley & Sons, West Sussex, UK.
3. Dworkin, M., S. Falkow, E. Rosenberg, K. H. Schleifer, and E. Stackerbrandt, eds. 2006. *The prokaryotes: Ecophysiology and biochemistry.* 3rd ed. Springer, Singapore.
4. Espinola, T. 1994. *Introduction to thermophysics.* Wm. C. Brown, Dubuque, IA.
5. Guillou, C., and J. F. Guespin-Michel. 1996. Evidence for two domains of growth temperature for the psychrotropic bacterium *Pseudomonas fluorescens* MF0. *Applied Environmental Microbiology* 62: 3319–3324.
6. Nelson, P. 2004. *Biological physics energy, information, life.* W. H. Freeman, New York.
7. Pauling, L. 1988. *General chemistry.* New York: Dover.
8. Phillips, R., J. Kondev, and J. Theriot. 2008. *Physical biology of the cell.* Garland Science, New York.
9. Schroeder, D. V. 2000. *An introduction to thermal physics.* Addison Wesley Longman, NY.
10. Sturge, M. D. 2003. *Statistical and thermal physics: Fundamentals and applications.* A. K. Peters, Natick, MA.

Chapter 6

1. Begley, C. M., and S. J. Kleis. 2000. The fluid dynamic and shear environments in the NASA/JSC RWPV bioreactor. *Biotechnology and Bioengineering* 70: 32–40.
2. Berg, H. C. 1993. *Random walks in biology.* Princeton University Press. Princeton, New Jersey.
3. Chaudhuri, J., and M. Al-Rubeai. 2005. *Bioreactors for tissue engineeering: Principles, design and operation.* Springer, Dordrecht, The Netherlands.

4. Cole, N. B., C. L. Smith, N. Sciaky, M. Terasaki, M. Edidin, and J. Lippinscott-Schwartz. 1996. Diffusional mobility of golgi proteins in membranes of living cells. *Science* 273: 797–801.
5. Glasstone, S., K. J. Laidler, and H. Eyring. 1941. *The theory of rate processes.* McGraw-Hill, New York.
6. Lovely, P. S., and F. W. Dahlquis. 1975. Statistical measures of bacterial motility and chemotaxis. *Journal of Theoretical Biology* 50: 477–496.
7. Purcell, E. M. 1997. Life at low Reynolds number. *American Journal of Physics* 45: 3–11.
8. Saffman, P. G., and M. Delbrück. 1975. Brownian motion in biological membranes. *Proceedings of the National Academy of Sciences* 72: 3111–3113.
9. Vogel, S. 1994. *Life in moving fluids.* Princeton University Press. Princeton, New Jersey.

Chapter 7

1. Glaser, R. 2001. *Biophysics.* Springer, Berlin, Germany.
2. Hallett, F. R., P. A. Speight, and R. H. Stinson. 1977. *Introductory biophysics.* Methuen Publications, Ontario, Canada.
3. Libby, P. 2002. Atherosclerosis: The new view. *Scientific American* 286: 46–55.
4. Rodriguez, R., K. Thompson, and M. Stewart. 2009. Lecture notes: Biophysics and Bio-imaging, Centenary College of Louisiana, Shreveport, LA, http://www.centenary.edu/biophysics/bphy304
5. Srivastava, P. K. 2005. *Elementary biophysics: An introduction.* Alpha Science International, Harrow, Middlesex, UK.
6. Tuszynski, J. A. 2003. *Introduction to molecular biophysics.* CRC series in pure and applied physics. CRC press, Boca Raton, Florida.

Chapter 8

1. Dan, D., and A. M. Jayannavar. 2002. Energetics of rocked inhomogeneous ratchets. *Physical Review* E 65: 037105-1-037105-5.
2. Fillingame, R. H. 2000. Getting to the bottom of the F_1–ATPase. *Nature Structural Biology* 7: 1002–1004.
3. Grimm, B., W. Rudiger, and H. Scheer. 1991. *Chlorophylls and bacteriochlorophylls: Biochemistry, biophysics, functions and applications.* Ed. B. Grimm, R. J. Porra, W. Rudiger, and H. Scheer. Springer, Dordrecht, The Netherlands.
4. Jülicher, F., A. Ajdari, and J. Prost. 1997. Modeling molecular motors. *Reviews of Modern Physics* 69: 1269–1281.

5. Junge, W., H. Sielaff, and S. Engelbrecht. 2009. Torque generation and elastic power transmission in the rotary F_0F_1–ATPase. *Nature* 459: 364–370.
6. Knight, A. E. 2009. *Single molecule biology.* Elsevier, San Diego, CA.
7. Magnasco, M. O. 1993. Forced thermal ratchets. *Physical Review Letters* 71: 1477–1481.
8. Meisenberg, G., and W. H. Simmons. 2006. *Principles of medical biochemistry.* 2nd ed. Mosby Elsevier, Philadelphia.
9. Miller, J. H. Jr., V. Vajrala, H. L. Infante, J. R. Claycomb, A. Palanisami, J. Fang, and G. T. Mercier. 2008. Physical mechanisms of biological molecular motors. *Physica B: Condensed Matter* 404: 503–506.
10. Nelson, P. 2004. *Biological physics energy, information, life.* W. H. Freeman, New York.
11. Plaxco, K. W. and M. Gross. 2006. *Astrobiology: A brief introduction.* The John Hopkins University Press, Baltimore, MD.
12. Rodriguez, R., K. Thompson, and M. Stewart. 2009. Lecture notes: Biophysics and Bio-imaging, Centenary College of Louisiana, Shreveport, LA, http://www.centenary.edu/biophysics/bphy304
13. Serdyuk, I. N., N. R. Zaccai, J. Zaccai, and G. Zaccai. 2007. *Methods in molecular biophysics: Structure, dynamics, function.* Cambridge University Press, New York.
14. Vale, R. D., and R. A. Milligan. 2000. The way things move: Looking under the hood of molecular motor proteins. *Science* 288: 88–95.
15. Xing, J., F. Bai, R. Berry, and G. Oster. 2006. Torque-speed relationship of the bacterial flagellar motor. *Proceedings of the National Academy of Sciences* 103: 1260–1265.
16. Xing, J., H. Wang, C. von Ballmoos, P. Dimroth, and G. Oster. 2004. Torque generation by the motor of the sodium ATPase. *Biophysical Journal* 87: 2148–2163.

Chapter 9

1. Claycomb, J. R., Q. Tran, V. Vajrala, and J. H. Miller, Jr. 2009. Harmonic analysis of neuronal membranes using SQUIDs. *IEEE Transactions on Applied Superconductivity* 19: 839–843.
2. Glaser, R. 2001. *Biophysics,* chap 3.5. Springer-Verlag, Berlin, Germany.
3. Grimnes, S., and Ø. G. Martinsen. 2000. *Bioimpedance and bioelectricity basics.* Academic Press, London.
4. Grosse, C., and H. P. Schwan. 1992. Cellular membrane potentials induced by alternating fields. *Biophysical Journal* 63: 1632–1642.

5. Kotnik, T., and D. Miklavčič. 2000. Theoretical evaluation of the distribution power dissipation in biological cells exposed to electric fields. *Bioelectromagnetics* 21: 385–394.
6. McShea, A., A. M. Woodward, and D. B. Kell. 1992. Non-linear dielectric properties of Rhodobacter capsulatus. *Bioelectrochemistry and Bioenergetics* 29: 205–214.
7. Nawarathna, D., J. R. Claycomb, G. Cardenas, V. Vajrala, D. Warmflash, J. Gardner, W. Widger, and J. H. Miller, Jr. 2007. SQUID-based biosensor for probing ion transporters in cell suspensions and tissue. *IEEE Transactions on Applied Superconductivity* 17: 812–815.
8. Prodan, E., C. Prodan, and J. H. Miller. 2008. The dielectric response of spherical live cells in suspension: An analytic solution. *Biophysics Journal* 95: 4174–4182.
9. *QuickField user's guide.* Tera Analysis. 2009. Svendborg, Denmark.
10. Rodriguez, R., K. Thompson, and Stewart M. 2009. *Lecture notes: Biophysics and bio-imaging.* Centenary College of Louisiana, Shreveport, LA, http://www.centenary.edu/biophysics/bphy304
11. Schwan, H. P. 1990. Alternating field evoked membrane potentials: Effects of membrane and surface conductance. Annual International Conference of the IEEE Engineering in Medicine and Biology Society 12: 1523–1524.
12. Schwan, H. P. 1994. Electrical properties of tissues and cell suspensions: Mechanisms and models. *Engineering in Medicine and Biology Society* 6: 70a–71a.
13. Woodward, A. M., and D. B. Kell. 1990. On the nonlinear dielectric properties of biological systems: Saccharomyces cerevisiae. *Bioelectrochemistry and Bioenergetics* 24: 83–100.

Chapter 10

1. ADInstruments LabTutor earthworm action potentials experiments. http://www.adinstruments.com
2. Claycomb, J. R. 2008. *Applied electromagnetics using QuickField™ and MATLAB®*. Jones and Bartlett, Sudbury, MA.
3. Claycomb, J. R., Q. Tran, V. Vajrala, and J. H. Miller, Jr. 2009. Harmonic analysis of neuronal membranes using SQUIDs. *IEEE Transactions on Applied Superconductivity* 19: 839–843.
4. Cotterill, R. 2003. *Biophysics: An introduction.* John Wiley & Sons, West Sussex.
5. Fitzhugh, R. 1961. Impulses and physiological states in theoretical models of nerve membrane. *Biophysical Journal* 1: 445–466.

6. Friedman, M. H. 2008. *Principles and models of biological transport.* Springer, New York.
7. Gulrajani, R. M. 1998. Bioelectricity and Biomagnetism. John Wiley & Sons, New York.
8. Hodgkin, A. L., and A. F. Huxley. 1952. A quantitative description of membrane current and its application to conduction and excitation in nerve. *Journal of Physiology* 117: 500–544.
9. Hodgkin, A. L., A. F. Huxley, and B. Katz. 1952. Measurement of current–voltage relations in the membrane of the giant axon of *Loligo. Journal of Physiology* 116: 424–448.
10. Lin, K. K. 2006. Entrainment and chaos in the pulse-driven Hodgkin–Huxley oscillator. *Society for Industrial and Applied Mathematics Journal on Applied Dynamical Systems* 5: 179–204.
11. Nelson, P. 2004. *Biological physics, energy, information, life.* W. H. Freeman and Company, New York.
12. Wallisch, P., M. Lusignan, M. Benayoun, T. I. Baker, A. S. Dickey, and N. G. Hatsopoulos. 2009. *Matlab® for neuroscientists: An introduction to scientific computing in Matlab®.* Elsevier Academic Press, Burlington, MA.

Chapter 11

1. Aranda, S., K. A. Riske, R. Lipowsky, and R. Dimova. 2008. Morphological transitions of vesicles induced by alternating electric fields. *Biophysical Journal* 95: L19–L21.
2. Brinckmann, E., ed. 2007. *Biology in space and life on earth.* Wiley-VCH, Weinheim, Germany.
3. Buckey, J. C. 2006. *Space physiology.* Oxford University Press, New York.
4. Claycomb, J. R. 2008. *Applied electromagnetics using QuickField™ and MATLAB®*, chap 12. Jones and Bartlett, Sudbury, MA.
5. Currey, J. D. 2002. *Bones: Structure and mechanics.* Princeton University Press, Princeton, NJ.
6. Davis, J. R., R. Johnson, J. Stepanek, and J. A. Fogarty. 2008. *Fundamentals of aerospace medicine.* 4th ed. Lippincott Williams & Wilkins, Philadelphia, PA.
7. Disalvo, E. A., and S. A. Simon, eds. 1995. *Permeability and stability of lipid bilayers.* CRC Press. A Taylor and Francis Group, Boca Raton, FL.
8. Glaser, R. 2001. *Biophysics*, chap 3. Springer-Verlag, Heidelberg, Germany.
9. Hianik, T., and V. I. Passechnik. 1995. *Bilayer lipid membranes: Structure and mechanical properties.* Kluwer Academic Publishers, Bratislava, Slovak Republic.

Chapter 12

1. Baule, G. M., and R. McFee. 1963. Detection of the magnetic field of the heart. *American Heart Journal* 66: 95–96.
2. Blakemore, R., and W. Hole. 1975. Magnetotactic bacteria. *Science* 190: 377–379.
3. Brazdeikis, A., Y. Y. Xue, and C. W. Chu. 2003. Non-invasive assessment of the heart function in unshielded clinical environment by SQUID gradiometry. *IEEE Transactions on Applied Superconductivity* 13: 385–388.
4. Clarke, J. 1986. SQUIDs, brains and gravity waves. *Physics Today* 39: 36–45.
5. Clarke, J., and A. I. Braginski, eds. 2006. The SQUID handbook: Applications of SQUIDs and SQUID systems. Wiley-VCH, Weinheim, Germany.
6. Claycomb, J. R., and J. H. Miller, Jr. 1999. Superconducting magnetic shields for SQUID applications. *Review of Scientific Instruments* 70: 4562–4568.
7. Claycomb, J. R., and J. H. Miller, Jr. 2006. Superconducting and high-permeability magnetic shields modeled for biomagnetism and nondestructive testing. *IEEE Transactions on Magnetics* 42: 1694–1702.
8. Claycomb, J. R., D. Nawarathna, V. Vajrala, and J. H. Miller, Jr. 2004. Impedance magnetocardiography measurements and modeling. *Journal of Applied Physics* 96: 7650–7654.
9. Gulrajani, R. M. 1998. *Bioelectricity and biomagnetism.* John Wiley & Sons, New York.
10. Jenks, W. G., S. S. H. Sadeghi, and J. P. Wikswo, Jr. 1997. SQUIDs for nondestructive evaluation. *Journal of Physics D: Applied Physics* 30: 293–323.
11. Johnsen, S., and K. J. Lohmann. 2008. Magnetoreception in animals. *Physics Today* 61: 29–35.
12. Lewis, M. J. 2003. Review of electromagnetic source investigations of the fetal heart. *Medical Engineering and Physics* 25: 801–810.
13. Padhye, N. S., and A. Brazdeikis. 2006. Change in complexity of fetal heart rate variability. *Proceedings of the 28th IEEE EMBS Annual International Conference,* August 2006, New York, 1796–1798.
14. STAR Cryoelectronics, Santa Fe, NM, http://www.starcryo.com
15. Tristan Technologies, Inc. San Diego, CA, http://www.tristantech.com
16. Verklan, M. T., N. S. Padhye, and A. Brazdeikis. 2006. Analysis of fetal heart rate variability obtained by magnetocardiography. *Journal of Perinatal and Neonatal Nursing* 20: 343–348.
17. Wajnberg, E., L. H. Salvo de Souza, H. G. P. Lins de Barros, and D. M. S. Esquivel. 1986. A study of magnetic properties of magnetotactic bacteria. *Biophysical Journal* 50: 451–455.

18. Wikswo, J. P. 1990. High resolution magnetic imaging: Cellular action currents and other applications. In *SQUID sensors: Fundamentals, fabrication and applications*, ed. H. Weinstock, 307–360. Kluwer Academic, Dordrecht, The Netherlands.
19. Wikswo, J. P. 2000. Applications of SQUID magnetometers to biomagnetism and nondestructive evaluation. In *Applications of superconductivity*, 139–228. H. Weinstock, ed. Kluwer Academic, Dordrecht.
20. Wikswo, J. P., J. P. Barach, and J. A. Freeman. 1980. Magnetic field of a nerve impulse: First measurements. *Science* 208: 53–55.

Chapter 13

1. Aihara, K., and G. Matsumoto. 1987. Forced oscillations and routes to chaos in the Hodgkin–Huxley axons and giant squid axons. In *Chaos in biological systems*, ed. H. Degn, A. V. Holden, and L. F. Olsen, NATO ASI Series 138, 121–131. Plenum Press, New York.
2. Baker, R. E., E. A. Gaffney, and P. K. Maini. 2008. Partial differential equations for self-organization in cellular and developmental biology. *Nonlinearity* 21: R251–R290.
3. Britton, N. F. 2003. *Essential Mathematical Biology.* Springer-Verlag, London.
4. Constantino, R. F., J. M. Crushing, B. Dennis, and R. A. Desharnais. 1995. Experimentally induced transitions in the dynamic behavior of insect populations. *Nature* 375: 227–230.
5. Cueda, S., and A. Sánchez. 2004. Nonlinear excitations in DNA: Aperiodic models versus actual genome sequences. *Physical Review* E 70: 051903-1-051903-8.
6. Englander, S. W., N. R. Kallenbach, A. J. Heeger, J. A. Krumhansl, and S. Litwin. 1980. Nature of the open state in long polynucleotide double helices: Possibility of soliton excitations. *Proceedings of the National Academy of Sciences* 77: 7222–7227.
7. Fall, C. P. 2002. *Computational cell biology*. Springer-Verlag, New York.
8. Feigenbaum, M. 1980. Universal behavior in nonlinear systems. *Los Alamos Science* 1: 4–27.
9. Jones, D. S., and B. D. Sleeman. 2003. *Differential equations and mathematical biology.* Chapman & Hall/CRC Press, Boca Raton, FL.
10. Kondo, S., and R. Asai. 1995. A reaction diffusion wave on the skin of the marine angelfish *Pomacanthus. Nature* 376: 765–768.
11. Maini, P. K., and H. G. Othmer. 2001. *Mathematical models for biological pattern formation*. Springer-Verlag, New York.
12. Strogatz, S. H. 1994. *Nonlinear dynamics and chaos: With applications to physics, biology, chemistry, and engineering*. Perseus Books, Cambridge, MA.

13. Yang, X-S. 2006. *An introduction to computational engineering with MATLAB.* Cambridge International Science Publishing, Cambridge, UK.
14. Zeeman, E. C. 1977. Differential equations for the heart beat and nerve impulse, *Catastrophe Theory*, E. C. Zeeman, ed., Addison-Westley, Reading, MA.

Chapter 14

1. Bak, P. 1996. *How nature works: The science of self-organized criticality.* Springer-Verlag, New York.
2. Bak, P., C. Tang, and K. Wiesenfeld. 1987. Self-organized criticality: An explanation of $1/f$ noise. *Physical Review Letters* 59: 381–384.
3. Bak, P., C. Tang, and K. Wiesenfeld. 1988. Self-organized criticality. *Physical Review* A 38: 364–374.
4. Bak, P., and K. Sneppen. 1993. Punctuated equilibrium and criticality in a simple model of evolution. *Physical Review Letters* 71: 4083–4086.
5. Baker, G. L., and J. P. Gollub. 1996. *Chaotic dynamics: An introduction.* 2nd ed. Cambridge University Press, New York.
6. Clarke, A. C. 1994. *Colors of infinity—Exploring the fractal universe.* Films for the Humanities and Sciences, Hamilton, NJ.
7. Claycomb, J. R., D. Nawarathna, V. Vajrala, and J. H. Miller, Jr. 2004. Power law behavior in chemical reactions. *Journal of Chemical Physics* 121: 12428–12430.
8. Claycomb, J. R., K. E. Bassler, J. H. Miller, Jr., M. Nersesyan, and D. Luss. 2001. Avalanche behavior in the dynamics of chemical reactions. *Physical Review Letters* 87: 178303–178306.
9. Claycomb, J. R., M. Nersesyan, D. Luss, and J. H. Miller, Jr. 2001. SQUID detection of magnetic fields produced by chemical reactions. *IEEE Transactions on Applied Superconductivity* 11: 863–866.
10. Dewey, T. G. 1997. *Fractals in molecular biophysics.* NewYork: Oxford University Press.
11. Eldredge, N., and S. J. Gould. 1972. Punctuated equilibria: An alternative to phyletic gradualism. In *Models in paleobiology*, ed. T. J. M. Schopf, 82–115. Freeman, Cooper, San Francisco, CA.
12. Gardner, M. 1970. Mathematical games: The fantastic combinations of John Conway's new solitaire game "life." *Scientific American* 223: 120–123.
13. Georgelin, Y., L. Poupard, R. Sartène, and J. C. Wallet. 1999. Experimental evidence for a power law in electroencephalographic α-wave dynamics. *European Physics Journal* B 2: 303–307.

14. Ghosh, S., A. Bhattacharjee, J. Banerjee, S. Manna, N. K. Bhatraju, M. K. Verma, and M. K. Das. 2009. "Electrical noise in cells, membranes and neurons." In *Complex dynamics in physiological systems: From heart to brain, understanding complex systems*, ed. S. K. Dana et al. Springer Science and Business Media B. V., New York.

15. Iannaccone, P. M., and M. K. Khokha. 1996. *Fractal geometry in biological systems: An analytical approach.* CRC Press, Boca Raton, FL.

16. Jensen, H. J. 2004. *Self-organized criticality emergent complex behavior in physical and biological systems.* Cambridge University Press, Cambridge, UK.

17. Losa, G. A., T. F. Nonnenmacher, D. Merlini, and E. R. Weibel, eds. 1998. *Fractals in biology and medicine,* I–IV. Birkhäuser Verlag, Basel, Switzerland.

18. Mandelbrot, B. B. 1982. *The fractal geometry of nature.* W. H. Freeman & Co, New York.

19. Raup, D. M. 1991. A kill curve for Phanerozoic marine species. *Paleobiology* 17: 37–48.

20. Sneppen, K., and G. Zocchi. 2005. *Physics in molecular biology.* Cambridge University Press, New York.

21. Sneppen, K., P. Bak, H. Flyvbjerg, and M. H. Jensen. 1995. Evolution as a self-organized critical phenomenon. *Proceedings of the National Academy of Sciences* 92: 5209–5213.

22. Styer, D. F. 2008. Entropy and evolution. *American Journal of Physics* 76: 1031–1033.

23. Vicsek, T., ed. *Fluctuations and scaling in biology.* Oxford University Press, New York.

24. Walleczek, J., ed. 2000. *Self-organized biological dynamics & nonlinear control.* Cambridge University Press, Cambridge, UK.

25. West, G. B., and J. H. Brown. 2004. Life's universal scaling laws. *Physics Today* 57: 36–42.

26. Zipf, G. K. 1949. *Human behavior and the principle of least effort: An introduction to human ecology*. Addison-Wesley Press, Cambridge, MA.

Chapter 15

1. Brinckmann, E., ed. 2007. *Biology in space and life on earth.* Wiley-VCH, Weinheim, Germany.

2. Davies, P. C. W. 2004. Multiverse cosmological models. *Modern physics letters* A 19: 727–744.

3. Fox, S. W. 1960. How did life begin? *Science* 132: 200–208.

4. Guth, A. H. 2007. Eternal inflation and its implications. *Journal of Physics A: Mathematical and Theoretical* 40: 6811–6826.
5. Jakosky, B. 2006. *Science, society, and the search for life in the universe.* University of Arizona Press, Tucson, AZ.
6. Kobakhidze, A., and L. Mersini-Houghton. 2007. Birth of the universe from the landscape of string theory. *European of Physics Journal C—Particles and Fields* 49: 869–876.
7. Livio, M., and M. J. Rees. 2005. Anthropic reasoning. *Science* 309: 1022–1023.
8. Miller, S. L. 1953. Production of amino acids under possible primitive Earth Conditions. *Science* 117: 528–529.
9. Miller, S. L., H. C. Urey. 1959. Organic compound synthesis on the primitive earth. *Science* 130: 245–251.
10. NASA astrobiology. http://astrobiology.nasa.gov/
11. Plaxco, K. W., and M. Gross. 2006. *Astrobiology: A brief introduction.* John Hopkins University Press, Baltimore, Maryland.
12. SETI. http://www.seti.org
13. Smolin, L. 2007. "Scientific alternatives to the anthropic principle." In *Universe or multiverse*, ed. B. Carr, Cambridge University Press, New York.
14. Tegmark, M. 2003. Parallel Universes. *Scientific American* May: 41–51.
15. Turok, N. and P. J. Steinhardt. 2008. *Endless Universe: Beyond the Big Bang.* Random House, New York.
16. Weiss, B. P., J. L. Kirschvink, F. J. Baudenbacher, H. Vali, N. T. Peters, F. A. Macdonald, and J. P. Wikswo. 2000. A low temperature transfer of ALH84001 from Mars to Earth. *Science* 27: 791–795.

Appendix 2

1. Hahn, B., and D. Valentine. 2007. *Essential MATLAB for engineers and scientists.* 3rd ed. Butterworth-Heinermann/Elsevier, Burlington, MA.
2. Higham, D. J., and N. J. Higham. 2005. *MATLAB Guide.* 2nd ed. SIAM. Philadelphia, PA.

Appendix 8

1. *QuickField user's guide*. 2009. Tera Analysis. Svendborg, Denmark.
2. QuickField Virtual Classroom. http://www.quickfield.com/demo/index.htm

Appendix 10

1. Harries, D. 1998. Solving the Poisson-Boltzmann equation for two parallel cylinders. *Langmuir* 14: 3149–3152.
2. Hsu, J-P., and B-T. Liu. 1996. Exact solution to the linearized Poisson-Boltzmann equation for spheroidal surfaces. *Journal of Colloid and Interface Science* 175: 785–788.
3. Moon, P., and D. E. Spencer. 2003. *Field theory handbook,* 2nd Ed, John T. Zubal Inc. Cleveland, OH.

Index

B

D

E

M

O

P

Q

W

Y

Z

CD Credits

Fluid-Structure Interaction Model of Active Eustachian Tube Function in Healthy Adult Patients

Courtesy of F. Sheer and S. Ghadiali.

Study of Compliance Mismatch within a Stented Artery

Courtesy of G. Coppola and K. Liu.

Hemodynamic Therapy of Middle Cerebral Artery Vasospasm Guided by a Multiphase Model of Oxygen Transport

Courtesy of S. Conrad, P. Chittiboina, and B. Guthikonda.

Validation of Measurement Strategies and Anisotropic Models Used in Electrical Reconstructions

Courtesy of R. Sadleir.

Simulating Microbubble Flows Using COMSOL Multiphysics

Courtesy of X. Chen and S. Ghadiali.

Modeling of Respiratory Lung Motion as a Contact Problem of Elasticity Theory

Courtesy of R. Werner, J. Ehrhardt, and H. Handels.

Coupled Fluid-Structural Analysis of Heart Mitral Valve

Courtesy of A. Avanzini and G. Donzella.

Virtual Thermal Ablation in the Head and Neck using COMSOL Multiphysics

Courtesy of U. Topaloglu, Y. Yan, P. Novak, P. Spring, J. Suen, and G. Shafirstein.

Numerical Modelling of the Diaphragmatic Signal in the Rib Cage

Courtesy of C. Konté, J. Caire, P. Gumery, H. Roux-Buisson, J. Bouteillon.

Simulating Hodgkin-Huxley-like Excitation using Comsol Multiphysics

Courtesy of J. Martinek, Y. Stickler, M. Reichel, and F. Rattay.

Modeling of snRNP Motion in the Nucleoplasm

Courtesy of M. Blaziková, J. Malínský, D. Stanek, and P. Herman.

Heat and Mass Transfer in Convective Drying Processes

Courtesy of C. Gavrila, A. Ghiaus, and I. Gruia.

Image Based-Mesh Generation for Realistic Simulation of the Transcranial Current Stimulation

Courtesy of R. Said, R. Cotton, P. Young, A. Datta, M. Elwassif, and M. Bikson.

Finite Element Analysis of Muscular Contractions from DC Pulses in the Liver

Courtesy of G. Long, D. Plescia, and P. Shires.

Electrical Stimulation of Brain using a realistic 3D Human Head Model: Improvement of Spatial Focality

Courtesy of A. Datta, M. Elwassif, and M. Bikson.

COMSOL Modeling of a Submarine Geothermal Chimney

Courtesy of M. Suárez and F. Samaniego.

Design of a High Field Gradient Electromagnet for Magnetic Drug Delivery to a Mouse Brain

Courtesy of I. Hoke, C. Dahmani, and T. Weyh.

CFD-based Evaluation of Drag Force on a Sphere Unsteadily Moving Perpendicularly toward a Solid Surface: a Simple Model of a Biological Spring, Vorticella Convallaria

Courtesy of S. Ryu and P. Matsudaira.

Magnetic Ratchet

Courtesy of A. Auge, F. Wittbracht, A. Weddemann, and A. Hütten.

COMSOL Multiphysics Simulations of Microfluidic Systems for Biomedical Applications

Courtesy of M. Dimaki, J. Moresco Lange, P. Vazquez, P. Shah, F. Okkels, and W. Svendsen.

Exploratory FEM-Based Multiphysics Oxygen Transport and Cell Viability Models for Isolated Pancreatic Islets

Courtesy of Peter Buchwald.

Solid Food Pasteurization by Ohmic Heating: Influence of Process Parameters

Courtesy of Markus Zell, Denis A. Cronin, Desmond J. Morgan, Francesco Marra, and James G. Lyng.

A Mean Field Approach to Many-particles Effects in Dielectrophoresis

Reprinted with permission from O. E. Nicotra, A. La Magna, and S. Coffa, *Applied Physics Letters*, 93, 193902 (2008). Copyright 2008, American Institute of Physics.

Fold-It screenshots

Courtesy of Animation Research Labs, University of Washington.